Lecture Notes in Business Information Processing 582

Series Editors

The series Lecture Notes in Computer Science (LNCS), including its subseries Lecture Notes in Artificial Intelligence (LNAI) and Lecture Notes in Bioinformatics (LNBI), has established itself as a medium for the publication of new developments in computer science and information technology research, teaching, and education.

LNCS enjoys close cooperation with the computer science R & D community, the series counts many renowned academics among its volume editors and paper authors, and collaborates with prestigious societies. Its mission is to serve this international community by providing an invaluable service, mainly focused on the publication of conference and workshop proceedings and postproceedings. LNCS commenced publication in 1973.

Ana Paula Cabral Seixas Costa ·
Sandro Radovanović · Guoqing Zhao ·
Uchitha Jayawickrama
Editors

Decision Support Systems XVI

Decision Support System Technology for Sustainable and Resilient Transitions

12th International Conference
on Decision Support System Technology, ICDSST 2026
Plymouth, UK, May 27–29, 2026, Proceedings

Springer

Editors
Ana Paula Cabral Seixas Costa
Universidade Federal de Pernambuco
Recife, Pernambuco, Brazil

Sandro Radovanović
University of Belgrade
Belgrade, Serbia

Guoqing Zhao
Swansea University
Swansea, UK

Uchitha Jayawickrama
Loughborough University
Loughborough, UK

ISSN 1865-1348 ISSN 1865-1356 (electronic)
Lecture Notes in Business Information Processing
ISBN 978-3-032-25145-9 ISBN 978-3-032-25146-6 (eBook)
https://doi.org/10.1007/978-3-032-25146-6

This Springer imprint is published by the registered company Springer Nature Switzerland AG
The registered company address is: Gewerbestrasse 11, 6330 Cham, Switzerland

Preface

The proceedings of this edition of the EWG-DSS Decision Support Systems published in the LNBIP series present a selection of reviewed and revised full papers from the twelfth International Conference on Decision Support System Technology (ICDSST 2026), held in Plymouth, UK, during May 27–29, 2026. The conference's main theme was: "Decision Support System Technology for Sustainable and Resilient Transitions" offering a platform to explore how DSS can drive and support a wide range of transition, green transition, digital transition, skills transition, societal transition, organization transition and supply chain transition.

The EWG-DSS series of International Conferences on Decision Support System Technology (ICDSST) consolidates the tradition of annual events organized by the EWG-DSS in offering a platform for European and international DSS communities, comprising the academic and industrial sectors, to present state-of-the-art DSS research and developments, to discuss current challenges that surround decision-making processes.

Building on this tradition, the scientific topic areas of ICDSST 2026 included:

- Decision systems and technology to support green transitions
- Decision systems and technology to support digital transitions
- Decision systems and technology to support workforce transitions
- Decision systems and technology to support society transitions
- Decision systems and technology to support organization transitions
- Decision systems and technology to support supply chain transitions
- Decision systems and technology to support a range of other relevant transitions
- Advances in decision making and decision support systems
- Decision-making integrated solutions within open data platforms
- Knowledge management & resource discovery for decision making
- Decision making using emerging technologies such as AI and digital twins
- Decision-making methods and processes
- Knowledge management and business intelligence
- Decision support for innovation and entrepreneurship
- Decision making in education
- Innovative decision-making approaches
- Big data analytics approaches for solving decision issues
- Big data visualization to support decision analysis
- Social network analysis for decision making
- Group & collaborative decision making
- Multi-attribute & multi-criteria decision making
- Decision tools, applications and business cases.

These topics allow exploration of how DSS can drive and support a wide range of transitions including green transition, digital transition, workforce transition, societal transition, organization transition and supply chain transition, to adapt to a fast-changing environment. The papers in this book share insights, methodologies, processes, models, tools, frameworks and applications that contribute to building resilient decision support systems to inform decision making.

Hence, this EWG-DSS LNBIP Springer edition has considered contributions of a "full-paper" format, selected through a single-blind paper reviewing process. In particular, at least two reviewers – members of the Program Committee – reviewed each submission through a rigorous two-staged process. The second stage involved the Guest Editors judging whether the revised versions were indeed addressing the issues that the reviewers raised. Papers that didn't address properly all the issues were either accepted to the conference but not included in this volume or were not accepted at all. Finally, we selected 27 out of 70 submissions, which corresponds to a 38% acceptance rate, to be included in this 12th EWG-DSS Springer LNBIP edition.

We proudly present the selected contributions, organized in five sections:

1. *Decision Support Systems and Models for Complex Transitions*: This session explores decision support systems and decision models designed to address complex societal and organizational transitions. The papers discuss applications related to sustainability, supply chains, food systems, workforce transformation and information security.
2. *Decision Support Systems and Models for Energy and Green Transition:* This session focuses on decision support systems and analytical models that facilitate energy and green transitions across different sectors. The papers address topics such as renewable energy adoption, hydrogen-based mobility, green productivity and sustainability practices in organizations.
3. *Decision Support Systems and Multicriteria Decision Making:* This session discusses methodological and theoretical advances in decision support systems and multicriteria decision making. The papers address topics such as normalization effects in MCDM methods, user behavior in visual decision support tools, and conceptual perspectives on decision quality. They also explore frameworks to enhance the adaptability of strategic decision support systems. Together, they contribute to improving the design, evaluation and effectiveness of DSS for complex decision contexts.
4. *Large Language Models and Data-Driven Decision Support:* This session explores the growing role of Large Language Models and data-driven approaches in decision support. The papers examine applications of LLMs in areas such as MCDA weight elicitation, resilient port management, business intelligence analysis and expert–AI comparative decision support.
5. *Decision Tools, Applications and Business Cases*: This session presents decision support tools applied to real-world operational and strategic problems across diverse sectors. The papers include applications in healthcare, transportation, supply chains, taxation, education, manufacturing and service operations. They demonstrate how optimization models, multicriteria methods and analytical tools can support more effective and evidence-based decisions, highlighting the practical impact of DSS in solving complex business challenges.

We would like to thank many people who greatly helped with the success of this LNBIP book. First, we would like to thank Springer for giving us the opportunity to guest edit this DSS book, and we especially wish to express our sincere gratitude to the Springer staff who have provided us with timely professional guidance and advice during the volume-editing process. Secondly, we need to thank all the authors for submitting their state-of-the-art work for consideration for this volume. Finally, we express our deep gratitude to the reviewers, who volunteered to assist in the improvement and the selection of the papers. We believe that this EWG-DSS Springer LNBIP volume has made a rigorous selection of high-quality papers addressing the conference theme. We hope that readers will enjoy the publication!

May 2026

Ana Paula Cabral Seixas Costa
Sandro Radovanović
Guoqing Zhao
Uchitha Jayawickrama

Organization

Honorary Chairs

Fatima Dargam	SimTech Simulation Technology & REACH Innovation, Austria
Isabelle Linden	University of Namur, Belgium
Jason Papathanasiou	University of Macedonia, Greece
Pascale Zarate	University Toulouse Capitole-IRIT, France

Conference General Chairs

Pavlos Delias	Democritus University of Thrace, Greece
Boris Delibašić	University of Belgrade, Serbia
Shaofeng Liu	University of Plymouth, UK

Local Organizing Committee

Huilan Chen	University of Plymouth, UK
Xiaoning Chen	University of Plymouth, UK
Dulekha Kasturiratne	University of Plymouth, UK
Xinyi Li	University of Plymouth, UK
Guoste Pivoraite	University of Plymouth, UK

Programme Committee Chairs

Ana Paula Cabral Seixas Costa	Universidade Federal de Pernambuco, Brazil
Uchitha Jayawickrama	Loughborough University, UK
Sandro Radovanović	University of Belgrade, Serbia
Guoqing Zhao	Swansea University, UK

Communication Chairs

Sergio Pedro Duarte	University of Porto, Portugal
Maite Escobar Urmeneta	University of Zaragoza, Spain

Uchitha Jayawickrama	Loughborough University, UK
Femi Olan	University of Essex, UK
Sandro Radovanović	University of Belgrade, Serbia

Program Committee

Carlos Antunes	University of Coimbra, Portugal
George Aretoulis	Aristotle University of Thessaloniki, Greece
Md Asaduzzaman	Staffordshire University, UK
Dragana Bečejski-Vujaklija	Serbian Society for Informatics, Serbia
Marko Bohanec	Jožef Stefan Institute, Slovenia
Ana Paula Cabral Seixas Costa	Federal University of Pernambuco, Brazil
HingKai Chan	Wenzhou-Kean University, China
Huilan Chen	University of Plymouth, UK
Christian Colot	University of Namur, Belgium
Fátima Dargam	SimTech Simulation Technology & REACH Innovation, Austria
Pavlos Delias	Democritus University of Thrace, Greece
Boris Delibašić	University of Belgrade, Serbia
Panagiota Digkoglou	University of Macedonia, Greece
Sérgio Pedro Duarte	University of Porto, Portugal
Sean Eom	Southeast Missouri State University, USA
María Teresa Escobar	University of Zaragoza, Spain
Oscar A. Forero	Agrosavia, Colombia
Jorge Freire de Sousa	Engineering University of Porto, Portugal
Julián Grigera	Universidad Nacional de La Plata, Argentina
Jiandong Hu	Henan Agricultural University, China
Ping Huang	Guizhou University of Commerce, China
River Huang	Paul Scherrer Institute, Switzerland
Lise Hunter	University of Plymouth, UK
Uchitha Jayawickrama	Loughborough University, UK
Daouda Kamissoko	IMT Mines Albi, France
Dulekha Kasturiratne	University of Plymouth, UK
Marc Kilgour	Wilfrid Laurier University, Canada
Kathrin Kirchner	Technical University of Denmark, Denmark
Emilio Larrodé	University of Zaragoza, Spain
Isabelle Linden	University of Namur, Belgium
Shaofeng Liu	University of Plymouth, UK
Antonio Lobo	Universidade do Porto, Porto
João Lourenço	Universidade de Lisboa, Portugal
Michael Madas	University of Macedonia, Greece

Francisco Marmolejo-Cossio	Boston College, USA
Nikolaos Matsatsinis	Technical University of Crete, Greece
José Maria Moreno-Jiménez	Zaragoza University, Spain
Ben C.K. Ngan	Worcester Polytechnic Institute, USA
Festus Oderanti	Liberty University, USA
Femi Olan	University of Essex, UK
Jason Papathanasiou	University of Macedonia, Greece
Gloria Phillips-Wren	Aristotle University of Thessaloniki, Greece
Sandro Radovanović	University of Belgrade, Serbia
Saeyeon Roh	University of Plymouth, UK
Ying Shen	Changchun University of Technology, China
Karim Soliman	Arab Academy for Science, Technology, and Maritime Transport, Egypt
Theodore Tarnanidis	University of Macedonia, Greece
Adiel Teixeira de Almeida	Federal University of Pernambuco, Brazil
Fernando Tricas Garcia	University of Zaragoza, Spain
Stelios Tsafarakis	Technical University of Crete, Greece
Alexis Tsoukias	Université Paris Dauphine-PSL, France
Alberto Turón	University of Zaragoza, Spain
Rudolf Vetschera	University of Vienna, Austria
Konstantinos Vergidis	University of Macedonia, Greece
Milan Vukićević	University of Belgrade, Serbia
Bo Wen	Swansea University, UK
Andy Wong	University of Strathclyde, UK
Xiaotian Xie	Newcastle University, UK
Jie Yan	East China University of Science and Technology, China
Guoqing Zhao	Swansea University, UK
Pascale Zaraté	IRIT/Toulouse University, France

Contents

Decision Tools, Applications and Business Cases

Decision Support Systems and Models for Complex Transitions

Research on the Influence Paths of Online Environmental Activities on Purchase Intention of Carbon-Labeled Products

Wu Hangyu, Li Ping, and Shen Ying(✉)

Changchun University of Technology, Changchun, JiLin 130102, China
shenying@ccut.edu.cn

Abstract. This study aims to boost sales of carbon-labeled products and alleviate the green burden on enterprises and consumers via efficient online promotion strategies. Using structural equation modeling and refined path analysis (parallel/serial mediation tests), it investigates the impact of Perception of Online Environmental Activities (POEA) on purchase intention, with Perception of Personal Image about Environmental (PPIE) as the key mediating factor. Data from 488 valid samples were processed via Box-Cox transformation and Min-Max normalization. Results show POEA exerts significant direct and indirect effects on purchase intention, while higher perceived frequency of Content Creation and Organize Activities has unexpectedly negative impacts—explained by Piaget's cognitive development theory. Government Responsibility Fulfillment (GRF) in the Perception of Fulfillment of Responsibility Status (PFRS) moderates some paths. The optimal model demonstrates excellent fit(χ^2/df = 1.640, RMSEA = 0.036, CFI = 0.997). This study suggests multidimensional promotion strategies and leveraging PPIE to enhance Collective Responsibility Attitude and Self-Categorization, providing practical references for improving the application effects of carbon labels and facilitating the green transition of markets.

Keywords: carbon label · purchase intention of eco-friendly products · Promotion Strategy · Consumer Behavior · influence path analysis

1 Introduction: To Promote Sales and Alleviate the Green Burden Caused by the Use of Carbon Labels

Carbon Label quantifies the greenhouse gas emissions of commodities in the whole life cycle in units of carbon dioxide equivalent, and informs consumers of the carbon information of the product in the form of a label. Although Carbon Label has been implemented for a long time in Europe and the United States, it is not long since the policy was formally decided to be piloted in China [1].

Undoubtedly, this measure will contribute to promoting sustainable and high-quality development in China. However, it is not difficult to find that although consumers can better identify the environmental protection level of products through Carbon Labels,

A. P. Cabral Seixas Costa et al. (Eds.): ICDSST 2026, LNBIP 582, pp. 3–17, 2026.
https://doi.org/10.1007/978-3-032-25146-6_1

due to the green premium caused by Carbon Labels [2], consumers will be forced to bear the additional consumption burden passed on by the increase in corporate costs.

In fact, many studies on Carbon Label have slightly different results due to different theoretical perspectives, data collection methods and measurement tools. In terms of individual factors, demographic characteristics such as age, income, gender and education level have all been proven to have a significant impact on consumers' willingness to purchase and pay [3].

From the perspective of game theory, Chen D.J. et al. believed that with the effective supervision and participation of the government, enterprises and consumers can take the green premium as an intermediary to maximize their own benefits and achieve the "win-win" situation expected for social green development [4]. However, the green premium itself is not the purpose of Carbon Labels' existence, and carbon labels should not be overused as a lever for corporate interests.

Nevertheless, the pressure on enterprises caused by technology investment and certification costs still deserves due attention, so is there a theoretical solution to relieve the pressure on both enterprises and consumers at the same time?

From the perspective of classical market equilibrium theory, this situation does exist. Carbon Label certification pushes up enterprises' costs, easily leads to higher product prices and shifts the supply curve upward. However, efficient promotion strategies can significantly boost demand and push the demand curve to the right. If the increase in demand is large enough, the new market equilibrium will result in increased production and slightly higher prices, with firms selling more to make up for reduced profits and consumers able to buy products at a reasonable premium.

So what constitutes an effective promotion strategy? Perhaps the prevailing era has already provided a definitive answer.

In China, residents spend an average of 5.5 h per day on the Internet, and the Internet penetration rate is as high as 92.9% [5]. Therefore, the core starting point of this study is to identify a pathway that boosts the sales of carbon-labeled products via online platforms—by shaping consumers' purchase intention (i.e., a low-carbon consumption decision-making process)—while alleviating the green cost burden incurred by the implementation of carbon labeling schemes. Carbon labeling and targeted online environmental promotion act as dual core mechanisms that underpin the transition of market supply systems and consumer consumption patterns toward low-carbon sustainability. As an exploratory inquiry, this study not only offers direct implications for policy advancement in China—the sample source country—but also provides actionable references for other nations seeking to further advance the eco-friendly transition of their markets.

2 Research Design: Structural Equation Models Pointing to Purchase Intentions

2.1 A Review of Relevant Research

Previous studies have examined the consumption of carbon-labeled products and their underlying mechanisms from diverse perspectives and with a range of methodologies. Drawing on the theory of planned behavior, Li Q. et al. argued that demographic factors give rise to substantial variations in consumers' purchase intentions toward carbon-labeled products [6]. From the dual perspectives of social norms and environmental concerns, Xiong H. et al. explored the significant impacts of social expectation discrepancies on consumers' decisions to purchase such products [7]. Based on a survey conducted across three Chinese cities, Wang J. et al. verified a positive correlation between low-carbon psychological constructs (e.g., cognitive level and sense of responsibility) and consumers' willingness to pay a premium [8].

In addition, Mei L. and Sun L. employed eye-tracking experiments to demonstrate that enhanced cognition of carbon labels exerts a markedly positive influence on participants' preference for carbon-labeled products [9]. Using structural equation modeling, they further confirmed that both carbon label cognition and carbon label promotion exert a direct and significant impact on consumers' purchase intentions toward low-carbon products [10].

Although many studies—including those cited above—have advocated for greater attention to online communication channels, more in-depth, targeted research on this topic remains scarce.

Given the lack of a systematic questionnaire template, this research will extensively incorporate a variety of classical hypotheses and theories related to the characteristics of online communication.

2.2 Perception of Online Environmental Activities: Browsing, Interaction, and Historical Behavior Perception

Perception of Online Environmental Activities (POEA) refers to users' subjective awareness and judgment of the environmentally conscious behaviors they conduct—either proactively or reactively—across diverse digital platforms.

From the perspective of platform users, user behaviors can be categorized into browsing and interaction, which generate a substantial volume of relevant data and thus form the platform's influence resources. Essentially, such data constitute a reflection of users' historical behavioral records. Both platforms and promoters can predict and profile users' characteristics through these historical records, thereby developing a generalized user classification system. The classification criteria selected will also be adjusted dynamically based on the specific objectives of promotion campaigns. Beyond demographic traits and browsing and interaction data, ritualized behavioral records—such as the organization and participation in relevant online activities—and consumption history (e.g., product purchase records) also serve as critical references for customer segmentation.

With the maturation of algorithm technology, the effect of the "Information Cocoon Room" has become increasingly pronounced [11]. Cass Sunstein first proposed the concept of the "Information Cocoon Room" in 2006, noting that long-term exposure to homogeneous information—driven by personal interest preferences, information selection habits, or algorithmic recommendation mechanisms—will lead to a phenomenon where individuals' cognitive horizons are confined within narrow boundaries. In recent years, the "information cocoon room" has been widely cited as a concept to critique the adverse effects of algorithmic and platform-based development; yet, it provides a valid entry point for collecting the online environmental behavior data required for this study.

Table 1. Specific Variables of POEA

Full name	Abrr.	Item description	Reference activities
Active search	AS	Proactively search for environmental protection-related content	Carbon neutrality knowledge, etc.
System recommendation	SR	Receive system-recommended environmental protection content	TikTok low-carbon topics, etc.
Interactive engagement	IE	Like and comment on environmental protection-related content	None
Share information	SI	Share environmental protection-related content	Repost low-carbon posters, etc.
Content creation	CC	Create and publish environmental protection-related content	Low-carbon life vlogs, etc.
Purchase history	PII	Purchase "carbon credit" or "carbon neutrality" products	Carbon-neutral milk, etc.
Public welfare donation	PWD	Participate in environmental crowdfunding or donations	Tencent Charity environmental protection projects, etc.
Environmental protection games	EPG	Participate in game-style environmental protection activities	Tree planting on Ant Forest, etc.
Environmental protection initiative	EPI	Participate in and support environmental protection policy initiatives	Sign electronic petitions, etc.
Organize activities	OA	Organize and launch environmental protection activities	Online low-carbon challenges, etc.

In this study, the self-report method is adopted for relevant measurements, whereby respondents' perceived evaluations of the frequency of their own online pro-environmental behaviors are used as a proxy for actual frequency data. The self-report method is widely employed in psychology to assess subjective experiences, including behavioral frequency and emotional states.

The rationale for the methodological approach adopted in this study requires further elaboration. Given the rapid updating and fragmentation of online information, attention resources are overextended, resulting in information overload; this in turn induces a general insensitivity among users toward low-frequency information. Conversely, information that achieves high levels of user perception typically reflects a protracted, systematic exposure process. By using participants' self-assessments and taking perceived intensity as a surrogate for actual frequency in the analytical model, this approach enables more accurate identification of genuinely impactful activities.

The specific variables corresponding to this factor are detailed in Table 1.

2.3 Perception of Personal Image About Environmental: Mediating Variables Based on Self-Evaluation

The research by Zeng Y.G. and Chen S., from a communication science perspective, provides a reasonable reference for the selection of mediating variables [12]. Their study clearly points out that ordinary people actively build an idealized public persona through selective content sharing, strengthen their self-distinctiveness via these personas, and achieve social interaction and value exchange under the algorithmic recommendation mechanism.

Table 2. Specific Variables of PPIE

Full name	Abbr.	Item description	Theory/prior research
Self-categorization	SC	I consider myself an active citizen practicing environmental protection.	Social identity theory [15]
Collective responsibility attitude	CRA	When others fail to comply with environmental norms, I feel responsible to remind them.	Personal norm activation theory [16]
Symbolic motivation	SM	My environmental behaviors are intended to show my values to others.	Symbolic consumption theory [17]
Relevant knowledge level	RKL	Before answering this questionnaire, did you truly understand the specific meaning and function of carbon labels?	Knowledge level moderator variable [11]
Moral satisfaction	MS	Purchasing carbon-labeled products makes me feel that I have contributed to environmental protection.	Theory of planned behavior - attitude and behavioral intention [18]

In this study, the term idealized public persona is operationalized as Perception of Personal Image about Environmental (PPIE). This transformation reflects the study's focus on respondents' perception of their own self-image, which aligns with the respondents' habitual ways of self-representation rather than using behaviorally defined labels imposed by external observers. It also clarifies that the study's survey questions are framed from the perspective of how respondents perceive their own image, rather than asking what kind of role they are playing—even though the underlying connotations of the two concepts are largely consistent.

Cognitive dissonance theory posits that individuals tend to maintain cognitive consistency. Specifically, when a discrepancy arises between their behaviors and cognitions, people experience a state of psychological tension and subsequently engage in behaviors aligned with their cognitions to alleviate such discomfort. However, whether individuals are still influenced by their self-image perceptions in the context of purchasing carbon-labeled products—and whether they will promote the purchase of such products to uphold their self-image—remains to be empirically tested and verified.

Owing to the lack of well-established scales for measuring self-image perceptions, this study was designed strictly on the basis of classical theories and certain prior research findings (see Table 2). In the final analysis, variables were included or excluded according to their factor loadings and the significance of the corresponding paths.

2.4 Perception of Fulfillment of Responsibility Status: A Supplementary Factor for Investigation

Beyond users' behaviors on online platforms themselves, the real-life environment is often regarded as a strong influencing factor [13]. Although this perspective falls outside the core scope of the present study, it is still deemed particularly valuable for consideration.

Based on the circle division specified in the social-ecological model proposed by American psychologist Urie Bronfenbrenner, this study designs the questionnaire items from three perspectives: government (Peripheral Circle) → family (Core Circle) → individual (Absolute Core Circle). Taking Perception of Fulfillment of Responsibility Status (PFRS) as the entry point, It assesses consumer attitudes toward the attribution of environmental responsibility. The specific items are set as follows (see Table 3).

Table 3. Specific Variables of PFRS

Full name	Abbr.	Item description	Circle
Government responsibility fulfillment	GRF	I believe environmental protection is mainly the government's responsibility.	Government
Family responsibility fulfillment	FRF	I believe my family members lack a strong sense of environmental protection.	Family
Individual responsibility fulfillment	IRF	I believe my environmental protection actions are insignificant.	Individual

2.5 Purchase Intention Measurement: Image Display Without Numerical Values

The measurement of purchase intention designed in this study is relatively clear, and the measurement is only determined from the perspective of behavioral intention. Before the question was asked, the questionnaire showed the respondents an introduction interface[1] (see Fig. 1) to help them recall the basic role of carbon labels.

The specific question is: "After understanding the concept of carbon label, suppose you encounter two similar products with the same price, but model A has A carbon label (indicating the carbon emissions), what is the likelihood that you will choose to buy model A?" When respondents choose a higher score, they believe that they have a higher purchase intention for products with carbon labels.

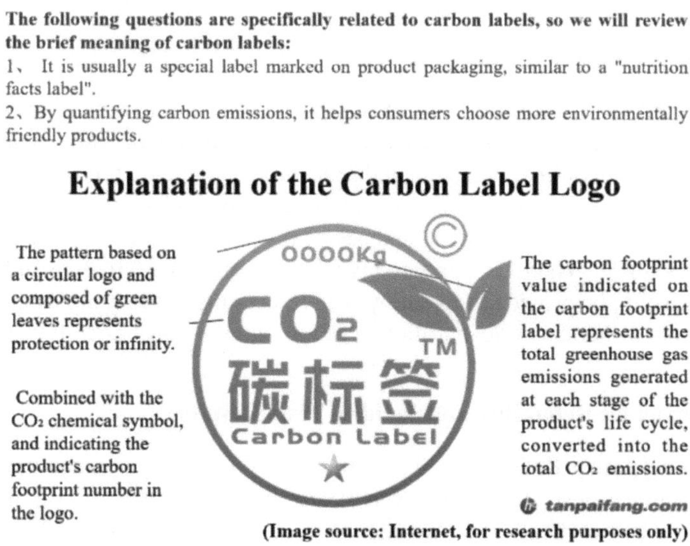

Fig. 1. Interface for Displaying Purchase Intention of Carbon-Labeled Products

2.6 Path Hypotheses Propositions

H1a: The POEA factor positively influences purchase intention for carbon-labeled products.

H1b: The POEA factor positively influences the PPIE factor.

H1c: The PPIE factor positively influences purchase intention for carbon-labeled products.

The above hypotheses were verified by Structural Equation Model (SEM). To compensate for the insufficiency of structural equation model in revealing the detailed correlation among observed variables, and in light of the realistic condition that the scales applied in this study have not undergone systematic development, the study further proposed the following path hypotheses for the relationships among observed variables:

H2a: Some variables in the PPIE factor mediate the impact of partial variables in the POEA factor on purchase intention.

[1] The original interface was in Chinese, and the interface in the text is the one translated into English.

H2b: Some variables in the POEA factor exert a direct effect on purchase intention.

The above analysis will derive a relational model among observed variables. Finally, based on the focus on the three variables in the PFRS factor, the study further proposes that:

H3: Partial variables in the PFRS factor moderate the effective paths proposed in Hypotheses H2.

3 Reliability Test, Validity Test and Data Preprocessing

3.1 Sample Collection Results

Research samples were collected by entrusting an online questionnaire collection platform named Wenjuan Network. After excluding the samples of respondents who had never participated in online environmental protection activities, a total of 488 valid samples were retained for the final analysis, and the sample precision was suitable for most exploratory surveys. The main data analysis of this study was conducted by using the software SPSSPRO. Where data processing was performed with Python and R, specific annotations would be provided separately. The test results of the samples indicated that the demographic characteristic distribution of the samples was basically reasonable.

3.2 Fundamental Reliability and Validity Tests Results

Internal Consistency Reliability. The standardized Cronbach's α coefficient derived from the combined test of all scales was 0.929, indicating that the overall reliability of the scales was excellent. The Cronbach's α coefficient test was conducted for each dimension of the questionnaire scales (see Table 4). All the coefficients were greater than 0.7, which confirmed that each measurement scale had good internal consistency.

Confirmatory Factor Analysis Validity. The overall validity of the scale yielded a KMO value of 0.943 and passed the Bartlett's test of sphericity, indicating excellent overall validity of the scale. The validity test results of each measurement scale (see Table 4) showed that both the POEA and PPIE factors had acceptable KMO values and passed the Bartlett's test of sphericity, confirming that the data were suitable for factor analysis.

However, when the three variables in the PFRS were combined into a single factor, its KMO value was significantly lower than those of the other factors. Therefore, the study concluded that this factor was not suitable for inclusion in factor analysis. Meanwhile, the low KMO score also verified that the variables classified based on the circle division in the social-ecological model were relatively independent of one another.

Table 4. Fundamental Reliability and Validity Tests Results.

Factor	Cronbach's α	Standardized Cronbach's α	KMO value	Bartlett's test of sphericity
POEA	0.936	0.935	0.949	0.000***
PPIE	0.774	0.778	0.815	0.000***
PFRS	0.819	0.819	0.699	0.000***

Note: *** indicates p < 0.001.

3.3 Results of Test-Retest Reliability Analysis

The Intraclass Correlation Coefficient (ICC) is a reliability coefficient index used to measure and evaluate inter-observer reliability and test-retest reliability. It can typically be applied in questionnaire surveys to assess the test-retest reliability of one subject across multiple samples over a period of time, or to conduct a consistency test of a group of subjects across multiple samples. An ICC value greater than 0.75 indicates good data reliability. The intraclass correlation coefficient was measured using SPSSPRO software, and the results (see Table 5) showed that the data of POEA, PPIE, and PFRS all had high intraclass reliability.

Table 5. Test-Retest Reliability Results

Factor	Intraclass Corr.	95% CI (L)	95% CI (U)	F	df_1	df_2	p
POEA ICC(1,1)	0.58	0.561	0.63	15.704	487	4392	0.000***
POEA ICC (1,k)	0.933	0.928	0.944	15.704	487	4392	0.000***
PPIE ICC(1,1)	0.394	0.364	0.451	4.425	487	1952	0.000***
PPIE ICC (1,k)	0.765	0.741	0.804	4.425	487	1952	0.000***
PFRS ICC (1,1)	0.585	0.555	0.645	5.518	487	976	0.000***
PFRS ICC(1,k)	0.809	0.789	0.845	5.518	487	976	0.000***

Note: *** indicates $p < 0.00$; ICC(1,1) = Single measurement, ICC(1,k) = Average measurement.

3.4 Results of Average Variance Extracted and Composite Reliability

The Average Variance Extracted (AVE) is also a statistical measure for examining the internal consistency of construct variables, while Composite Reliability (CR) reflects whether all items within each latent variable consistently explain that latent variable. Generally, an AVE value higher than 0.5 or a CR value higher than 0.7 indicates high convergent validity, and only one of the two metrics needs to be examined. According to the results (see Table 6), the composite reliability among the variables of each scale was relatively high.

Table 6. AVE and CR Results.

Factor	AVE value	CR value	Item no.
POEA	0.629	0.942	10
PPIE	0.42	0.777	5
PFRS	0.624	0.829	3

3.5 Results of Construct Validity Test

Based on the factor loading coefficient table of the scale (see Table 7), the measurement items of each factor were significant at the statistical level, thus rejecting the null hypothesis. With the exception of the two variables (MS and EPG), the standardized factor loadings of all other variables were greater than 0.6. Variables MS and EPG were therefore deleted for subsequent analyses of the study.

Table 7. Construct Validity Test Results

Factor	Var.	Unstandardized loading	Standardized loading	z	S.E.	p
POEA	AS	1	0.87	–	–	–
	SR	0.738	0.719	19.151	0.039	0.000***
	IE	0.844	0.793	22.386	0.038	0.000***
	SI	0.92	0.831	24.321	0.038	0.000***
	CC	1.095	0.844	25.039	0.044	0.000***
	PH	0.879	0.772	21.429	0.041	0.000***
	EPG	0.379	0.426	9.802	0.039	0.000***
	EPI	0.933	0.806	23.055	0.04	0.000***
	OA	1.143	0.837	24.666	0.046	0.000***
	PWD	0.855	0.778	21.67	0.039	0.000***
PPIE	SC	1	0.654	–	–	–
	CRA	1.142	0.673	12.731	0.09	0.000***
	SM	1.143	0.604	11.612	0.098	0.000***
	RKL	1.484	0.757	13.99	0.106	0.000***
	MS	0.826	0.491	9.681	0.085	0.000***

Note: *** indicates $p < 0.001$; – represents fixed parameter with no value.

3.6 Data Preprocessing

Description of Data Distribution Characteristics. The Shapiro-Wilk (S-W) test results indicated that all data were significant, suggesting that the sample did not strictly conform to a normal distribution. Meanwhile, most variables exhibited high skewness, with absolute values close to or even exceeding 1. Given that all variables showed a left-skewed distribution and a mode of 6, it can be concluded that consumers generally tended to select higher values.

Data Treating. With reference to the validation and recommendations of the method proposed by Nwakuya, this study ultimately adopted the Box-Cox method for data transformation [14].

To ensure the consistency of data dimensions, this study further performed Min-Max normalization on the transformed data.

After the processing, the characteristics of data distribution were significantly improved. Although the S-W test results remained statistically significant for all variables, the skewness values were markedly reduced. Combined with the distribution plots, the transformed data could be reasonably accepted as approximately following a normal distribution. Meanwhile, this study verified that the factor loadings of the transformed data remained unchanged, thus the transformed results were adopted for subsequent analyses.

4 Results of Model Construction

4.1 Results of Structural Equation Model

Based on the hypotheses related to H1, the screened factor variables were incorporated into the structural equation model for testing (see Fig. 2). The results showed that Hypothesis H1a was not supported, while Hypotheses H1b and H1c were supported. According to the impact coefficients, the path from POEA to Purchase Intention was not significant. Therefore, the actual effective path was "POEA→PPIE→Purchase Intention", indicating that PPIE exerted a full mediating effect in the relationship between POEA and Purchase Intention. Combined with the model fit evaluation results (see Table 8), this path can be regarded as acceptable.

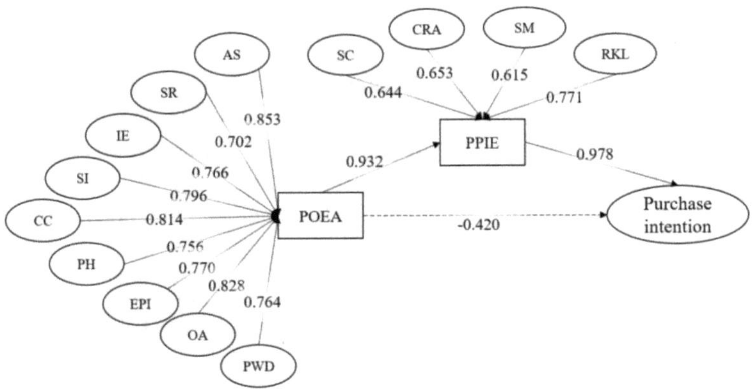

Fig. 2. Structural Equation Model Diagram

Table 8. Model Fit Evaluation Results of SEM

Fit index	Optimal range	Actual score	Fit index	Optimal range	Actual score
χ^2/df	<3	4.551	GFI	>0.9	0.922
RMSEA	<0.08	0.085	RMR	<0.05	0.003
CFI	>0.9	0.938	NFI	>0.9	0.922
NNFI	>0.9	0.923			

4.2 Variable Extraction: A Path Model Was Constructed with Variables as the Minimum Observation Units

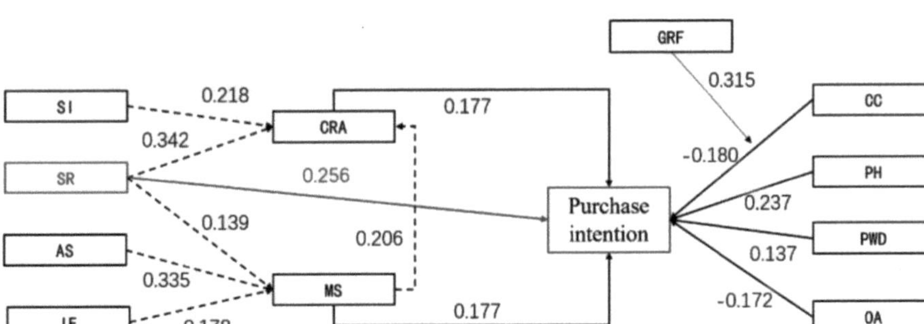

Fig. 3. Variable Path Model Diagram

Table 9. Model Fit Evaluation Results of Optimal Path Model

Fit index	Optimal range	Actual score	Fit index	Optimal range	Actual score
χ^2/df	<3	1.640	GFI	>0.9	0.973
RMSEA	<0.08	0.036	RMR	<0.05	0.001
CFI	>0.9	0.997	NFI	>0.9	0.992
NNFI	>0.9	0.987			

Based on the results of the total effect analysis via the structural equation model, the mediating effect of PPIE in the model was generally verified. However, the structural equation model is essentially a path analysis that integrates variables into latent factors by means of the maximum likelihood estimation method. It evaluates the structural relationships among latent variables, thus providing overall guiding significance. In contrast, it fails to fully reveal the relationships among observed variables.

Therefore, this study introduced more refined path analysis methods to explore the path relationships among variables—parallel mediation test and serial mediation test—and ultimately obtained the model with the best data fit (see Fig. 3).[2] Variables in the PFRS were incorporated into the model. The model results were extremely satisfactory, exceeding expectations (see Table 9).

5 Conclusion

This study takes improving the sales of products with carbon labels through efficient online promotion strategies, and further alleviating the burdens on enterprises and consumers in the low-carbon transition, as its starting point. Using path analysis methods such as structural equation modeling, it concludes that Perception of Online Environmental Activities has a significant positive effect on increasing product purchase intention (i.e., their low-carbon consumption decision-making process) through both direct and indirect pathways, with Perception of Personal Image about Environmental serving as the key mediating factor.

The results indicate that higher perceived frequency of Content Creation and environmental activities leads to greater resistance, and the perceived frequency of Organize Activities also negatively impacts purchase intention toward carbon-labeled products. Although these two negative impact paths are counterintuitive, this study argues they can be explained by Piaget's theory of cognitive development stages. This theory posits that individuals gradually construct a systematic understanding of things through assimilation and accommodation mechanisms, and consumers' cognition of products becomes more systematic as information accumulates and experience grows. As cognitive levels improve, consumers develop more complex evaluation systems, shifting their product requirements from a single attribute to a multidimensional, comprehensive assessment.

[2] The coefficient of the path from CC to Purchase Intention affected by CRF is independent, rather than the direct impact of CRF on Purchase Intention.

In the questionnaire measurement of this study, respondents were only asked to express their purchase intention for products with the single attribute of a carbon label. In contrast, those who organize environmental activities frequently and environmental content creators generally exhibit higher-than-average sensitivity to environmental topics. With increased practice, they express skepticism and caution toward products with only a carbon label, leading to relatively lower ratings. Conversely, general consumers with less in-depth practice tend to be more radical and provide higher purchase intention scores under the influence of social supervision mechanisms, ultimately resulting in the counterintuitive findings of this data analysis. This suggests that enterprises and governments should adopt multidimensional feature combinations in promoting carbon-labeled products, coordinate promotion with related or similar concepts, or expand the practical value of carbon labels.

Based on the core findings, this study further argues that the facilitating effect of Perception of Personal Image about Environmental on purchase intention should be leveraged effectively, particularly in developing detailed strategies to drive consumers' Collective Responsibility Attitude and Self-Categorization. This will amplify the impact of users' participation in basic environmental activities on platforms (including: System Recommendation, Share Information, Active Search, and Interactive and Engagement). Additionally, efforts should be made to encourage more potential consumers to participate in environmental public welfare initiatives, thereby increasing purchase intention through practical activities.

References

1. The State Council of the People's Republic of China: Notice of the State Council on Issuing the Action Plan for Carbon Peaking Before 2030 (in Chinese). https://www.gov.cn/zhengce/content/2021-10/26
2. Zhang, S.Y.: Research on the Formation Mechanism of Green Premium Payment Willingness and Green Business Models (in Chinese). Beijing University of Posts and Telecommunications (2022). https://doi.org/10.26969/d.cnki.gbydu.2022.000857
3. Zhao, R.: Opportunities and challenges of carbon label practice from the perspective of stakeholder response (in Chinese). J. Sichuan Normal Univ. (Nat. Sci. Ed.). **44**(06), 711–719+708 (2021). https://doi.org/10.3969/j.issn.1001-8395.2021.06.001
4. Chen, D.J., Gao, G.R.: How green premium promotes industrial green transformation: from the perspective of evolutionary game (in Chinese). World Surv. Res. **1**, 42–56 (2025). https://doi.org/10.13778/j.cnki.11-3705/c.2025.01.004
5. National Bureau of Statistics of the People's Republic of China: Communiqué of the Third National Time Use Survey (in Chinese). https://www.stats.gov.cn/sj/zxfb/202410
6. Li, Q., Long, R., Chen, H.: Empirical study of the willingness of consumers to purchase low-carbon products by considering carbon labels: a case study. J. Clean. Prod. **161**, 1237–1250 (2017). https://doi.org/10.1016/j.jclepro.2017.04.154
7. Xiong, H., Hu, W., Xu, M., et al.: Revisiting heterogenous social desirability bias in consumer willingness to pay for food carbon label: social norms and environmental concerns. Food Policy. **128**, 102690 (2024). https://doi.org/10.1016/j.foodpol.2024.102690
8. Wang, J., Meng, Y.: Research on the premium payment level and influencing factors of carbon-labeled products (in Chinese). J. Mudanjiang Normal Univ. (Soc. Sci. Ed.). **6**, 37–45 (2022). https://doi.org/10.3969/j.issn.1003-6121.2022.06.004

9. Mei, L., Sun, L.F., Zhang, P., et al.: The impact of carbon label information dissemination on consumers' green purchase behavior: Evidence from eye-tracking experiments (in Chinese). For. Econ. **45**(05), 75–96 (2023). https://doi.org/10.13843/j.cnki.lyjj.20230712.001
10. Mei, L., Sun, L.F., Li, W., et al.: Research on the path of carbon labels' influence on low-carbon purchase intention: a moderated mediation effect based on pro-social behavior (in Chinese). Chin. J. Environ. Manag. **15**(05), 117–128 (2023). https://doi.org/10.16868/j.cnki.1674-6252.2023.05.117
11. Liu, Y.T., Wang, X.M.: Reflection and regulation of information cocoon effect under algorithm recommendation mechanism: taking Toutiao as an example (in Chinese). Media. **13**, 38–40 (2024)
12. Zeng, Y.G., Chen, S.: Digital survival and persona display: research on self-image construction in platform society (in Chinese). Future Commun. **30**(1), 19–27 (2023). https://doi.org/10.13628/j.cnki.zjcmxb.2023.01.006
13. Liu, X.Q.: Irrational online public opinion field and its challenges and suggestions for social governance (in Chinese). Social Gov. **7**, 85–91+10 (2022). https://doi.org/10.16775/j.cnki.10-1285/d.2022.07.008
14. Nwakuya, M.T., Nwabueze, C.C.: Application of box-cox transformation as a corrective measure to heteroscedasticity using an economic data. Am. J. Math. Stat. **8**, 8–12 (2018). https://doi.org/10.5923/j.ajms.20180801.02
15. Tajfel, H.: The social identity theory of intergroup behavior. In: Austin, W.G., Worchel, S. (Eds.) The social Psychology of Intergroup Relations, pp. 33–47. Brooks/Cole (1979)
16. Schwartz, S.H.: Normative influences on altruism. In: Berkowitz, L. (Ed.) Advances in Experimental Social Psychology, Vol. 10, pp. 221–279. Academic Press (1977)
17. Griskevicius, V., Tybur, J.M., Van den Bergh, B.: Going green to be seen: status, reputation, and conspicuous conservation. J. Pers. Soc. Psychol. **98**(3), 392–404 (2010). https://doi.org/10.1037/a0017346
18. Ajzen, I.: The theory of planned behavior. Organ. Behav. Hum. Decis. Processes **50**(2), 179–211 (1991). https://doi.org/10.1016/0749-5978(91)90020-T

Decision Support System Technology for Sustainable Workforce Transition in Human-Centric Last-Mile Supply Chains

Karim Soliman[1,2](✉)

[1] University of Business and Technology, Jeddah, Saudi Arabia
k.soliman@ubt.edu.sa
[2] Arab Academy for Science, Technology, and Maritime Transport, Cairo, Egypt
karim.mohamed@aast.edu

Abstract. Decision support system (DSS) technologies are increasingly integrated into last-mile supply chains, thereby reshaping the generation, communication, and execution of operational decisions. Although these technologies present their own efficiency gains, they also present massive workforce- transition issues, especially in the context of seeking sustainable and resilience-based transformation. Although DSS has a central influence in such transitions, the empirical literature has limited the understanding of how such systems enable workforce preparedness and competence in times of change. This research discusses the role of DSS technology in promoting sustainable workforce transition in human-centric last-mile supply chains using Saudi Arabia as the empirical context. A quantitative research design was chosen, and the data were gathered among professionals who were involved in last-mile operations. The relationships between DSS technology and workforce-transition readiness (WTR) and the relationship between DSS technology and workforce-transition capability (WTC) were measured using the Partial Least Squares Structural Equation Modeling (PLS-SEM). Findings indicate that DSS technology significantly enhances WTR and WTC, which in turn contribute to social sustainability (SS) and operational resilience (OR). Additionally, WTR preparedness emerges as a key mechanism linking DSS technology to SS outcomes. This research contributes to DSS research by proposing a transition-based and human-centric concept using DSS in last-mile supply chains. It positions WTR as a central explanatory mechanism, thereby moving beyond the traditional view of DSS as a passively adopted technology.

Keywords: Decision Support Systems · Workforce Transition · Human-Centric Supply Chains · Last-Mile Logistics · Sustainability

1 Introduction

The digitalization of supply chain operations accelerated by new technology has led to a tremendous adoption of DSS technologies to improve planning, coordination, and controlling operations, especially in the realm of last-mile logistics. These technologies

A. P. Cabral Seixas Costa et al. (Eds.): ICDSST 2026, LNBIP 582, pp. 18–32, 2026.
https://doi.org/10.1007/978-3-032-25146-6_2

increasingly affect the way in which decisions are generated, evaluated, and implemented, and the impact on operational efficiency is not the only influence on the roles and responsibilities and interaction of the workforce. In last-mile supply chains - characterized by labor intensity, time sensitivity and direct customer interaction - DSS technologies have become central mechanisms to manage complexity and uncertainty [1, 2]. Although previous studies have discussed extensively the application of DSS in optimizing routing, demand forecasting, and improving performance in the supply chain, most of the existing literature has been technology-centric or performance-oriented. Comparatively less attention has been given to the human implications of DSS-mediated transformation, and how such systems alter the dynamics of the workforce [3, 4]. The integration of data-driven analytics, automation, and real-time decision infrastructures changes decision authority, task allocation, and needed skill profiles, and thus creates significant challenges for workforce transition. These challenges go beyond the system adoption and include employees' willingness to engage with the DSS outputs, their trust in the recommendations of these systems, and their ability to adapt to the reconfigured decision-making processes and organizational support mechanisms [5]. Recent developments in supply chain digitalization and Industry 5.0 focus on the importance of human-centered decision environments, arguing that sustainable and resilient transformation cannot be reached by means of techno-logical advancement only [6, 7]. Rather, DSS technologies need to be designed and implemented in ways that actively facilitate workforce transition, including initiatives to reskill workforces, to redefine roles, and to promote fairness and transparency in decision processes. Nevertheless, empirical DSS research has often treated factors related to workforce as contextual or secondary factors rather than central mechanisms through which technological transformation yields long-term sustainability out-comes [8]. At the same time, sustainability in supply chains is increasingly conceptualized as a multidimensional construct that covers not only environmental performance but SS and OR [9]. Within the scope of last mile logistics, SS is closely linked with job continuity, fair workload distribution as well as skill development opportunities, whereas OR is the capacity for uninterrupted continuity of service under conditions of disruption and operational variability [10]. DSS technologies may contribute to both dimensions by allowing for informed, transparent, and adaptive decision making. However, the degree to which these technologies are linked with sustainable outcomes is very much down to their ability to support effective workforce transition at times of change [11]. This problem is especially relevant in the Saudi Arabian context, where the rapid growth in e-commerce and logistics services and the country's digital transformation initiatives have accelerated the implementation of data-driven DSS technologies in last-mile operations [12]. With more organizations in the Saudi logistics sector relying on advanced analytics and decision infrastructures, it's becoming important to know how these technologies relate to workforce transition and realize sustainable and resilient supply chain transformation. Despite this practical relevance, empirical studies investigating DSS-enabled workforce transition in this context are still scarce. To fill these gaps, this study focuses on exploring the role of DSS technology in sustainable workforce transition in human-centric last-mile supply chains in Saudi Arabia. Specifically, it conceptualizes WTR and WTC as central explanatory mechanisms to which DSS technology is associated to achieve social sustainability and operational resilience outcomes. Theoretically, this study contributes to the advancement of DSS research in three aspects. First, it changes the analytical focus from technology adoption to workforce transition processes, which extends traditional DSS perspectives that focus more on system functionality and decision performance [13]. Second, it integrates socio-technical and sustainability-oriented logics through a conceptual integration of workforce transition as the mechanism of linking digital decision infrastructural to long-term social and resilience results. Third, it presents WTR not just as an attitudinal construct, but as a process-based construct that measures cognitive and behavior preparedness throughout digital transformation. By developing a transition-oriented DSS perspective, the study moves beyond static paradigms of adoption and adds to a more dynamic and human-incentive understanding of the relation between decision technologies and sustainable transformation within labor-intensive supply chain contexts [13].

2 Literature Review and Hypotheses Development

2.1 Decision Support System Technology in Last-Mile Supply Chains

DSS are designed to support decision makers by bringing together data, analytical models, and easy-to-use interfaces that facilitate the semi-structured and unstructured decision making processes [1, 2, 14]. In the field of supply chain management, DSS technologies have made a journey from static planning utilities to dynamic and data-driven platforms with the potential for real-time and predictive decision-making [3]. In the case of last mile supply chains, these technologies are increasingly used to solve routing complexity, demand volatility, service level issues and operational uncertainty [15]. It is known that the last mile is generally considered as one of the most complex and labor-intensive parts of the supply chain, where the decisions related to task allocation, delivery sequencing, and resource utilization require quick and frequent resolving [16]. DSS technologies help to make these decisions easier with analytics-driven insights, scenario evaluations, and coordination across operational actors. Empirical studies have revealed that the use of DSS in last mile logistics improves operational efficiency and responsiveness; however, these research studies mainly focus on operational performance outcomes while overlooking workforce related implications [17]. From the point of view of DSS research, recent scholars highlighted the need to move away from tool-centric conceptualizations and to understand DSS as decision infrastructures that shape the understanding, trustworthiness, and enactment of decisions by human actors [13]. This paradigm shift is particularly relevant in last-mile contexts, where workers are in direct contact with DSS outputs, and where the quality of decisions made depends on the effective collaboration of humans and systems, rather than on automation.

2.2 Workforce Transition and Human-Centric Decision Support

Workforce transition refers to the process by which employees adapt to changes in roles, decision power and skill requirements caused by technological and organizational transformation [5]. Contrary to technology adoption, which focuses on a preliminary acceptance or utilization, workforce transition emphasizes on sustained adaptation, capability development, and role reconfiguration across the organizational lifecycle. Within environments that are supported by DSS, workforce transition is intricately connected with the redistribution of decision authority between human agents and computational systems [6]. In the context of Industry 5.0 human-centric DSS has emerged as a key principle to advocate technologies for augmenting, not replacing, human decision-making processes [6]. Under this paradigm, the expectation is that DSS technologies will enable transparency, explainability and active user participation, building trust and creating effective engagement. Empirical studies of human-automation interaction show that employees' tendency to use DSS output is dependent on their understanding of system logic and their perceived ability to intervene when necessary [18]. Furthermore, the existing literature identifies WTR and WTC as significant dimensions of workforce transition. Moreover, WTR reflects employee psychological and cognitive preparedness to interact with DSS-driven decision processes, whereas transition capability represents employee skill sets, training and adaptive capacity relative to novel workflows [19]. Despite their recognized value, these constructs have not been well-developed in DSS research, in which workforce variables are often seen as contextual moderators, rather than as central mechanisms.

2.3 Sustainability as a Transition Outcome in Last-Mile Supply Chains

Sustainability in supply chains has traditionally been conceptualized regarding environmental performance. Recent research, however, supports a more comprehensive framework that includes SS and OR [9]. SS is inclusive of equitable labor practices, continuity of employment and opportunities for skill development, whereas OR refers to a supply chain's ability to maintain service continuity in the face of disruption [10]. In last mile logistics workforce stability and adaptability are indispensable to both dimensions. Elevated turnover rates, skills mismatches and resistance to changes imposed by DSS can undermine the reliability of service and increase susceptibility to disruption [20]. DSS technologies have the potential to enable sustainable results by enabling informed and adaptive decision-making; however, the effectiveness of these tools depends on the successful management of workforce transition processes [11]. [21] argues that sustainability should be considered an emergent outcome of the transition processes and not a direct consequence of technology being deployed. This perspective implies that DSS technologies help contribute to sustainability indirectly by increasing workforce readiness and ability during times of change. Nonetheless, there is a lack of empirical evidence that supports this mechanism, especially within last mile supply chain contexts.

2.4 Hypotheses Development

Building on DSS theory and literature on workforce transition, this study conceptualizes DSS technology as an enabler for sustainable workforce transition within last-mile supply chains. DSS technology is expected to support workforce transition through increased transparency of decisions, analytical support, and coordination, thereby strengthening both transition readiness and capability.

H1: *DSS technology is positively associated with WTR.*

DSS technologies provide structured and predictive information as well as decision rationale, which has been shown to reduce uncertainty and enhance user confidence in technology-mediated environments [1, 18]. By clarifying task expectations and supporting informed decision-making, DSS is expected to strengthen workers' preparedness to participate in digitally supported decision processes.

H2: *DSS technology is positively associated with WTC.*

By supporting training, experiential learning, and adaptive work processes, DSS technologies complement employees' ability to operate effectively in digitally mediated decision environments [5, 19]. As digital transformation requires ongoing skill development and workflow adjustment, DSS-supported environments are expected to be positively associated with WTC.

H3: *WTR is positively associated with SS outcomes in last-mile supply chains.*

Employees who are prepared to engage in DSS-driven decision-making processes are more likely to experience clarity in roles, perceived fairness, and job continuity, which are core components of SS in labor-intensive supply chains [9, 20]. Workforce preparedness may therefore be associated with the maintenance of socially sustainable conditions during technological change.

H4: *WTC is positively associated with operational resilience.*

Adaptive and skilled workforces demonstrate greater responsiveness to disruption and operational variability, thereby reinforcing continuity of operations and resilience [10, 16]. When employees possess the capability to interpret and implement DSS-supported recommendations OR is more likely to be sustained under dynamic conditions. Therefore, WTC is expected to be positively associated with operational resilience in last-mile supply chains [13, 21].

H5: *Workforce transition readiness mediates the relationship between DSS technology and sustainability outcomes in last-mile supply chains.*

3 Conceptual Framework

This study adopts a transition-oriented perspective on DSS to explain how DSS technology relates to sustainable outcomes in last-mile supply chains. Traditional DSS research has predominantly conceptualized these systems as technological artefacts designed to enhance decision quality and performance efficiency [1, 13]. While such perspectives provide important insights into system design and adoption, they offer limited explanation of how DSS technologies reshape workforce structures and long-term organizational sustainability. In contrast, the present framework conceptualizes DSS technology as a decision infrastructure that reconfigures information flows, authority structures, and role expectations. Within this view, sustainable outcomes do not emerge directly from system deployment but are associated with workforce transition processes that mediate technological change. By explicitly modeling workforce transition readiness and capability as central mechanisms, the framework extends socio-technical reasoning by identifying how digital decision environments interact with human adaptation processes to shape sustainability outcomes.

As shown in Fig. 1, data-driven DSS technology is conceptualized as an exogenous construct that reflects the extent to which data-driven DSS, analytics, and decision

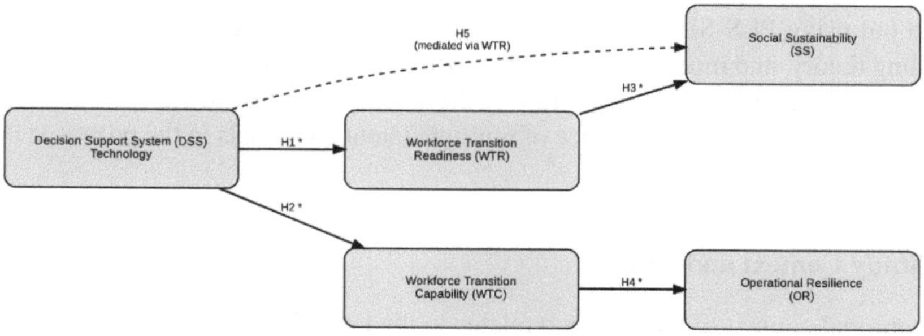

Fig. 1. The conceptual framework

guidance are embedded in the last mile operations. Workforce transition is modelled as two complementary mechanisms - WTR and WTC. Transition readiness describes preparedness, openness and willingness of workers to engage with DSS supported decision processes, while transition capability describes the skills, training and adaptive capacity needed to operationalize novel workflows and decision roles [5, 19]. Moreover, the framework conceptualizes sustainability because of successful workforce transition, instead of as a direct result of DSS adoption. Two dimensions of sustainability are highlighted, SS and OR. SS indicates job continuity, fairness and stability in the workforce - all factors of salience in labor-intensive last-mile contexts [9, 20]. On the other hand, OR refers to the ability of last-mile supply chains to ensure continuity of service and respond to disruptions in an adaptive and informed manner through decision making [10, 16]. Importantly, the framework assumes that WTR is a mediating mechanism between DSS technology and SS outcomes. This mediation reflects a process logic that suggests DSS technologies contribute to sustainability only when workers are adequately prepared to engage, trust and act on DSS outputs, while WTC complements this mechanism by enhancing the workforce capability to execute DSS-supported decisions under dynamic and uncertain operating conditions. By combining DSS technology, workforce transition mechanisms and sustainability outcomes in one framework, this research contributes towards a human-centric and transition-based conceptualization of DSS in last mile supply chains while considering the human, organizational, and sustainability dimensions in transitional contexts [13, 21].

4 Research Methodology

4.1 Research Design

This research follows a quantitative cross section research design to empirically test the proposed conceptual framework and related hypotheses. A survey-based methodology was chosen, owing to its appropriateness to the measurement of perceptual, behavioral and organizational dimensions relevant to the use of DSS and workforce transition mechanisms. This design is consistent with the extant scholarship on DSS and supply chain management that focuses on technology-enable decision making and sustainability-related outcomes using structural modelling techniques [13, 21]. Empirical analysis was carried out using PLS-SEM that is well suited to prediction-oriented research, such as extending theory, and models that include mediation effects. Furthermore, this approach is also suitable to the exploratory nature of the workforce transition in DSS-enable last-mile supply chains with the presence of multiple latent constructs in the proposed model [22, 23].

4.2 Study Context and Sample

The empirical context of the research relates to the last mile supply chain operations in Saudi Arabia where the digital transformation initiatives and workforce transitions are increasingly being experienced. Last-mile logistics in Saudi Arabia is marked by high labor intensity, time critical and often stressful delivery requirements, and an increasing reliance on decision support technologies based on data. This environment provides an appropriate environment for an analysis of the degree to which DSS technology enables workforce transitions and maintains desirable out-comes. The target population included operational and supervisory personnel involved in last-mile logistics operations in Saudi Arabia including dispatchers, route planners, delivery coordinators, and frontline supervisors. A purposive sampling strategy was utilized because respondents were to have direct experience with DSS-supported decision processes. Professional networks, logistics companies, and industry associations were contacted to distribute the survey electronically. These respondents are perfect to establish the DDS implementations as well as workforce transition dynamics. Data was obtained from a structured questionnaire, and this data was distributed electronically. Participation was totally voluntary and confidentiality of respondents was guaranteed. A total of 420 questionnaire was distributed through electronic means with 318 responses received, and 284 valid questionnaires were retained after excluding incomplete questionnaire and checking for inconsistency. This sample size is adequate based on minimum recommendations for PLS - SEM under the complexity of the proposed research model [22, 23].

4.3 Measurement Instrument

All constructs were measured using multi-item reflective scales adapted from established literature. To ensure content validity, measurement items were drawn from prior validated studies and contextually adjusted to reflect DSS-supported decision environments in last-mile supply chains. All items were measured using a five-point Likert scale ranging from 1 ("strongly disagree") to 5 ("strongly agree"). DSS technology was measured using six items adapted from decision support and digital decision infrastructure literature [1, 13, 14]. The items captured the extent of data-driven decision support, transparency of system rationale, real-time data integration, analytical capability, and coordination support. A representative item is: "Our organization uses data-driven systems to support operational decision-making." Minor wording adaptations were made to contextualize the scale to last-mile logistics. WTR was assessed using five items adapted from digital transformation and human–automation interaction research [5, 18]. These items assessed the preparedness of employees, their openness and willingness to participate in DSS-supported decision processes. A sample item is: "I feel prepared to work in

DSS-supported decision environments." Items were context refined to represent operational settings. WTC was measured from five items based on digital skills and adaptive capability literature [5, 19]. The scale measured the perceived skills, training adequacy, and the ability to adapt the workflow to DSS-enabled processes of the employees. A representative item is: "I have the digital skills necessary to effectively use DSS." Minor wording changes were made for context. SS was operationalized with six items based on research on sustainable supply chain [9, 20]. These items reflected job continuity, fairness, work-force stability, and employee well-being in DSS-enabled operations. An example item is: "DSS-supported processes contribute to job continuity." OR was measured with the help of five items derived from the supply chain resilience literature [10, 16]. The items measured service continuity, ability to respond to disruptions and ability to adapt operationally. A sample item is: "Our operations ensure service continuity in the face of disruptions." All adaptations were restricted to contextual changes in wording to make them applicable to the last-mile supply chain operations and still maintain the original conceptual meaning of the scales.

4.4 Data Analysis Technique

To empirically test the proposed framework and research hypotheses, the non-parametric PLS-SEM technique was employed [22, 23]. This methodological approach is appropriate for prediction-oriented research, particularly when modeling multiple latent constructs and mediation effects in emerging research contexts. It is also suitable given the study's objective of maximizing the explained variance of endogenous constructs [22]. SmartPLS 4 software was used, following the recommended two-step analytical procedure [22, 23]. First, the measurement model was evaluated to assess reliability and validity. Second, the structural model was examined to test the hypothesized relationships and mediation effects using bootstrapping procedures.

The measurement model was assessed prior to structural analysis to ensure the reliability and validity of the latent constructs. Internal consistency reliability was evaluated using Cronbach's alpha (α) and composite reliability (CR), while convergent validity was assessed using the average variance extracted (AVE). All constructs satisfied the recommended thresholds for reliability and convergent validity [22, 23]. Discriminant validity was examined using the heterotrait–monotrait ratio (HTMT), a stringent criterion in PLS-SEM analysis. All HTMT values were below the established cut-off levels, indicating adequate discriminant validity among the constructs [22, 23]. Following confirmation of measurement model adequacy, the structural model was evaluated to test the hypothesized relationships. Collinearity among predictor constructs was assessed using inner variance inflation factor (VIF) values, all of which were below recommended thresholds, indicating that collinearity did not bias the estimated path coefficients [23]. The explanatory power of the model was examined using coefficients of determination (R^2), and predictive relevance was assessed using the Stone–Geisser Q^2 statistic obtained through blindfolding. The results indicated acceptable explanatory power and predictive relevance across the endogenous constructs [22]. Effect sizes (f^2) were also calculated to evaluate the substantive contribution of exogenous constructs to their respective endogenous variables [22]. The direct hypotheses (H1–H4) were tested using standardized path coefficients obtained through bootstrapping with 5,000 resamples. Statistical significance was determined based on p-values and confidence intervals that did not include zero [23]. The mediating effect hypothesized in H5 was examined using bootstrapped indirect effects with bias-corrected confidence intervals. Mediation was considered supported when the indirect effect was statistically significant, and the nature of the mediation was assessed by comparing direct and indirect effects [22]. All analyses were conducted using SmartPLS with default settings. Given the cross-sectional and self-reported nature of the data, several procedural and statistical remedies were implemented to mitigate potential common method bias (CMB). Procedurally, respondent anonymity and confidentiality were assured, scale items were carefully worded to reduce ambiguity, and predictor and criterion constructs were presented in separate sections of the questionnaire to minimize evaluation apprehension. Statistically, Harman's single-factor test was conducted, and the first factor accounted for less than 50% of the total variance, suggesting that CMB is unlikely to pose a serious concern. In addition, full collinearity variance inflation factor (VIF) values were examined and found to be below the recommended threshold of 3.3, providing further evidence that common method bias does not significantly threaten the validity of the results.

5 Results

5.1 Sample Description and Screening

A total of 318 responses were collected. After excluding incomplete questionnaires, responses with excessive missing values, and cases exhibiting patterned response behavior, 284 valid observations were retained for analysis. The demographic and professional characteristics of the respondents are summarized in Table 1.

Table 1. Sample characteristics

Characteristic	Category	n	%
Gender	Male	208	73.2
	Female	76	26.8
Age	18–24	38	13.4
	25–34	142	50.0
	35–44	84	29.6
	45+	20	7.0
Role	Dispatcher/Planner	91	32.0
	Delivery coordinator	122	43.0
	Supervisor/Manager	71	25.0
Experience	<2 years	48	16.9
	2–5 years	139	48.9
	>5 years	97	34.2

5.2 Measurement Model Assessment

5.2.1 Reliability and Convergent Validity

Internal consistency reliability and convergent validity were assessed using Cronbach's alpha, CR, and AVE. As shown in Table 2, all constructs satisfied the recommended thresholds, indicating adequate reliability and convergent validity.

Table 2. Reliability and convergent validity

Construct	Items	Cronbach's α	CR	AVE
DSS	6	0.881	0.911	0.632
WTR	5	0.869	0.903	0.645
WTC	5	0.852	0.891	0.618
SS	6	0.892	0.920	0.658
OR	5	0.874	0.908	0.635

5.2.2 Discriminant Validity

Discriminant validity was assessed using the HTMT ratio. All HTMT values were below the recommended thresholds, supporting adequate discriminant validity among the constructs (Table 3).

Table 3. HTMT ratios.

	DSS	WTR	WTC	SS	OR
DSS	–				
WTR	0.628	–			
WTC	0.592	0.714	–		
SS	0.551	0.749	0.688	–	
OR	0.518	0.671	0.732	0.659	–

5.3 Structural Model Assessment

5.3.1 Collinearity and Predictive Power

Collinearity was assessed using inner variance inflation factor values, which were below recommended thresholds for all predictor constructs. The explanatory power and predictive relevance of the model were evaluated using R^2 and Q^2 values. As shown in Table 4, the model demonstrates moderate to substantial explanatory power and satisfactory predictive relevance.

Table 4. Model predictive power

Endogenous construct	R^2	Q^2 (optional)	Strength
WTR	0.395	0.286	Moderate
WTC	0.351	0.248	Moderate
SS	0.562	0.381	Substantial
OR	0.498	0.339	Moderate–substantial

5.3.2 Hypotheses Testing (Direct Effects)

The direct hypotheses (H1–H4) were tested using standardized path coefficients obtained through bootstrapping with 5,000 resamples. The results are reported in Table 5.

Table 5. Structural paths and hypothesis testing

Hypothesis	Path	β	t	p	Decision
H1	DSS → WTR	0.629	12.18	<0.001	Supported
H2	DSS → WTC	0.592	11.03	<0.001	Supported
H3	WTR → SS	0.501	9.14	<0.001	Supported
H4	WTC → OR	0.529	9.87	<0.001	Supported

As shown in Table 5, all hypothesized direct relationships are positive and statistically significant, providing strong empirical support for H1–H4. DSS technology shows a substantial and statistically significant association with both WTR and WTC, highlighting its role in shaping workforce adaptation in last-mile supply chains. In turn, WTR is positively associated with SS outcomes, while WTC contributes to OR. Collectively, these findings support the proposed transition-oriented framework by demonstrating that workforce-related mechanisms serve as key pathways linking DSS technology to sustainability and resilience outcomes.

5.4 Mediation Test (H5)

The mediating role of WTR in the relationship between DSS technology and SS (H5) was explored by using bootstrap indirect effects along with 5,000 resamples. Statistical significance of the mediation effect was determined based on confidence intervals for the bias-corrected difference in the effect size that did not include zero. The results of the mediation analysis are shown in Table 6.

Table 6. Mediation analysis (bootstrapped indirect effect)

Hypothesis	Indirect path	Indirect effect (β)	t	p	95% CI LL	95% CI UL	Decision
H5	DSS → WTR → SS	0.315	7.92	<0.001	0.238	0.396	Supported

As indicated in Table 6, the indirect effect of DSS technology on the SS through WTR is positive and statistically significant. This finding offers empirical evidence for H5 and confirms that WTR is an important mediating mechanism between DSS technology and SS outcomes. Moreover, the result underpins the transition-oriented perspective advanced in this study, indicating that the sustainability benefits of DSS technology are realized primarily through its capacity to enhance workforce preparedness for DSS-supported decision processes, rather than through direct technological effects alone.

6 Discussion

This research contributes to the DSS literature by empirically examining how DSS technology is associated with sustainability-oriented outcomes in last-mile supply chains through workforce transition mechanisms. Consistent with prior work emphasizing analytical support and decision transparency in shaping decision behavior [1, 13], the results indicate that DSS technology is positively associated with WTR and WTC, supporting a transition-oriented, human-centric perspective on DSS deployment. The findings suggest that DSS-enabled decision environments are linked not only to changes in operational decision processes but also to workforce preparedness to engage with data-driven decision practices. By strengthening WTR, DSS technology appears to be associated with reduced uncertainty surrounding technology-mediated change and higher readiness to rely on DSS-supported recommendations. In parallel, the positive association between DSS and WTC aligns with the view that digital transformation involves sustained capability development rather than a one-time adoption event [5, 18]. Regarding sustainability outcomes, the results indicate that WTR and WTC play distinct yet complementary roles. The positive relationship between WTR and SS is consistent with arguments in the social sustainability literature that emphasize workforce preparedness, perceived fairness, and role clarity as critical elements in labor-intensive supply chains [9, 19]. When employees demonstrate readiness to interact with DSS-supported decision processes, social outcomes such as job continuity and workforce stability appear more likely to be maintained during periods of operational change. These findings offer nuance to concerns that digital decision technologies per se undermine SS, and instead that the risks may be conditional on inadequate workforce preparation rather than DSS deployment per se. Similarly, the positive link between WTC and OR fits with cycles of resilience research which identifies human adaptability as a key element of resilient supply chains [10, 15]. While DSS may increase situational awareness and coordination, the findings suggest that OR is closely associated with the workforce's ability to operationalize DSS-supported decisions under dynamic and uncertain conditions. This view emphasizes the importance of considering the concept of resilience not as a strictly technological characteristic, but as an out-come created by the interaction between technologies for decision support and workforce capabilities. Most importantly, the mediation analysis is supportive of a transition-oriented logic of DSS. The statistically significant indirect effect of DSS on SS through WTR implies that workforce readiness plays an important role as a mechanism relating DSS technology and sustainability outcomes. This interpretation is consistent with conceptualizations that present sustainability because of organizational transition processes and not as a direct result of the implementation of technology [20]. By empirically identifying this mediating pathway, the study helps to move existing DSS literature beyond static adoption models and provides a more process-oriented understanding of the relationship between DSS-enabled transformation and socially sustainable outcomes in last mile supply chains. Collectively, these findings point to the central role of workforce transition in shaping the sustainability implications of DSS. Rather than replacing human judgement, DSS technology seems to run the risk of reshaping decision environments in ways that can support workforce readiness and capability when appropriately managed. This human-centered interpretation answers calls for DSS research to integrate technological, organizational, and sustainability dimensions in transition contexts [13, 20], and gives a more nuanced account of how DSS can be associated with sustainable and resilient transformation in last-mile supply chains.

7 Conclusion, Recommendations, and Limitations

This study focused on the association of DSS technology with the sustainability-related outcomes in the last mile supply chains through the workforce transition mechanisms. By taking a transition-oriented decision support perspective, the results indicate that the sustainability and resilience implications of DSS are accounted for largely through WTR and WTC and not as a direct result of system deployment. In doing so, the study goes further in a human-centric interpretation of DSS-enabled transformation, in which workforce transition processes are central to understanding when and how DSS relates to socially sustainable and resilient last-mile operations. From a theoretical perspective, this research adds to the literatures on DSS by repurposing DSS technology as a transition-enabling decision infrastructure instead of a stationary decision tool. Traditional DSS research work has been focused mainly on the system functionality, adoption and performance improvement. However, this workplaces WTR and WTC as key explanatory mechanisms between DSS technology and SS and OR. By specifically modeling workforce transition as the means to link sustainability outcomes with DSS deployment, the study brings together DSS theory with the socio-technical and sustainability perspectives. This transition-oriented approach builds on the existing paradigms of DSS in three ways: it puts the focus on dynamic workforce adaptation instead of technological implementation per se. From a managerial perspective, the findings emphasize the need to match DSS implementation with conscious workforce transition strategies. Organizations implementing DSS technologies in last-mile operations should promote WTR actively, e.g. by increasing the transparency of decisions, increasing trust in system recommendations, and involving employees in DSS-supported decision processes. At the same time, strengthening WTC by structured training, digital skills and adaptive workflow design may facilitate improvements in OR and help make organizations more responsive to disruptions. Managers are there-fore encouraged to take human-centered approaches to DSS implementation that explicitly address workforce transition processes as part of broader sustainability and resilience strategies. Despite the contributions that it has made, several limitations should be acknowledged. First, the cross-sectional research design reduces the possibilities to capture the dynamic evolution of WTR and WTC over time. Second, the use of self-reported measures may be prone to perceptual bias. Third, the empirical focus on last-mile supply chains in Saudi Arabia might limit the potential for generalizability to different institutional or geo-graphical situations. Future research could use longitudinal research designs, include objective indicators of DSS use and operational performance and test the proposed framework in different supply chain environments to further refine understanding of DSS-enabled workforce transition and sustainability outcomes.

References

1. Power, D.J.: Decision Support Systems: Concepts and Resources for Managers. Greenwood Publishing Group, Westport (2002)
2. Eboigbe, E.O., Farayola, O.A., Olatoye, F.O., Nnabugwu, O.C., Daraojimba, C.: Business intelligence transformation through AI and data analytics. Eng. Sci. Technol. J. **4**(5), 285–307 (2023)
3. Malekzadeh, M., Torabi, S.A., Sheikhalishahi, M.: A supply chain digital twin for resilience analysis under immediate and long-term disruptions: a case study in plastic parts' manufacturing. Int. J. Prod. Res. **63**(24), 10445–10467 (2025)
4. Cerchione, R., Centobelli, P., Borin, E., Usai, A., Oropallo, E.: The WISED knowledge-creating company: rethinking SECI model in light of the digital transition. J. Knowl. Manag. **28**(10), 2997–3022 (2024)
5. Gong, C., Ribiere, V.: Understanding the role of organizational agility in the context of digital transformation: an integrative literature review. VINE J. Inf. Knowl. Manag. Syst. **55**(2), 351–378 (2025)
6. Castagnoli, R., Cugno, M., Maroncelli, S., Cugno, A.: A new research agenda for human-centric manufacturing: a systematic literature review. IEEE Trans. Eng. Manag. **71**, 15236–15253 (2024)
7. dos Santos Filho, V.H., de Resende, L.M., Pontes, J.: Development of a theoretical model for digital risks arising from the implementation of industry 4.0 (TMR-I4. 0). Future Internet. **16**(6), 215 (2024)
8. Fosso Wamba, S., Guthrie, C., Queiroz, M.M., Minner, S.: ChatGPT and generative artificial intelligence: an exploratory study of key benefits and challenges in operations and supply chain management. Int. J. Prod. Res. **62**(16), 5676–5696 (2024)
9. Fritz, M.M., Ruel, S.: What does "sustainable supply chain management" really mean? A contribution to bridging the gap between research, education and practice. Int. J. Logist. Manag. **35**(2), 332–363 (2024)
10. Yanamandra, R.: Building competitive supply chains: a way through supply chain resilience. In: Emerging Trends in Decision Sciences and Business Operations, pp. 1–23. Routledge, India (2022)
11. Grieves, M., Vickers, J.: Digital twin: mitigating unpredictable, undesirable emergent behavior in complex systems. In: Kahlen, F.J., Flumerfelt, S., Alves, A. (eds.) Transdisciplinary Perspectives on Complex Systems, pp. 85–113. Springer, Cham (2017)
12. Alshammari, M., Al-Kandari, A., Al-Mutairi, N.: Digital transformation in logistics services: evidence from Saudi Arabia. Sustainability. **13**(18), 10207 (2021)

13. Onwujekwe, G., Weistroffer, H.R.: Intelligent decision support systems: an analysis of the literature and a framework for development. Inf. Syst. Front. **27**, 1–32 (2025)
14. Soliman, K., El-Barkouky, N.: An e-government procurement decision support system model for public–private partnership projects in Egypt. In: Proceedings of the International Conference on Decision Support System Technology (ICDSST); 2020 May 18–20; Cham, Switzerland, pp. 112–124. Springer International Publishing, Cham (2020)
15. Jazemi, R., Alidadiani, E., Ahn, K., Jang, J.: A review of literature on vehicle routing problems of last-mile delivery in urban areas. Appl. Sci. **13**(24), 13015 (2023)
16. Soliman, K., Bahareth, A.: Statistical analysis of green third-party logistics (3PL) and supply chain performance in Saudi E-commerce last-mile systems. J. Stat. Appl. Pro. **14**, 681–700 (2025)
17. Moradi, N., Mafakheri, F., Wang, C.: A systematic review of sustainable ground-based last-mile delivery of parcels: insights from operations research. Vehicles. **7**(4), 121 (2025)
18. Wörle, J., Metz, B.: Misuse or abuse of automation? Exploring drivers' intentions to nap during automated driving. Transport. Res. F: Traffic Psychol. Behav. **99**, 460–472 (2023)
19. González-Pérez, L.I., Ramírez-Montoya, M.S.: Components of education 4.0 in 21st century skills frameworks: systematic review. Sustainability. **14**(3), 1493 (2022)
20. Grote, G., Kolbe, M., Waller, M.J.: The dual role of adaptive coordination in teams: balancing flexibility and stability. Organ. Psychol. Rev. **8**(4), 257–277 (2018)
21. Kohle, F.H.: Green, clean and sustainable: transforming education in film, TV, and media integrating the triple bottom line into the film, TV, and media value chain in a Dutch Applied Sciences University F. Kohle, PhD, June 8, 2022. Media Pract. Educ. **23**(4), 365–387 (2022)
22. Hair, J.F.: A Primer on Partial Least Squares Structural Equation Modeling (PLS-SEM). Sage (2014)
23. Hair, J.F., Risher, J.J., Sarstedt, M., Ringle, C.M.: When to use and how to report the results of PLS-SEM. Eur. Bus. Rev. **31**(1), 2–4 (2019)

Decision Support for Alternative Protein Transitions by Assessing Cultured Meat Acceptance Risk State Under Misinformation Belief Endorsement

Guoste Pivoraite[1(✉)], Shaofeng Liu[1], Saeyeon Roh[1], and Guoqing Zhao[2]

[1] Plymouth Business School, University of Plymouth, Plymouth PL4 8AA, UK
guoste.pivoraite@plymouth.ac.uk

[2] School of Management, Swansea University, Swansea SA1 8EN, UK

Abstract. Food system transitions involve contested claims, evolving governance, and uncertain public legitimacy. For pre-commercialised technologies such as cultured meat (CM), acceptance conditions can constrain transition progress before stable markets exist, especially when public evaluation is shaped through digital information environments. This paper develops an evidence-informed decision support framework to assess CM acceptance vulnerability under misinformation belief endorsement. The framework defines endorsement as misinformation-related claims being accepted and used as a basis for judgement, distinct from simple exposure. It situates endorsement within an information environment shaped by predispositions and credibility signals, and links endorsement patterns to an acceptance risk state that summarises whether acceptance conditions appear relatively stable or increasingly fragile. For decision use, endorsement is organised into three content domains, societal engineering, health and biotech conspiracies, and market and policy deception, and each domain is linked to acceptance risk cues and broad response classes. It can help firms, regulators, and transition stakeholders align engagement, governance signalling, and communication choices to the dominant acceptance vulnerabilities under contested information conditions.

Keywords: Alternative protein transitions · Decision support · Cultured Meat · Information environment

1 Introduction

Food system transitions are increasingly understood as complex and contested processes beyond linear technological substitutions. Decisions about which innovations to support, scale, or delay are often made under conditions of uncertainty, incomplete evidence, and competing social expectations [11, 25]. In this context, transition pathways are shaped not only by technical feasibility and economic viability, but also by public legitimacy, trust in institutions, and perceptions of fairness and control [5, 8].

Alternative proteins have been positioned as one potential component of future food system transitions, yet their role seems to be neither fixed nor uncontested. Cultured meat

A. P. Cabral Seixas Costa et al. (Eds.): ICDSST 2026, LNBIP 582, pp. 33–50, 2026.
https://doi.org/10.1007/978-3-032-25146-6_3

(CM), in particular, remains a pre-commercialised technology, with ongoing uncertainty surrounding production scale, costs, regulatory trajectories, and sustainability outcomes [16, 23]. While it is often discussed in relation to environmental or ethical potential, existing evidence indicates that consumer acceptance represents a critical constraint, especially in early stages where direct experience is limited and public understanding is largely mediated through information environments [9, 28].

Research on CM acceptance has identified a wide range of influencing factors, including perceived naturalness, safety concerns, trust in producers and regulators, and affective responses such as disgust or unease [9, 18]. Comparative work across alternative protein categories further shows that acceptance drivers and barriers are not uniform, suggesting that transition decisions cannot rely on generic assumptions about "alternative protein consumers" [2, 14]. However, much of this literature has focused on explaining or predicting attitudes and intentions, with more limited attention to supporting decision-making under uncertainty.

At the same time, decision-makers in early transition contexts face an additional challenge where public perceptions are increasingly shaped within information environments characterised by misinformation, selective exposure, and distrust of institutions. Recent work has shown that misinformation effects are often overstated or poorly specified, and that simple notions of exposure obscure the importance of belief endorsement and meaning-making processes [1]. Nevertheless, evidence suggests that misinformation should not be ignored, particularly when it reinforces existing distrust or conspiracy-oriented worldviews that are resistant to factual correction [7, 26]. Reviews of audience resilience further indicate that susceptibility and resistance to misinformation are shaped by interacting individual, social, and contextual factors [13]. Information deficits alone offer an incomplete account. For pre-commercialised technologies such as CM, these dynamics can be especially consequential. When products are not yet widely available, public understanding can be largely constructed through narratives, claims, and counter-claims circulating in digital and media environments. As a result, endorsement of misinformation-related beliefs can translate into reputational risks, legitimacy challenges, and strategic uncertainty for firms and ecosystem actors long before market behaviour can be observed [11, 16]. From a transition perspective, acceptance therefore functions both as an outcome to be measured later, and as a decision-relevant risk that should be interpreted and managed in real time.

Despite this, a gap remains between consumer acceptance research and the needs of transition decision-makers. While existing studies document acceptance determinants, they rarely offer structured ways to interpret misinformation belief endorsement as a signal for decision-making, or to connect such signals to appropriate response options. This gap is particularly evident in contexts where acceptance concerns intersect with broader narratives of societal control, health risks, or institutional deception, which are difficult to address through conventional information provision alone [9, 26].

The paper develops a decision-support framework for cultured meat transition decision-making by assessing acceptance vulnerability under misinformation belief endorsement. The framework draws on two prior research outputs including a conceptual synthesis and framework paper grounded in systematic literature review, and a Twitter/X discourse analysis used for qualitative, exploratory grounding [20, 21]. The proposed framework conceptualises acceptance risk as a decision-relevant state reflecting the combined influence of knowledge, perceptions, and attitudes as they are shaped within the information environment. Rather than offering population-level behavioural prediction, the framework is presented as a risk-sensing and response-alignment heuristic for early transition decision moments. The aim is structured interpretation of belief endorsement patterns and their implications for acceptance vulnerability at early-stage transition decision points. It complements existing risk registers and stakeholder mapping tools by distinguishing between different endorsement logics (e.g., control, harm, deception) and linking them to distinct acceptance risk cues and response-option classes. The framework is not intended to delegitimise good-faith critique. It focuses on endorsement of demonstrably false or misleading claims and on how such endorsement dynamics reshape credibility constraints for decision-makers.

The paper makes three contributions. First, it reframes consumer acceptance as a decision-relevant risk in transition contexts, alongside its role as a downstream behavioural outcome. Second, it provides a structured way to connect misinformation belief endorsement to acceptance risk cues and corresponding response-option classes that can inform decision-making under uncertainty. Third, it offers a decision-support framework designed to be usable in multidisciplinary transition discussions where legitimacy, trust, and evidence uncertainty interact.

2 Literature Review and Conceptual Foundations

Cultured meat has been positioned as one possible pathway within broader protein transitions, but the direction and pace of such transitions are shaped by more than technological readiness. Decisions are made under conditions of contested sustainability claims, evolving governance, and uncertain social legitimacy, where acceptance becomes a material source of transition risk going beyond a downstream "marketing" concern [11, 16, 23]. In this setting, consumer acceptance is best understood as dynamic and information-contingent, particularly because beliefs about cultured meat can be formed and reinforced in digital information environments where misleading claims, distrust cues, and conspiratorial narratives circulate and stabilise [1, 15, 19]. Importantly, contested transition debates may also include legitimate critique and unresolved evidence questions where the challenge for decision support is to distinguish these from endorsement of demonstrably false or misleading claims without treating all dissent as misinformation.

2.1 Protein Transitions as Multi-Actor Decisions Under Uncertainty

Protein transition decisions are typically framed as systemic changes involving multiple actors (industry, civil society, regulators, and consumers) and interacting policy and market dynamics [6, 8, 11]. Within this framing, governance is often discussed through "policy mix" approaches that recognise that transitions involve interacting instruments and unintended repercussions, including distributional and legitimacy effects [11]. Questions of inclusion and legitimacy have been emphasised as particularly salient in food-system change, because transitions can be contested not only on technical grounds but also in terms of procedural fairness and voice [8].

Across transition scholarship, decision-making is characterised by uncertainty (about outcomes and legitimacy), multiplicity (stakeholders and goals), and contestation (over what counts as sustainable and acceptable change). This matters for decision support because different framings of systemic change imply different expectations about speed, disruption, and institutional reconfiguration [5], while firm and sector-level change has also been described as constrained by structural lock-ins and conflicting pressures [25]. For CM specifically, "just transition" critiques highlight that alternative protein innovation can reproduce unsustainable political-economic dynamics if justice and power are not explicitly addressed, reinforcing that legitimacy conditions can become as consequential as technical feasibility [16].

Uncertainty is especially pronounced for CM because the technology remains pre-commercialised in most settings, and evidence on sustainability performance and scalability is still developing. Recent synthesis suggests that sustainability discussions remain fragmented across environmental assessment, social and health implications, and techno-economic feasibility, creating a knowledge context that can be interpreted selectively by different actors [23]. Under these conditions, decision-makers face a dual uncertainty including uncertainty about what the technology can deliver, and uncertainty about how it will be interpreted and accepted within the public sphere.

2.2 Acceptance of Cultured Meat and Adjacent Substitutes

A substantial body of consumer research has examined determinants of acceptance, often operationalised as attitudes, willingness to try, or purchase intentions. Recent reviews converge on a familiar set of drivers and barriers such as perceived naturalness/unnaturalness, perceived health and safety risk, disgust-related responses, trust in institutions and producers, perceived benefits (e.g., animal welfare), and cost/value perceptions [9, 18, 28]. The literature also indicates variability across contexts and populations, suggesting that acceptance is not a stable trait but a contingent judgement shaped by cultural and informational conditions [24, 28].

Comparative research across alternative protein categories is particularly relevant for transition decisions because choices are rarely binary (meat vs. cultured meat); instead, portfolios are considered (plant-based, mycoprotein, insects, cultivated meat). Evidence from a nationally representative study in Singapore compared acceptance and drivers across several alternative proteins, reinforcing the point that "novel protein acceptance" is not a single phenomenon and that barriers differ by product type [14]. Preference structures also complicate adoption pathways, including how consumers categorise costs and benefits and respond to perceived externalities, implying that consumer response cannot be inferred from general pro-sustainability attitudes alone where product categories carry distinct symbolic meanings and perceived risks [2].

Communication effects are also mixed and often smaller than implied in policy and advocacy debates. Experimental evidence in Australia showed that fact-based environmental messaging did not shift attitudes or intentions toward cultivated seafood, suggesting that "more information" or sustainability claims may be insufficient when deeper concerns (trust, perceived unnaturalness, risk) dominate evaluation [3]. However, social cues and source dynamics can matter as influencer endorsements perceived as authentic and competent have been shown to affect willingness to buy cultured meat, indicating that informational social influence may shape attitude formation in high-uncertainty contexts [22]. Together, these findings point to a limitation in the acceptance literature for transition decision support: acceptance is frequently treated as an outcome of individual preferences, while information environments, trust cues, and belief formation processes are not consistently translated into decision-relevant signals [3, 18, 28].

Finally, several streams of research suggest that aversion can be partly driven by evolved threat-management responses, including disgust and contamination concerns, which may be difficult to shift through factual correction alone [10, 28]. This further supports the need to move beyond "information deficit" assumptions when considering decision support for acceptance risk.

2.3 Information Environments, Misinformation, and Belief Formation

Misinformation research provides conceptual and methodological cautions for integrating misinformation into acceptance assessment. Misinformation is not a singular, easily measured exposure as definitions vary and public concern can exceed what is empirically demonstrated in some contexts, creating risks of over-claiming effects or mis-specifying mechanisms [1]. At the same time, arguments that misinformation can be ignored have been challenged on conceptual and evidential grounds, including the claim that low prevalence in some measurements does not imply negligible harm, especially when effects are concentrated in high-stakes domains or among vulnerable groups [7]. For transition decision making, the implication is that neither alarmism nor dismissal is adequate. A mechanism-oriented approach is needed that clarifies how particular belief forms become credible and how they shape evaluations in specific contexts. This also implies that acceptance-related controversy should not be treated as "misinformation" by default as the decision-relevant question is when claims become endorsed as judgement anchors and thereby constrain credibility and legitimacy.

The concept of an information environment is increasingly used to capture the broader ecosystem shaping belief formation, including sources, signals, social networks, and institutional responses. In health policy literature, the information environment has been argued to function as an independent social determinant, influencing beliefs and behaviours in ways analogous to other structural determinants [19]. Empirical work in communication further shows that properties of digital information environments (including processing dynamics and presumed influence) can shape behavioural responses to misinformation, reinforcing the relevance of environment level conditions beyond individual traits [27]. A systematic synthesis also suggests that resilience or vulnerability to disinformation is multi-level, spanning individual dispositions, media use, trust, cultural context, and environmental factors, while noting that audience resilience is often invoked without consistent definitional clarity [13]. These points align with transition contexts where distrust in institutions and perceived manipulation can be central features of discourse around food technologies.

A further foundation concerns predispositions that shape whether misinformation is interpreted as plausible and worth endorsing [4, 26]. Technology-mediated processes can reinforce conspiracy mindsets, suggesting that belief systems can become self stabilising and hard to correct [26]. For assessment purposes, this supports treating endorsement as the proximal mechanism of interest because exposure or knowledge alone may not explain acceptance outcomes. Food-specific scholarship similarly highlights that misleading information, established misconceptions, and "fake news" can shape perceived healthiness and acceptability, interacting with longstanding psychological and social drivers rather than operating as a simple additive effect [15].

2.4 Research Gaps

Despite this evidence base, translation into decision-useful interpretation at concrete transition moments remains limited. Consumer acceptance research has generated substantial evidence on factors associated with attitudes, willingness, and intentions, yet it remains primarily organised around explanatory outcomes instead of structured guidance for decision teams [18, 28].

This means that where communication effects are examined, results can be mixed, indicating that acceptance risk cannot be assumed to shift reliably through generic information provision alone [3]. Second, because CM remains pre-commercialised, sustainability evidence is still developing across environmental, social, and techno-economic dimensions, creating uncertainty that can interact with trust and legitimacy concerns in public debate [16, 23]. Third, misinformation scholarship increasingly emphasises the need for conceptual clarity and mechanism specification, distinguishing endorsement processes from vague notions of exposure, yet these mechanisms do not seem to be consistently integrated into alternative protein acceptance discussions in ways that could be directly usable for transition decision support [1, 7]. Existing decision tools such as risk registers or stakeholder mapping may record "concern" but they typically do not distinguish endorsement logics that imply different acceptance vulnerabilities and require different credible response options. Finally, transition scholarship highlights legitimacy, inclusion, and systemic constraints, but consumer-facing information dynamics are not always incorporated into transition decision framings despite their potential to destabilise legitimacy and slow adoption pathways [8, 19, 25].

3 Evidence Base and Framework Development Approach

3.1 Evidence Base

The framework was developed using two main evidence sources, supplemented by targeted literature aligned with the conference focus. First, it draws on prior conceptual work that synthesised the CM acceptance literature and organised it into a decision-relevant framework intended for early-stage commercialisation and transition contexts [20]. That work established acceptance as multi-dimensional, extending beyond product attributes to include affective and contextual influences that shape evaluation when direct experience is limited [20]. In the present paper, this conceptual foundation is carried forward, with greater emphasis placed on the role of information conditions and belief formation in high-uncertainty settings, reflecting recent scholarship in misinformation and information-environment research [1, 19].

Second, the framework is informed by a Twitter/X discourse analysis undertaken as part of the wider research project [21]. Twitter/X was treated as a naturally occurring space in which claims, doubts, and legitimacy challenges around cultured meat are articulated and contested in real time. Topic modelling was used to structure a large volume of posts, followed by qualitative interpretation to examine recurring claim types and justificatory patterns in everyday language [21]. Topic clusters were then grouped into higher-order topic families and interpreted iteratively to derive the three endorsement domains used in this paper. This evidence was not treated as representative of public prevalence and it was used to identify salient belief patterns that plausibly shape how CM is evaluated under contested information conditions. To support transparency without reproducing raw posts, Table 1 in Sect. 5 provides paraphrased endorsement markers that illustrate the domain logic.

In addition, the framework was refined through targeted engagement with recent literature emphasising transitions as contested, multi-actor processes where legitimacy and trust shape feasible pathways, and where decision-making often proceeds under uncertainty [8, 11, 25]. Together, these inputs supported an evidence-informed approach in which well-established acceptance determinants were retained, while discourse-grounded belief patterns were used to foreground information-environment dynamics that can destabilise acceptance conditions.

3.2 Development Process

Framework development followed an iterative conceptual development approach rather than empirical model testing. It was designed for early transition decisions around CM, where decision teams may often interpret evolving public narratives and legitimacy pressures without relying on mature market indicators. For that reason, the framework is oriented toward interpreting acceptance vulnerability as a decision-relevant risk state, with explanatory and predictive modelling reserved for subsequent work. The framework retains a simple structure to support practical use. It treats the information environment as the context shaping what is regarded as credible and actionable, and it focuses on misinformation belief endorsement as the pathway through which that context can translate into acceptance vulnerability [1, 19]. Endorsement is organised into three content domains including societal engineering beliefs, health and biotech conspiracy beliefs, and market and policy deception beliefs, based on recurring discourse patterns and supported by literature emphasising that vulnerability and resistance to misleading information reflect interacting individual, social, and contextual factors [13]. The domains are treated as analytically distinct because they differ in their dominant evaluative logic (autonomy/control, harm/contamination, and governance illegitimacy/capture), even where specific topics may overlap. To ensure that the framework is decision-relevant, these endorsement patterns are linked to acceptance risk cues and broad response directions, consistent with work treating misinformation as a practical organisational risk issue [17].

3.3 Scope and Analytical Limits

The framework is presented as a conceptual decision-support structure. Illustrative claim types are paraphrased and used to clarify the logic without reporting prevalence estimates or reproducing raw posts. The intention is to provide minimal but decision-relevant transparency on how endorsement logics were structured, while avoiding claims about population prevalence or effect size. Detailed discourse reporting, measurement development, and empirical testing are reserved for subsequent outputs.

4 The Decision-Support Framework

4.1 Framework Overview

The proposed framework is intended for early transition contexts where public evaluation is shaped through information environments and where decision teams must interpret emerging signals without relying on mature market indicators or behavioural data [11, 19, 25]. Figure 1 summarises the framework logic. The framework treats misinformation belief endorsement as the key mechanism linking the information environment to acceptance vulnerability. Endorsement is defined as misinformation-related claims being accepted and used as a basis for judgement (e.g., repeated, defended, or treated as credible), distinct from simple exposure to misleading content [1, 7]. Misinformation research has shown that exposure cannot be treated as equivalent to belief or downstream impact without clarifying what is accepted and how it anchors evaluation [1].

Belief endorsement is shaped within an information environment that includes individual and social predispositions affecting credibility judgements in uncertain contexts, and sources and signals that shape perceived plausibility and legitimacy [13, 19, 27]. The framework's decision-relevant outcome is an acceptance risk state, used here as a structured summary of whether acceptance conditions appear relatively stable or increasingly fragile. In this paper, "risk state" refers to vulnerability in acceptance conditions that can affect legitimacy, stakeholder relationships, and feasibility of transition progress even before widespread product access exists [16, 23]. The construct is interpretive, and is intended to support structured judgement under uncertainty, without estimating behavioural probabilities.

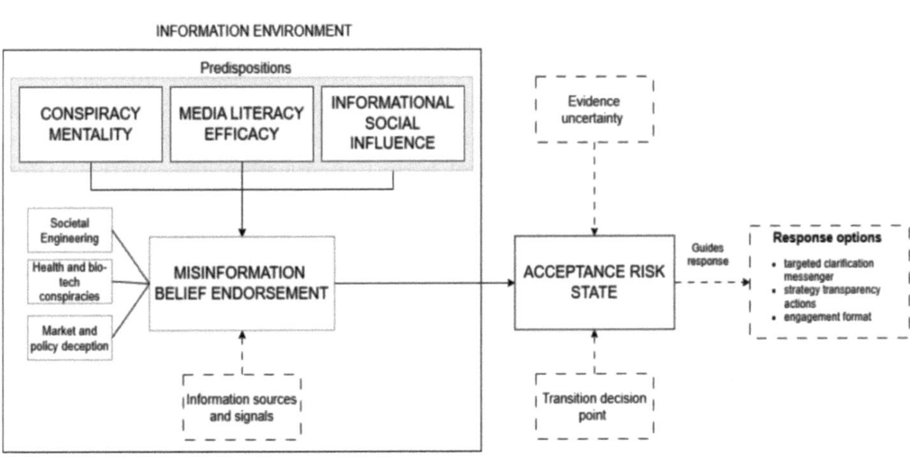

Fig. 1. Conceptual framework

The framework is intended to support structured interpretation of endorsement patterns as signals of emerging acceptance vulnerability under contested information conditions.

4.2 Core Components and Definitions

Information Environment (Predispositions). Predispositions are treated as influences that shape how information is interpreted and whether certain claims feel plausible or worth repeating in high-uncertainty contexts. In technology related domains, conspiracy-linked belief dynamics can become self-reinforcing through social and informational mechanisms, which can shift evaluation away from evidence appraisal and toward distrust based interpretations [26]. Wider synthesis also indicates that vulnerability or resistance to misleading information reflects interacting individual, social, and contextual influences, with influences extending beyond information deficits alone [13]. For this paper, three predisposition types are retained as conceptually relevant which include conspiracy-oriented interpretations, confidence in evaluating online information quality, and reliance on social proof under uncertainty [12, 13, 27].

Information Environment (Sources and Signals). In this framework it refers to the channels and credibility cues through which claims circulate (e.g., perceived source credibility, social proof signals, and the wider platform context). Research on digital information environments indicates that behavioural and evaluative responses to misinformation are shaped by processing dynamics and environmental properties, as well as by message content [27]. Related arguments also frame the information environment as a structuring condition shaping what is regarded as credible and actionable [19].

Misinformation Belief Endorsement. It is organised into three domains used in this paper as a practical structure for interpreting claim logics observed in CM discourse and aligning response thinking. These elements are not treated as exhaustive of scepticism, nor are they intended to classify all criticism as misinformation, but as distinct endorsement patterns with different implications for acceptance vulnerability:

- Societal engineering beliefs- CM framed as coercive societal control or imposed change, aligning with broader technology conspiracy dynamics and legitimacy concerns [16, 26].
- Health and biotech conspiracy beliefs- CM framed as inherently unsafe, contaminating, or deliberately harmful. Food related misleading information has been discussed as shaping perceived healthiness and acceptability, particularly where scientific uncertainty exists and narratives simplify complexity [15, 23]. Disgust and contamination intuitions are also highlighted as barriers that may be difficult to shift through factual correction alone [10].
- Market and policy deception beliefs- CM approvals, subsidies, and labelling framed as corrupt, captured, or designed to mislead. This connects directly to institutional trust and legitimacy conditions that are salient in transition governance contexts [11, 19].

Acceptance Risk State. The acceptance risk state summarises whether acceptance conditions appear increasingly fragile under prevailing endorsement patterns. It consolidates acceptance concerns repeatedly identified in the CM and alternative protein literature (e.g., perceived risk, trust, disgust-related responses, legitimacy concerns) into a decision-oriented construct suitable for early stage contexts where market behaviour cannot yet be observed [9, 18, 28]. The purpose of the construct is interpretive, supporting decision teams in recognising when evaluation appears to be shifting toward entrenched distrust, categorical rejection, or intensified legitimacy contestation in ways that may constrain transition feasibility. For practical interpretation, acceptance conditions can be observed along a spectrum. They may appear:

- relatively stable, where concerns remain responsive to discussion and evidence;
- increasingly sensitive, where endorsement patterns consolidate and trust signals weaken;
- fragile, where distrust becomes entrenched and rejection is framed in categorical or identity-linked terms.

The distinctions are qualitative and intended to inform judgement about when closer monitoring or strategic recalibration may be necessary.

4.3 Contextual Elements Supporting Transition Decision Use

To reflect how transition decisions are made under uncertainty and legitimacy pressure, the framework includes contextual elements that shape how endorsement signals can be interpreted and acted upon. Evidence uncertainty is treated as a background condition because sustainability and feasibility assessments remain contested and developing for cultured meat, which creates space for selective interpretation in public debate [23]. The framework also recognises that credibility is channel-contingent including sources, platform cues, and social signals influence both how claims travel and how responses are received [19, 27]. In addition, acceptance vulnerability can intensify at specific decision moments, such as controversies, regulatory milestones, or market entry debates, when legitimacy judgements become especially salient for governance and strategy [11, 25]. Finally, endorsement patterns are linked to response classes as decision-support prompts. The response classes are not presented as tested interventions, consistent with risk-management perspectives that treat misinformation as an organisational risk to be managed under uncertainty [7, 17].

5 Applying the Framework at Decision Moments

CM transition decisions are often taken at moments when public evaluation is being shaped by contested claims, uneven trust, and fast-moving online narratives. In these situations, decision teams need a structured way to interpret which misinformation related claims are being treated as credible (i.e. endorsed) and what that implies for acceptance vulnerability and response choices. The framework can be used to support structured interpretation since it focuses on endorsement and links dominant endorsement logics to acceptance risk cues and response directions that remain feasible under prevailing credibility constraints [1, 7, 13, 19].

5.1 Practical Steps for Decision Use

In practice, the framework can be applied through a short workflow as follows:

- Monitor and capture recurring claims from relevant areas (mainstream coverage, stakeholder conversations, regulator-facing debate, and major social platforms), focusing on claims that are repeated, defended, or treated as credible [1, 7]. In resource-constrained settings, this can be done through periodic scans around key events (e.g., consultation windows, announcements) instead of continuous monitoring.
- Classify endorsement logic using the three belief domains (societal engineering, health/biotech conspiracies, market/policy deception). Multi-classification can be expected when claims blend logics [13].
- Assess acceptance vulnerability by identifying what the dominant logic implies considering autonomy threat and resistance, harm certainty and contamination intuitions, or governance illegitimacy and capture assumptions [7, 19].
- Select response direction as a class of action aligned to the credibility constraints of the moment (e.g. messenger choice, governance signals, oversight transparency), avoiding default reliance on generic fact correction when the dominant logic is not evidence-responsive [1, 7].
- Record and review whether endorsement patterns shift over time to support disciplined interpretation and response alignment under uncertainty [1, 13].

5.2 Linking Belief Endorsement Domains to Acceptance Risk Cues and Response Directions

Table 1 translates the three belief domains into decision-relevant cues and response directions.

Endorsement cues are paraphrased from the Twitter/X corpus and reflect recurring narrative markers [21]. The response directions are framed as credibility oriented classes of action presented without intervention evaluation evidence, consistent with the need to avoid over specifying effects in a pre-commercialised context [1, 7, 15]. Examples are included to clarify practical meaning.

The logic outlined in Table 1 can be observed in high-salience regulatory or consultation periods. For instance, when recurring and defended claims frame cultured meat as part of an "elite" or externally imposed project linked to loss of autonomy, such endorsements fall within the societal engineering domain. In this configuration, acceptance vulnerability is indicated less by product-specific safety concerns and more by heightened autonomy reactance and symbolic resistance. Under these conditions, responses centred on detailed technical reassurance are unlikely to address the dominant evaluative logic. Greater emphasis is instead placed on autonomy-respecting engagement, visible procedural fairness, and the use of locally trusted intermediaries, consistent with the response classes summarised in Table 1. Ongoing review of endorsement patterns would then indicate whether evaluation shifts toward conditional concerns or remains anchored in control-based distrust. The table is intended to support structured judgement in contexts where fixed intervention rules are not available. Its value lies in differentiating endorsement logics so that response alignment reflects prevailing credibility conditions.

5.3 Using the Framework in Common Decision Moments

Before applying the framework, mixed endorsement patterns should be expected, particularly when controversies blend health risk claims with broader distrust of governance. In such cases, classification should prioritise the dominant evaluative logic shaping acceptance vulnerability, while noting secondary logics that may require parallel response emphases. Use may also differ by actor type. For example, startups may have limited legitimacy capital and therefore rely more heavily on third-party scrutiny and credible intermediaries, incumbents may face stronger expectations of transparency and accountability, while regulators may prioritise procedural fairness, clear rationale, and visible independence to prevent capture interpretations

The framework has a potential to be most useful when endorsement patterns increase around high salience decision contexts. During fast-moving controversies, efficient classification may help teams avoid mismatched responses (e.g. detailed safety rebuttals when the dominant logic is perceived as an autonomy threat, or branding reassurance when the dominant logic is capture and deception) [7, 27]. Around regulatory milestones

Table 1. Endorsement-to-response mapping

Endorsement domain	Illustrative endorsement markers from Twitter/X	Acceptance risk cues for decision teams	Response emphasis (classes, not evaluated interventions)
Societal engineering	CM framed as an "elite/globalist" project linked to WEF, "Great Reset", Agenda 2030, coercion/mandates, "control", "depopulation", and paired with "eat bugs" narratives; personified blame toward powerful actors (e.g. tech figures)	High reactance and identity-based refusal. Evaluation shifts from product evidence to autonomy threat and symbolic resistance. Distrust becomes self-protective, not just informational	Autonomy-respecting engagement (choice-preserving framing, avoid coercive tone. E.g., emphasising voluntariness, opt-in pilots). Procedural fairness and voice signals (credible participation formats, stakeholder inclusion. E.g., open consultation formats chaired by trusted intermediaries). Messenger strategy that reduces "elite" cueing and emphasises locally trusted intermediaries (e.g., local public health figures, community-based organisations).

(continued)

Table 1. *(continued)*

Endorsement domain	Illustrative endorsement markers from Twitter/X	Acceptance risk cues for decision teams	Response emphasis (classes, not evaluated interventions)
Health and biotech conspiracies	Claims that the product is grown from "cancer cells", causes cancer, contains mRNA, is GMO/engineered, is "toxic/chemical", "Frankenfood", "contaminating", "untested", "lab rat" framing	Harm certainty and contamination intuitions Disgust and perceived unnaturalness intensify. Precautionary demands (ban/stop) replace conditional openness. Correction attempts may be read as minimization.	Oversight visibility and process transparency (what testing/monitoring exists and how decisions are made. E.g. publishing clear testing pathways and decision criteria). Uncertainty-competent safety communication (clear boundaries of what is known/unknown. E.g. explicitly stating where evidence is still developing). Credible third-party scrutiny rather than brand-led reassurance (e.g. independent scientific review, regulator-facing summaries).
Market and policy deception	Claims that FDA/USDA approvals are corrupt/captured, "bought", bribed; "follow the money" frames (subsidies/taxpayer funding); "labelling to dupe consumers"; bans/procurement restrictions framed as political collusion	Governance legitimacy breakdown. Suspicion toward regulators, standards, and firms. Reputational fragility and partnership hesitation. Even accurate information can be discounted as captured.	Governance credibility actions (traceable compliance processes, clarity on standards, conflicts, funding, and decision pathways. E.g. transparent conflict of interest disclosures, audit-ready documentation). Transparent regulatory engagement posture (e.g. open publication of consultation responses and rationale). Accountability signaling that reduces capture interpretations (e.g. clear delineation of roles between firms, regulators, and assessors).

and consultation periods, market/policy deception narratives tend to become more consequential because they target approval legitimacy including governance and process credibility therefore become central [11, 19]. During market-entry and naming/labelling debates, contamination and perceived consumer deception frames can escalate acceptance vulnerability even when the underlying technical evidence has not changed. More credible responses combine uncertainty-competent explanation with clear oversight signals, and not treating naming as a purely commercial branding choice [10, 15].

6 Contribution to Knowledge and Decision Support for Sustainable and Resilient Transitions

The framework contributes to decision support for early stage CM transitions by positioning acceptance vulnerability as an interpretive condition shaping commitments under uncertainty. In contested transition contexts, firms and governance actors often need to make commitments (e.g. investment, engagement, regulatory positioning, partnership-building) under uncertainty and legitimacy pressure, while public meaning-making can shift faster than the underlying technical evidence. A resilience oriented approach to decision support therefore benefits from structured interpretation of emerging vulnerabilities, consistent with transition literature emphasising contestation, legitimacy, and governance conditions as determinants of whether pathways accelerate or stall [11, 16, 25]. Continued uncertainty in sustainability and feasibility assessments further intensifies these interpretive pressures in public debate [23].

A second contribution concerns how misinformation is incorporated into acceptance assessment. The framework centres belief endorsement as the mechanism linking contested information conditions to acceptance vulnerability. This responds to conceptual critiques that misinformation research becomes weak when it blurs exposure, belief, and downstream effects without specifying what is accepted and how it is used in evaluation [1]. By focusing on claims that are accepted, repeated, defended, or used as judgement anchors, the framework links misinformation dynamics to acceptance vulnerability without making prevalence claims. Such an approach is consistent with the argument that misinformation should not be dismissed simply because measured exposure or average effects may be uneven, especially when it reinforces distrust or entrenched interpretive frames that resist conventional correction [7]. It also aligns with evidence that conspiracy-linked belief systems can become self-reinforcing and identity-protective, shaping evaluation of emerging technologies [26]. The framework does not treat all scepticism or normative disagreement as misinformation. It focuses on recurring endorsement patterns that attribute intentional harm, coercive control, or institutional corruption in ways that alter credibility conditions.

The decision support contribution lies in translating these points into a problem-structuring framework that can be used at decision moments. The framework supports early warning and sensemaking by helping teams identify which endorsement logics are salient, what kind of acceptance vulnerability they imply, and which broad response directions are more credible under prevailing trust conditions. This draws on an information environment perspective in which belief formation is shaped by interacting individual, social, and contextual factors, rather than information deficits alone [13]. It also reflects evidence that digital information environments influence behavioural responses through processing dynamics and social signals, making environment-level conditions part of the decision context [27]. Related literature further argues that information environments can operate as structural determinants that shape what is regarded as credible and actionable, reinforcing the need for decision support that is calibrated to credibility constraints [19].

Finally, the framework's differentiation of endorsement logics supports adaptive decision-making in line with sustainable and resilient transitions which require responses to heterogeneous legitimacy challenges, not a single communication solution. By organising endorsement into distinct belief domains and linking them to acceptance risk cues and response-option classes, the paper provides a structured basis for judgement under uncertainty while remaining consistent with the evidential limits of a pre-commercialised context. Figure 1 summarises the conceptual logic and Table 1 operationalises this differentiation into domain-specific risk cues and response classes. In doing so, it offers a contribution as a decision support framework intended to strengthen interpretation and response alignment in the information conditions that increasingly shape transition feasibility.

7 Conclusion

This paper developed an evidence informed decision support framework for assessing Cultured Meat acceptance vulnerability under misinformation belief endorsement. In pre-commercialised transition contexts, decision teams often need structured interpretation of emerging acceptance vulnerability before population-level behavioural prediction is feasable. Therefore, the framework uses an acceptance risk state as a decision-oriented summary of whether acceptance conditions appear relatively stable or increasingly fragile. It positions the information environment as the context shaping whether misinformation related claims are endorsed and treats endorsement as the mechanism through which contested information conditions can translate into heightened acceptance vulnerability. To support use at transition decision points, the paper also provided illustrative mapping from endorsement domains to acceptance risk cues and broad response classes, intended to support judgement and guide response alignment, without prescribing tested interventions. Figure 1 presents the conceptual structure of this logic, and Table 1 translates it into a domain-specific endorsement-to-response mapping tool.

Several limitations follow from the scope and evidence base. The framework is conceptual and designed for decision support under uncertainty, so it does not provide prevalence estimates, validated effect sizes, or behavioural prediction claims. The empirical grounding is qualitative and exploratory, drawing on Twitter/X discourse to identify recurring endorsement logics. This offers insight into how claims are framed and defended in naturally occurring debate, but it is not representative of the wider public and cannot be used to infer population-level attitudes. Also, the mapping between endorsement patterns, acceptance risk cues, and response classes remains provisional as well, reflecting structured synthesis and discourse-informed plausibility rather than evaluation evidence on what responses are effective across audiences or settings. In addition, the acceptance risk state compresses diverse acceptance processes into a single decision-oriented construct and does not capture the full variation likely to exist across consumer segments and cultural contexts. Although operational clarification has been provided in terms of indicative stability and fragility conditions, further empirical refinement would be required for systematic monitoring or comparative application.

Next steps focus on strengthening empirical foundations and decision relevance. Further work could look at refining the endorsement domains through discourse analysis across additional platforms, including how narratives cluster, co-occur, and intensify around key decision points. In addition, measurement development and primary data collection can be used to test the proposed pathway from information-environment predispositions to endorsement and to acceptance vulnerability. Such work would allow future research to examine whether the framework functions primarily as an interpretive heuristic, a risk-sensing device, or a precursor to more formalised modelling approaches within transition decision support.

Acknowledgments. This research was conducted as part of a Doctoral Training Account (DTA) funded by the University of Plymouth.

Disclosure of Interests. The authors declare no competing interests relevant to the content of this article.

References

1. Altay, S., Berriche, M., Acerbi, A.: Misinformation on misinformation: conceptual and methodological challenges. Soc. Media + Soc. **9**(1), 1–13 (2023). https://doi.org/10.1177/20563051221150412
2. Andersson, H., Ouvrard, B.: Not on my plate! Challenges to promote meat substitutes. Ecol. Econ. **240**, 108795 (2026). https://doi.org/10.1016/j.ecolecon.2025.108795
3. Cook, S.J., Cardilini, A.P.A., Hayley, A., Francis, P.: Fact-based environmental messaging did not influence Australians' attitudes and intentions towards cultivated seafood. Food Qual. Prefer. **125**, 105514 (2025). https://doi.org/10.1016/j.foodqual.2025.105514
4. Cosgrove, T.: When education fails: narcissism, uniqueness, and need for closure in conspiracy beliefs and misinformation. Personal. Individ. Differ. **251**, 113567 (2026). https://doi.org/10.1016/j.paid.2025.113567
5. Dalimier, J., Maréchal, K., Dendoncker, N.: Transition and transformation: a cross-fertilizing approach to complement the study of systemic and sustainable changes. Environ. Innov. Soc. Trans. **58**, 101059 (2026). https://doi.org/10.1016/j.eist.2025.101059
6. da Silveira, A.B., Abib, G., Reis, G.G.: On what basis do cooperatives engage in the cultivated meat market? Insights from Brazilian cooperatives. J. Bus. Res. **204**, 115816 (2026). https://doi.org/10.1016/j.jbusres.2025.115816
7. Ecker, U.K.H., Lewandowsky, S., Cook, J., et al.: Why misinformation must not be ignored. Am. Psychol. **80**(6), 867–878 (2025). https://doi.org/10.1037/amp0001448
8. Giesbers, E., Mattijssen, T.J.M., Leeuwis, C.: Citizen participation in food systems transitions: how inclusive should it be? Environ. Innov. Soc. Trans. **56**, 100986 (2025). https://doi.org/10.1016/j.eist.2025.100986
9. Hanan, F.A., Karim, S.A.B., Aziz, Y.A., Ishak, F.A.C., Sumarjan, N.: Consumers' cultured meat perception and acceptance determinants: a systematic review and future research agenda. Int. J. Consum. Stud. **48**, e13088 (2024). https://doi.org/10.1111/ijcs.13088
10. Havermans, R.C.: Future foods refusal: how our behavioral immune system impedes the acceptance of cultured meat. Trends Food Sci. Technol. **163**, 105179 (2025). https://doi.org/10.1016/j.tifs.2025.105179

11. Hundscheid, L., Voigt, C., Bergthaler, D., Plank, C., Wurzinger, M., Melcher, A.H.: Policy mix for the sustainable protein transition in Austria. Environ. Innov. Soc. Trans. **51**, 100819 (2024). https://doi.org/10.1016/j.eist.2024.100819
12. Katsiroumpa, A. et al.: The online misinformation susceptibility scale: development and initial validation. Healthcare. **13**, 2252 (2025). https://doi.org/10.3390/healthcare13172252
13. Kont, J., Elving, W., Broersma, M., Bozdağ, Ç.: What makes audiences resilient to disinformation? Communications. **50**(2), 534–555 (2025). https://doi.org/10.1515/commun-2023-0078
14. Lwin, M.O., Rawski, S.L., Boland, M.: Feeding the future: a comparison of drivers and barriers towards consumers' acceptance of alternative proteins in Singapore. Food Qual. Prefer. **137**, 105761 (2026). https://doi.org/10.1016/j.foodqual.2025.105761
15. Melios, S., Asimakopoulou, A.A., Greene, C.M., Crofton, E., Grasso, S.: Food-related fake news, misleading information, established misconceptions, and food choice. Curr. Opin. Food Sci. **63**, 101309 (2025). https://doi.org/10.1016/j.cofs.2025.101309
16. Moritz, J., Murphy, J., Bryant, C.: Prospects of justice for cellular agriculture: a just transition or reinvesting in unsustainability? Food Ethics. **9**, 22 (2024). https://doi.org/10.1007/s41055-024-00156-8
17. Noronha, E., Varela da Costa, J., Mira da Silva, M.: A multivocal grey literature review on fake news and risk management. Discov. Comput. **29**, 21 (2026). https://doi.org/10.1007/s10791-025-09902-w
18. Onwezen, M.C., Dagevos, H.: A meta-review of consumer behaviour studies on meat reduction and alternative protein acceptance. Food Qual. Prefer. **114**, 105067 (2024). https://doi.org/10.1016/j.foodqual.2023.105067
19. Palmer, A., Gorman, S.: Misinformation, trust, and health: the case for information environment as a major independent social determinant of health. Soc. Sci. Med. **381**, 118272 (2025). https://doi.org/10.1016/j.socscimed.2025.118272
20. Pivoraite, G., Liu, S., Roh, S., Zhao, G.: Framework for understanding consumer perceptions and attitudes to support decisions on cultured meat: a theoretical approach and future directions. In: Duarte, S.P., Lobo, A., Delibašić, B., Kamissoko, D. (eds.) Decision Support Systems XIV. Human-Centric Group Decision, Negotiation and Decision Support Systems for Societal Transitions. ICDSST 2024 Lecture Notes in Business Information Processing, vol. 506, pp. 109–125. Springer, Cham (2024). https://doi.org/10.1007/978-3-031-59376-5_9
21. Pivoraite, G., Liu, S., Roh, S., Zhao, G.: Assessing consumer perceptions of cultured meat using social media discourse to support decision-making under uncertainty. Lecture Notes in Business Information Processing. (2025). Retrieved from: https://pearl.plymouth.ac.uk/pbs-research/674
22. Polli Leite, F., Septianto, F., Pontes, N.: "Meat" the influencers: crafting authentic endorsements that drive willingness to buy cultured meat. Appetite. **199**, 107401 (2024). https://doi.org/10.1016/j.appet.2024.107401
23. Rana, R.L., Bux, C., Tommaso, V.: The sustainability nexus of cultured meat. Food Res. Int. **227**, 118245 (2026). https://doi.org/10.1016/j.foodres.2025.118245
24. Simeone, M., Verneau, F.: Lab-grown bites: what drives or stops the world from eating cultured meat? A global perspective on developed and developing countries. Br. Food J. **128**(3), 920–938 (2025). https://doi.org/10.1108/BFJ-03-2025-0355
25. Strom-Andersen, N., Szulecka, J., Bugge, M.M., Forsberg, E.-M.: Navigating sustainability paradoxes. Res. Policy. **55**, 105407 (2026). https://doi.org/10.1016/j.respol.2025.105407
26. Trang, S., Kraemer, T., Trenz, M., Weiger, W.H.: Deeper down the rabbit hole: how technology conspiracy beliefs emerge. Inf. Syst. Res. **36**(2), 709–735 (2025). https://doi.org/10.1287/isre.2022.0494

27. Wei, R., Lo, V.-H., Zhang, X., Lu, M., Qiu, J.L.: Examining the impact of digital information environments, information processing, and presumed influence on behavioral responses to COVID-19 misinformation in Asia. New Media Soc. **27**(9), 5257–5280 (2025). https://doi.org/10.1177/14614448241252391
28. Yu, Y., Wassmann, B., Lanz, M., Siegrist, M.: Willingness to consume cultured meat: a meta-analysis. Trends Food Sci. Technol. **164**, 105226 (2025). https://doi.org/10.1016/j.tifs.2025.105226

Sustainable Community Value Networks: A Decision Support Model for Food Sharing Community Transitions

Xinyi Li(✉), Shaofeng Liu, Dulekha Kasturiratne, and Huilan Chen

University of Plymouth, Plymouth PL4 8AA, UK
xinyi.li@plymouth.ac.uk

Abstract. Recent global events such as the COVID-19 pandemic, climate disruption, ongoing living crisis and inflation have increasingly driven individuals and families into the critical state of food insecurity. In response, food sharing initiatives such as food sharing organisations, charities, communities and hubs have been established across many countries. One of the key decision-making challenges faced by food sharing communities is to build sustainable community value networks that can create long-lasting community value and enhance operational efficiency. This paper addresses this challenge by investigating how sustainable community value can be systematically generated and used to support decision-making within food sharing communities through a structured value network model. Drawing upon a comprehensive Systematic Literature and Practice Review (SLPR), this study conceptualises and develops the Sustainable Community Value Networks (SCVN) framework by integrating insights from both academic literature and real-world practice. The SCVN framework functions as a decision support model to guide sustainable and resilient community and society transitions, offering practical recommendations and strategic insights for decision-makers within food sharing communities to enhance adaptive capacity and strengthen the overall resilience of food sharing systems.

Keywords: Decision Support Systems · Food Sharing Communities · Community Value Networks · Resilient Transitions · Social Innovation · Sustainable Community Development · Systematic Literature and Practice Review

Introduction

Hunger remains both a profound violation of human dignity and a major barrier to social, economic, and political progress (FAO Council, 2010). Recent global disruptions—ranging from the COVID-19 pandemic and climate instability to the ongoing cost-of-living crisis and inflation—have intensified levels of food insecurity among individuals and families worldwide (UN World Food Programme, 2023; FAO et al., 2024). In response, countries such as the UK, Italy, Germany, and Denmark have launched a variety of food sharing initiatives, including community-led food projects and ICT-enabled food sharing platforms, which are often perceived as a collective action against issues such as food crisis by re-distributing surplus food to population in need and prevent food

A. P. Cabral Seixas Costa et al. (Eds.): ICDSST 2026, LNBIP 582, pp. 51–63, 2026.
https://doi.org/10.1007/978-3-032-25146-6_4

waste (Berns et al., 2023; Mackenzie & Davies, 2019; Lombardi & Costantino, 2020; Schanes & Stagl, 2019). Together, these efforts have fostered various food sharing communities that supported disadvantage groups at both local and national levels, offering vital services such as food access, recipes sharing, and health advice (Davies et al., 2019; Harvey et al., 2020). Such communities have demonstrated success in enhancing food crisis support and reducing food waste, also contribute to foster food sustainability and democracy (Jaeggi & Gurven, 2013, Mackenzie & Davies, 2019; Mazzucchelli et al., 2021; Zhao et al., 2023).

However, one of the key decision-making issues faced by food sharing communities is how to build sustainable community value networks that can guide effective action, ensure resource efficiency, and enhance long-term community resilience. Community value, as a specific form of social value, emerges through collective activities and shared practices within communities, and plays a significant role in sustaining community networks and society transition (Mair & Martí, 2006; Pret & Carter, 2017; Santos, 2012; Shaw & Carter, 2007). Although value network models are well-developed and widely applied in common communities such as business and education, the domain of food sharing remains relatively underexplored, with limited empirical evidence on how sustainable community value can be systematically generated and leveraged in decision-making processes (Lombardi & Costantino, 2020; Morone et al., 2018). Thus, there is an urgent need for research aimed at exploring how to create sustainable community value networks in food sharing communities to enhance operational efficiency and support resilient, sustainable transitions in an increasingly dynamic environment. To address this gap, this study aims to investigate how to create a sustainable community value network in food sharing communities. Drawing on a Systematic Literature and Practice Review (SLPR) on current literature and practical initiatives, this study identifies the key components of the sustainable community value network. The framework offers valuable insights and efficient decision-making recommendations for food sharing communities and strengthens the capacity and resilience of food sharing communities to meet the growing demand for sustainability.

1 Methodology: Systematic Literature & Practice Review

A Systematic Literature Review (SLR) was identified as the most appropriate methodological approach for this study due to its structured and transparent nature. Unlike traditional narrative reviews, the SLR provides a rigorous process for synthesising and consolidating dispersed knowledge from existing scholarly work, enabling the development of new theoretical insights and contributions (Tranfield et al., 2003). To reflect the applied dimension of this research, the review also incorporated evidence from real-world food sharing initiatives, including both ongoing and completed projects. These practical sources include project reports (e.g. CULTIVATE), publications (e.g. FoodSHIFT2030), project databases (e.g. SHARECITY100), and official websites of food sharing organisations (e.g. Feeding Devon, Feeding Britain, FareShare, etc.). The integration of academic and practice-based sources forms the basis of a Systematic Literature and Practice Review (SLPR), which offers both conceptual and evidence-based grounding for the development of sustainable community value networks. The SLPR process is organised into five phases, each of which is described in detail below (see Fig. 1):

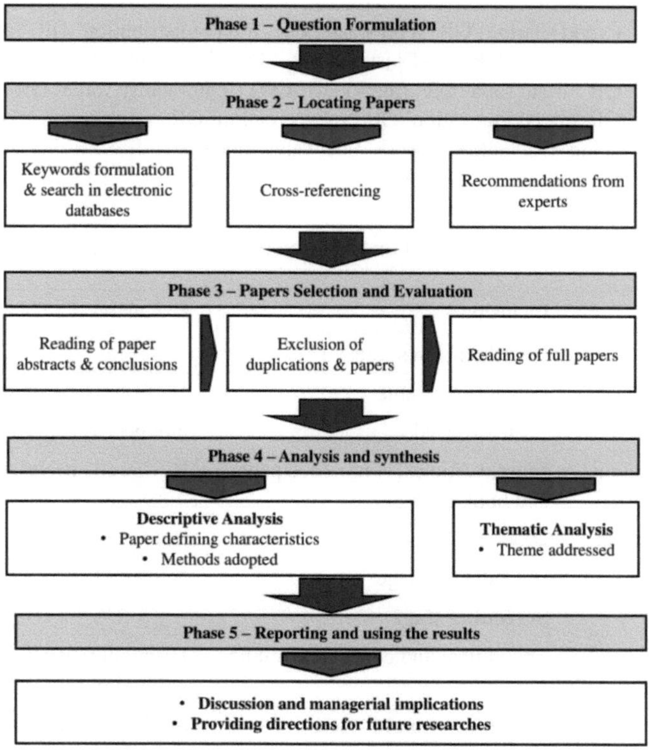

Fig. 1. SLPR methodology adopted in the study

The first stage of the SLR involved the formulation of precise, well-defined research questions to ensure conceptual clarity and minimise ambiguity throughout the review process (Hohenstein et al., 2015):

– *RQ1.* What are the main components of sustainable community value networks, and how are these components interconnected?

The second stage focused on identifying and retrieving relevant, high-quality academic literature to build a comprehensive set of contributions aligned with the research objectives (Denyer & Tranfield, 2009). A selection of trusted academic and project databases, namely Web of Science, Scopus, ScienceDirect, Taylor & Francis Online, Wiley Online Library, and CORDIS, was used due to their global recognition and relevance to business and food sharing research. A set of predefined keywords (e.g. "food sharing", "community value", and related terms) was employed as the search criteria and applied across titles, abstracts, and keyword fields (see Table 1).

Following the initial database search, a set of clearly defined inclusion and exclusion criteria was applied to guide the selection and evaluation of relevant studies (see Table 2).

Table 1. Databases and search strings

Databases	Web of Science, Scopus, ScienceDirect, Taylor & Francis, Wiley Online Library, CORDIS
Search strings	('food sharing' OR 'food waste' OR 'food redistribution' OR 'food democracy' OR 'food sustainability' OR 'food bank' OR 'sharing economy' OR 'urban food' OR 'free food') AND ('community value' OR 'social value' OR 'social innovation' OR 'value creation' OR 'value generate' OR 'create value' OR 'value network' OR 'value model')

Table 2. Criteria for inclusion or exclusion studies

Criteria	Inclusion	Exclusion
	Identification stage	
Availability	Full-text available	No full-text available
Peer-review	Peer-reviewed	Not peer-reviewed
Type of publication	Journal articles, conference papers and book chapters	Reviewed articles, posters, etc.
Year of publication	2005–2026	Before 2005
Language	English Language	Non-English
	Screening stage	
Types of study	Theoretical and conceptual studies	Discussion papers, overviews, opinion papers, etc.
Relevance	Studies relevant to research questions (e.g. contain keywords, same/similar research focus, etc.)	Studies that are not relevant to research questions

These criteria considered factors such as publication year, type, thematic relevance, and overall methodological quality. This rigorous screening process ensured that only studies directly aligned with the research questions and demonstrating sufficient academic robustness were retained, thereby improving the reliability and validity of the review. Papers that failed to meet the criteria or appeared as duplicates were excluded.

The screening process involved multiple stages, including an initial review of titles, abstracts, and introductions, followed by a full-text assessment of the remaining articles. Additional relevant sources were identified through backward and forward citation tracking, as well as expert recommendations from both academic researchers and practitioners in food sharing organisations. In total, 42 sources including peer-reviewed journal articles, conference proceedings, and practical project outputs were selected for final inclusion. The full literature identification and selection process is summarised in Fig. 2.

The subsequent stage involved data extraction and synthesis of findings. Key information was systematically extracted from the selected sources to address the research

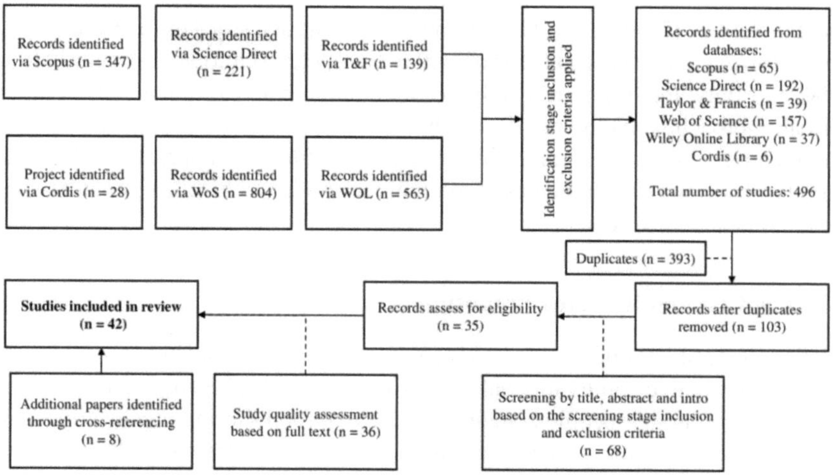

Fig. 2. Paper Selection flowchart

question, identify recurring themes, and highlight critical gaps in the existing literature. This process enabled a comprehensive understanding of the core components of sustainable community value networks in food sharing communities, as well as the interconnections between them. Drawing on the insights generated through the SLPR, the following section presents the key findings from both academic literature and practical initiatives. These findings form the foundation for the development of a conceptual framework designed to demonstrate how sustainable and resilient community value can be generated, delivered, captured, and reinvested within food sharing communities.

2 Key Findings: Descriptive and Thematic Analysis

Based on the SLPR findings, the results are presented in two parts. The first section provides a descriptive analysis, offering a detailed overview of the selected studies with particular attention to their core characteristics and methodological approaches. The second section presents a thematic analysis, identifying key themes and outlining the principal components that define the concept of community value within food sharing communities, as revealed through the review process.

2.1 General Overview of Analysed Papers

This study review examined 42 papers published between 2010 and 2024 across 28 journals, including journals specific to disciplines such as business, management, sustainability, and environment. The plurality of our selected papers were published in *Sustainability* (4), *Geoforum* (3), and *Journal of Cleaner Production* (2), highlighting the strong connection between community value creation in food sharing communities and topics in research areas like food sustainability and social innovation.

In terms of the authors' affiliation, seven papers were led by researchers affiliated with institutions in Ireland. In a number of other publications and projects, the first author is based in the UK (7), Italy (5), Netherlands (3) or Sweden (3). This distribution aligns with the growing interest in food sharing within European countries, particularly Ireland, the UK, and Italy. Regarding research methods, the majority of papers (12) employed empirical approaches, including 10 that based on case studies and 2 that used semi-structured interviews. Analytical models (6) and mixed-method approaches (5) were less frequently observed. This indicates a predominant focus on practical, context-specific studies in this area of research.

2.2 Define Community Value Propositions

According to the SLPR findings, defining community value propositions is a critical step in creating sustainable community value (Davies et al., 2018; Morone et al., 2018). This involves identifying the specific benefits or value that food sharing organisations, initiatives, and projects deliver to their communities. Previous literature and practices have identified three key community value propositions: improve food crisis support, foster social connectivity and promote a healthy lifestyle (Morone et al., 2018; SHARECITY, 2022; UNWFP, 2023). Local food sharing communities, often formed through activities organised by food sharing organisations, are frequently centred around these propositions. For instance, food sharing organisations such as Feeding Devon (2024) and Feeding Britain (2024) claimed in their impact reports that their primary objectives are to benefit their members by "providing emergency food assistance," "promoting healthy lifestyle," "fostering social connections," and, ultimately, aiming to enhance members' perceived health and overall well-being.

Firstly, a key priority for food sharing communities is addressing food crises by providing individuals and families experiencing food insecurity with critical access to food (Lombardi & Costantino, 2020; UNWFP, 2023). Additionally, it encompasses the sharing of kitchen spaces, food preparation, storage equipment, as well as techniques for growing, processing, and cooking food (Davies, 2016; Davies et al., 2018). Beyond addressing food crises, food sharing communities also aim to promote healthier lifestyles by providing members with low-cost and nutritious food, as well as cooking and nutrition activities such as 'Healthy Eating Project' and 'Healthy Holidays' (Feeding Britain, 2024; Feeding Devon, 2024). These efforts align with broader propositions to enhance the well-being of individuals and communities. By combining access to healthy food with community-driven initiatives, food sharing projects not only nourish their members but also encourage healthier habits and sustainable practices (Makov et al., 2020). Furthermore, food sharing communities play a vital role in fostering social connections, serve as platforms for mutual support, collaboration and interaction (Ganglbauer et al., 2014; Parker et al., 2010). FareShare (2024) reports that 83% of their members feel closer to their community, which indicates how food sharing communities enables individuals with shared values to meet, interact, and build connections (Harvey et al., 2020). As Ganglbauer et al. (2014) and Lozano (2007) highlight, local food sharing communities serve as "social networking sites" that contribute to the development of strong social ties and a sense of belonging and support among members. Together, these three components – improving food crisis support, fostering social connectivity, and promoting a healthy lifestyle – form the foundation of community value propositions and contribute to the broader impact of food sharing communities (Berns et al., 2023; Schanes & Stagl, 2019).

2.3 Map Community Value Beneficiaries

Locating community value beneficiaries is another key element to be taken into consideration in creating long-lasting community value. Disadvantaged groups represent the primary beneficiaries of community value within food sharing communities, making their identification a critical component in fostering sustainable and long-lasting community value (FareShare, 2024; Morrow, 2019). Existing literature consistently highlights the vital role food sharing communities play in addressing food insecurity and inequality, particularly among vulnerable populations (Davies et al., 2019; Harvey et al., 2020). These groups often include low-income households, single-parent families, refugees, and elderly individuals, especially in areas marked by poverty and social disadvantage, such as rural or economically deprived neighbourhoods (FAO et al., 2024).

For instance, FareShare (2024) reports that their efforts benefit a wide range of vulnerable individuals, with 75% of beneficiaries being families with children, 73% living on low or no income, and 60% being elder adults. Similarly, Feeding Devon (2024) notes that their community kitchens and food banks predominantly serve low-income families facing financial instability, single parents experiencing delays in receiving benefits, and grandparents raising grandchildren. These examples underscore that disadvantaged groups are not only the most frequent participants in food sharing communities but also the primary recipients of the social and economic value generated by such initiatives (Berns et al., 2021).

2.4 Community Value Creation and Delivery

The creation and delivery of community value are central to the development of sustainable community value networks, as they ensure long-term benefits for members of food-sharing communities. Existing literature and practices have identified three critical aspects of sustainable community value creation and delivery: financial and budgeting advice, nutrition and cooking skills, and health and wellbeing management (Davies et al., 2019; Veen, 2019). These dimensions highlight how food-sharing organisations go beyond merely distributing food to address the broader needs of their beneficiaries. For example, many organisations facilitate cooking and nutrition workshops and encourage the exchange of recipes using ingredients sourced from food-sharing activities (Davies, 2016; Rut et al., 2021). Feeding Devon (2024), for instance, provides "meal-kit" food boxes containing pre-portioned ingredients, accompanied by recipe cards and video links to cooking demonstrations. Additionally, members can participate in real-time cooking classes to enhance their culinary skills, promoting healthier eating habits and nutritional literacy.

Beyond food provision, food-sharing organisations often deliver wraparound services to address the root causes of food insecurity and poverty (SHARECITY, 2022). These services include welfare rights advice, financial literacy programs, health checks, and mental health support action plans (Feeding Britain, 2024). Together, these initiatives enable members to better tackle poverty, gain essential skills to improve their employability, and strengthen their overall resilience, fostering opportunities for greater economic and social inclusion (Makov et al., 2020; Mazzucchelli et al., 2021).

2.5 Community Value Measure and Feedback

One critical challenge in establishing sustainable community value networks lies in effectively measuring the community value created and effectively integrating this feedback into food sharing community (Meshulam et al., 2023). Previous literature and practices provided three vital criteria for evaluating value and guiding continuous feedback on the sustainable community network, including personal growth, skill and knowledge development, community cohesion, engagement, and empowerment (Marovelli, 2019; Morrow, 2019). As Lave and Wenger (1991) and Fox (2000) note, learning is inherently social and can be cultivated through active participation in community practices. In food-sharing contexts, activities such as food-related information sharing workshops and wraparound advice services enable community members to enhance their skills and knowledge through shared experiences (Donald & Ford, 2023; Rigg et al., 2021).

Research has demonstrated that community value significantly contributes to social value creation in multiple ways, including fostering community cohesion, community engagement and empowerment (Santos, 2012; Wenger et al., 2011). Community cohesion is closely tied to the development of human and social capital, evidenced by the establishment of social relationships, networks, and the creation of positive spillover effects within the community (Hietschold et al., 2023). Community engagement focuses on fostering inclusive participation, collaborative efforts, and active involvement among community members. This dynamic participation supports the equitable distribution of resources and the development of inclusive practices that enable communities to thrive collectively (Morrow, 2019; Mackenzie & Davies, 2019; Rut et al., 2021). Meanwhile, community empowerment involves both the recognition of opportunities and the empowerment and acknowledgment of community members, which strengthens their sense of identity and involvement (Berns et al., 2023; Donald & Ford, 2023; Rut et al., 2021).

By systematically measuring these dimensions of community value, food-sharing communities can gather actionable evidence to refine their services, set clearer goals, and evolve their value propositions (Berns et al., 2023; Davies et al., 2018). This feedback loop provides an evidence-driven foundation for future directions, enabling communities to better tackle challenges, address beneficiaries' needs, and sustain long-term community value and impact.

2.6 The Links Between Community Value Network Components

By analysing and synthesising these key components of a sustainable community value network, the links between each component also revealed and formed a series of cohesive and interconnected processes. The process begins with defining the key value propositions by identifying the primary goals and needs of food-sharing communities. These value propositions establish a foundation for determining the objectives that food-sharing organisations aim to achieve, such as addressing food insecurity, fostering social connections, and promoting healthier lifestyles (Davies et al., 2018; Feeding Britain, 2024).

Building on these propositions, value beneficiaries can be identified, enabling organisations to pinpoint the target groups—often disadvantaged populations—that stand to benefit most from these initiatives (FAO et al., 2024). This step informs the design and implementation of appropriate channels for community value creation and delivery, ensuring that interventions are tailored to meet the specific needs of beneficiaries.

Once community value has been created and delivered through these channels, establishing robust measurement mechanisms becomes crucial to evaluate the outcomes effectively. These mechanisms assess how well the initiatives have addressed the intended goals, such as improving individual well-being, fostering community cohesion, and enhancing social inclusion (Davies et al., 2019; Rut et al., 2021). Feedback gathered through these evaluation processes can then be reintegrated into the system, allowing food-sharing communities to refine their value propositions and improve their delivery strategies. This iterative feedback loop not only strengthens the alignment between services and community needs but also enhances the efficiency and impact of food-sharing activities.

By prioritising the cyclical creation, capture, measurement, and reinvestment of community value through this closed-loop approach, food-sharing communities can achieve greater adaptability and sustainability. This process supports food-sharing organisations in making informed decisions on how best to design and refine their interventions, ultimately bolstering their resilience and ensuring long-lasting benefits for the communities they serve. Through this interconnected system, sustainable community value networks can evolve to address emerging challenges while continuously enhancing their contributions to social value creation.

3 Discussion: The Conceptual Framework - SCVN Model

Drawing on a review of current literature and practical initiatives, the 'Sustainable Community Value Networks' conceptual framework has been developed to provide a comprehensive and structured approach to build sustainable community value networks in food sharing community (as shown in Fig. 3). This SCVN framework integrates four key components, including *define community value propositions*, *map community value beneficiaries*, *community value creation and delivery*, and *community value measurement and feedback*, each with detailed sub elements (Morrow, 2019; Rut et al., 2021; Santos, 2012; Wenger et al., 2011).

The SCVN framework operates as a closed-loop cycle, beginning with the identification of key community value propositions, such as improving food crisis support, fostering social connectivity, and promoting a healthy lifestyle. These propositions guide the process of mapping beneficiaries, ensuring that disadvantaged groups such as low-income families, elderly individuals, and single parents receive targeted support. Once the beneficiaries are identified, the framework facilitates community value creation through three theme-driven delivery channels, including financial and budgeting, nutrition and cooking, health and wellbeing management. The final component, community value measurement and feedback, plays a crucial role in assessing impact and capturing insights on both individual and community levels. At individual level, criteria include personal growth and fulfilment, as well as knowledge and skill development. At the community level, criteria encompass community cohesion, engagement and empowerment.

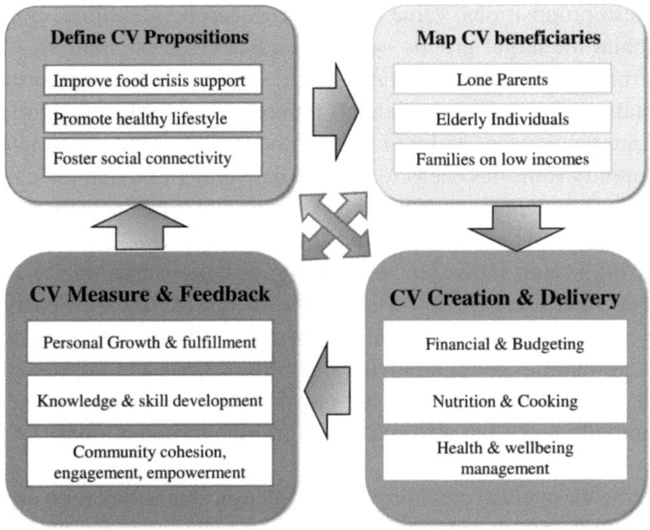

Fig. 3. SCVN Conceptual Framework

The evaluated outcomes are then reinvested and feedback into the framework, refining value propositions and improving delivery mechanisms, thereby forming a sustainable and adaptive cycle.

By structuring these interconnected components into a closed-loop system, the SCVN framework ensures that community value is continuously created, measured, and reinvested to ensure long-lasting impact. The iterative nature of the SCVN model allows food sharing organisations to refine their strategies based on evaluated outcomes, ensuring that activities and initiatives remain relevant and effective in addressing the evolving needs of beneficiaries. This structured approach not only enhances the operational efficiency and sustainability of food sharing communities but also strengthens their capacity to foster resilient transitions in today's changing environment. Ultimately, SCVN serves as a guiding model for food sharing organisations to systematically generate, capture, and reinvest community value, supporting the creation of long-lasting and impactful community value networks.

4 Contribution to New Knowledge

This study offers a significant contribution to both theory and practice by proposing the Sustainable Community Value Networks (SCVN) framework, a structured decision support model designed to strengthen sustainable and resilient transitions in food sharing communities. The framework integrates four key components: community value propositions, beneficiary mapping, value creation and delivery, and measurement and feedback—into a closed-loop system. By conceptualising the interdependencies between these elements, the SCVN model advances current understanding of how communities can generate, manage, and sustain value in response to complex and evolving challenges.

In addition to its theoretical relevance, the SCVN framework provides practical value for decision-makers, including community leaders, practitioners, and policymakers. It supports evidence-based planning and resource allocation, enabling food sharing communities to design more targeted, effective interventions. In doing so, the framework contributes to more adaptive, resilient, and sustainable systems of support for vulnerable populations. Ultimately, it enhances decision-making processes that are central to managing community transitions in an increasingly uncertain global landscape.

5 Conclusion

This paper has presented the results of a Systematic Literature and Practice Review (SLPR) to investigate how sustainable community value networks can be developed to support decision-making within food sharing communities. A total of 42 academic and practical sources were reviewed to identify critical components and patterns, culminating in the development of the SCVN conceptual framework. This model integrates key elements, including the definition of community value, identification of beneficiaries, value creation and delivery mechanisms, and feedback systems into a coherent, closed-loop decision support tool.

The SCVN framework provides a practical tool to guide community actors in planning and implementing effective interventions that contribute to sustainable and resilient community transitions. It equips decision-makers with a structured approach to navigating uncertainty, adapting to change, and maintaining long-term community well-being. As food sharing initiatives face increasing pressure to respond to social, economic, and environmental disruptions, the SCVN framework provides both the strategic clarity and practical guidance needed to support inclusive, future-ready community development.

Acknowledgements. The work reported in this paper was conducted as part of a University Research Studentship (URS) funded by the University of Plymouth.

Reference[1]s

Berns, K., Rossitto, C., Tholander, J.: Queuing for waste: sociotechnical interactions within a food sharing community. In: CHI'21 Making Waves, Combining Strengths, 10. Online Virtual Conference (Originally Yokohama, Japan). Association for Computing Machinery (2021)

Berns, K., Rossitto, C., Tholander, J.: Learning from other communities: organising collective action in a grassroots food-sharing initiative. Comput. Supported Coop. Work. **32**(4), 951–999 (2023)

Davies, A.: SHARECITY Sustainability of City-Based Food Sharing: Typologies of Food Sharing. ShareCity Working Paper No. 1. Department of Geography, School of Natural Sciences, Trinity College Dublin, Dublin (2016)

Davies, A., Weymes, M., Mackenzie, S.: Communicating goals and impacts of food-sharing. Urban Agricult. Mag. **34**, 38–40 (2018)

[1] **Note**: not all 43 papers included in the SLPR but only selected key references are listed here because of page limit

Davies, A.R., Cretella, A., Franck, V.: Food sharing initiatives and food democracy: practice and policy in three European cities. Polit. Governance. **7**(4), 8–20 (2019)
Denyer, D., Tranfield, D.: Producing a systematic review. In: Buchanan, D., Bryman, A. (eds.) The Sage Handbook of Organizational Research Methods, Chapter 39, pp. 671–689. Sage Publications Ltd, London (2009)
Donald, W.E., Ford, N.: Fostering social mobility and employability: the case for peer learning. Teach. High. Educ. **28**(3), 672–678 (2023)
FAO Council: The Right to Food. FAO (2010)
FAO, IFAD, UNICEF, WFP & WHO: The State of Food Security and Nutrition in the World 2024: Financing to End Hunger, Food Insecurity and Malnutrition in all its Forms. FAO, Rome (2024)
FareShare: Fareshare's Impact Report 2024: Transforming Surplus Food into Support. FareShare (2024)
Feeding Devon: Health Eating Project. Feeding Devon (2024)
Feeding Britain: Feeding Hope: The Impact of Affordable Food Clubs on Nutrition and Wellbeing. Feeding Britain (2024)
Fox, S.: Communities of practice, Foucault and actor-network theory. J. Manag. Stud. **37**(6), 853–868 (2000)
Ganglbauer, E., Fitzpatrick, G., Subasi, Ö., Güldenpfennig, F.: Think globally, act locally: a case study of a free food sharing community and social networking. In: Proceedings of the 17th ACM Conference on Computer Supported Cooperative Work & Social Computing. CSCW'14: Computer Supported Cooperative Work, pp. 911–921. Association for Computing Machinery, New York (2014)
Harvey, J., Smith, A., Goulding, J., Branco Illodo, I.: Food sharing, redistribution, and waste reduction via mobile applications: a social network analysis. Ind. Mark. Manag. **88**, 437–448 (2020)
Hietschold, N., Voegtlin, C., Scherer, A.G., Gehman, J.: Pathways to social value and social change: an integrative review of the social entrepreneurship literature. Int. J. Manag. Rev. **25**(3), 564–586 (2023)
Hohenstein, N., Feisel, E., Hartmann, E., Giunipero, L.: Research on the phenomenon of supply chain resilience: a systematic review and paths for further investigation. Int. J. Phys. Distrib. Logist. Manag. **45**(1–2), 90–117 (2015)
Jaeggi, A.V., Gurven, M.: Natural co-operators: food sharing in humans and other primates. Evol. Anthropol. Issues, News, Rev. **22**(4), 186–195 (2013)
Lave, J., Wenger, E.: Situated Learning: Legitimate Peripheral Participation. Cambridge University Press, Cambridge, MA (1991)
Lombardi, M., Costantino, M.: A social innovation model for reducing food waste: the case study of an Italian non-profit organization. Adm. Sci. **10**, 45 (2020)
Lozano, R.: Collaboration as a pathway for sustainability. Sustain. Dev. **15**(6), 370–381 (2007)
Mackenzie, S.G., Davies, A.R.: SHARE IT: co-designing a sustainability impact assessment framework for urban food sharing initiatives. Environ. Impact Assess. Rev. **79**, 106300 (2019)
Makov, T., Shepon, A., Krones, J., Gupta, C., Chertow, M.: Social and environmental analysis of food waste abatement via the peer-to-peer sharing economy. Nat. Commun. **11**(1), 1156 (2020)
Mair, J., Martí, I.: Social entrepreneurship research: a source of explanation, prediction, and delight. J. World Bus. **41**(1), 36–44 (2006)
Marovelli, B.: Cooking and eating together in London: food sharing initiatives as collective spaces of encounter. Geoforum. **99**, 190–201 (2019)
Mazzucchelli, A., Gurioli, M., Graziano, D., Quacquarelli, B., Aouina-Mejri, C.: How to fight against food waste in the digital era: key factors for a successful food sharing platform. J. Bus. Res. **124**, 47–58 (2021)

Meshulam, T., Font-Vivanco, D., Blass, V., Makov, T.: Sharing economy rebound: the case of peer-to-peer sharing of food waste. J. Ind. Ecol. **27**(3), 882–895 (2023)
Morone, P., Pasquale, M.F., Enrica, I., Morone, A.: Does food sharing lead to food waste reduction? An experimental analysis to assess challenges and opportunities of a new consumption model. J. Clean. Prod. **185**, 749–760 (2018)
Morrow, O.: Community self-organizing and the urban food commons in Berlin and New York. Sustainability. **11**(13), 36–41 (2019)
Parker, A., Landry, B.M., Grinter, R.E.: Characteristics of shared health reflections in a local community. In: Proceedings of CSCW '10, pp. 435–444. ACM, New York (2010)
Pret, T., Carter, S.: The importance of 'fitting in': collaboration and social value creation in response to community norms and expectations. Entrep. Reg. Dev. **29**(7–8), 639–667 (2017)
Rigg, C., Coughlan, P., O'Leary, D., Coghlan, D.: A practice perspective on knowledge, learning and innovation - insights from an EU network of small food producers. Entrep. Reg. Dev. **33**(7–8), 621–640 (2021)
Rut, M., Davies, A.R., Ng, H.: Participating in food waste transitions: exploring surplus food redistribution in Singapore through the ecologies of participation framework. J. Environ. Policy Plann. **23**(1), 34–47 (2021)
Santos, F.M.: A positive theory of social entrepreneurship. J. Bus. Ethics. **111**(3), 335–351 (2012)
Schanes, K., Stagl, S.: Food waste fighters: what motivates people to engage in food sharing? J. Clean. Prod. **211**, 1491–1501 (2019)
Shaw, E., Carter, S.: Social entrepreneurship: theoretical antecedents and empirical analysis of entrepreneurial processes and outcomes. J. Small Bus. Enterp. Dev. **14**(3), 418–434 (2007)
SHARECITY: Sustainability Impacts of Food Sharing in Smart Dublin 8. SHARECITY (2022)
United Nations World Food Programme: Global Report on Food Crises: Number of People Facing Acute Food Insecurity Rose to 258 Million in 58 Countries in 2022. United Nations World Food Programme (2023)
Veen, E.J.: Fostering community values through meal sharing with strangers. Sustainability. **11**(7), 2121 (2019)
Tranfield, D., Denyer, D., Smart, P.: Towards a methodology for developing evidence-informed management knowledge by means of systematic review. Br. J. Manag. **14**(3), 207–222 (2003)
Wenger, E., Trayner, B., Laat, D.: Promoting and Assessing Value Creation in Communities and Networks: A Conceptual Framework. Ruud de Moor Centrum Raporten, Open Universiteit, Netherlands (2011)
Zhao, G., Liu, S., Wang, Y., Lopez, C., Ong, A., Chen, X.: Reducing food waste from social innovation perspective: a review of measures, research gaps and future directions. Int. Food Agribus. Manag. Rev. **26**(2), 199–223 (2023)

A Decision-Making for Understanding Risk Propagation and Resilience Capabilities in AFSCs Based on Agrifood Transition: A Conceptual Framework

Xiaoning Chen[1(✉)], Shaofeng Liu[1], Huilan Chen[1], and Dulekha Kasturiratne[2]

[1] Plymouth Business School, University of Plymouth, Plymouth PL4 8AA, UK
xiaoning.chen@plymouth.ac.uk

[2] Henley Business School, University of Reading, Reading RG6 6UD, UK

Abstract. Driven by the combined influence of global climate change, geopolitical conflicts, and public health crises, the agrifood transition system is undergoing a profound transformation from traditional production and consumption patterns toward low-carbon, digitalized, and sustainable models. However, agrifood supply chains (AFSCs) face high levels of uncertainty during this transition, as risks often spread rapidly across different stages, triggering cascading effects and threatening food safety and supply chain stability. Although existing studies have examined risk factors and resilience in agrifood supply chains, there is still a lack of a systematic conceptual framework addressing risk propagation pathways and their relationships with resilience capabilities in the context of agrifood system transition. To fill this gap, the present study proposes a conceptual framework for investigating risk factors, characterizing risk propagation pathways, and resilience capabilities within agrifood transition systems. Starting with the identification of risks, risk propagation pathways, and resilience capabilities, the framework illustrates how risks are identified and propagated within AFSCs. Furthermore, it conceptualizes resilience capabilities across three dimensions, pre-shock, during shock and after shock, to explain how different capabilities mitigate risks, reduce their negative impacts, and support AFSCs during agrifood transition. The contribution of this study lies in developing a comprehensive framework that systematically links risks, their propagation pathways, and resilience capabilities used to mitigate these risks, thereby facilitating the sustainable transition of agrifood systems.

Keywords: Agrifood supply chain · risk factors · risk propagation · resilience capabilities · Agrifood transition

1 Introduction

The agrifood system also as known as food system. [73] defined the system is a way that people gain and consume food in space and time. Agrifood system integrates all elements related to produce, processing, distribution and consumption as well the output of these activities, and it also encompasses social economic and environmental outcomes

A. P. Cabral Seixas Costa et al. (Eds.): ICDSST 2026, LNBIP 582, pp. 64–81, 2026.
https://doi.org/10.1007/978-3-032-25146-6_5

[53]. Agrifood systems play a crucial role in achieving sustainable development goals, as it provides important sources of nutrition, employment opportunities and economic prospects for numerous individuals and the global population [84]. On average, agriculture produces approximately 237 million tons of food each year. This includes 195 million tons of tubers, fruits, grains, roots and vegetables, 11 million tons of meat, and 2.1 billion liters of milk. Agriculture provides livelihoods for over 2.5 billion people, most of whom live in low-income and middle-income countries, and it continues to play a key role in promoting global development [44]. Due to the influence of factors such as globalization and technological innovation, agrifood system have been continuously evolving and becoming increasingly complex [25, 34].

Currently, agrifood systems face multiple problems and issues, such as an increase in food demand, the climate change, scarcity of natural resources, loos of biodiversity, as well as consumers' increasing demands for health and sustainable [40]. These challenges not only represent external pressures but also may be transformed into specific risks through the interdependent and complex structures within agrifood system [69]. Specifically, risks can propagate through the stages of production, processing, logistics, retail, and consumption, generating cumulative or amplifying effects [26]. While agrifood transition not only involves adjustment in agricultural production methods and food consumption patterns, but also derives the restructuring of processing, logistics, retail and other stages [72, 95]. Especially, the innovation and systematic transition play an essential role in reshaping the agrifood system and preventing various risks [42]. As proposed by *GLOPAN(Glob.Panel Agric. Food Syst. Nutr.)* [48], the innovation and transition of the system not only aims to drive the system toward a healthy and sustainable direction, but also the transition measures can help fundamentally address the various risks currently faced by the agrifood system.

Agrifood supply chain (AFSCs), as one of key links in agrifood systems, play an essential role in ensuring food security [111]. The potential of AFSCs has been widely recognized in the global policy discussions regarding the need for a sustainable transition of agrifood system [11, 17, 32, 107]. AFSCs cover the entire farm-to-fork process, bringing together farmers, processors, wholesalers, retailers, and consumers [75]. These diverse stakeholders share common goals such as ensuring food security, sustaining economic development, and supporting livelihoods [2, 5]. Currently, AFSCs faces one of the major long-term challenges in agrifood transition, namely the sustainability of AFSCs. The pursuit of sustainability in AFSCs is driven by several critical factors. The sector's rapid expansion (Joshi et al., 2023), essential for meeting rising global food demands, has concurrently led to significant environmental degradation, resource scarcity, and heightened social inequities [9]. These challenges highlight the imperative for a systemic transition towards sustainable practices, which is vital for preserving ecological integrity and ensuring equitable resource distribution and food security [46]. In addition, the growing trend of globalization and the continuous enhancement of innovation capabilities have increased the complexity of AFSCs, as well as strengthened the interconnectedness between suppliers and processors and the interdependencies among different practitioners within AFSCs [97]. This kind of highly interconnected structure implies risks tend to spread across different sectors of the system, resulting in widespread risk propagation across AFSCs [4, 58, 80]. Meanwhile, AFSCs inherently exhibits characteristics such as strong seasonality, high perishability, long production cycles, and strict temperature control requirements [65], which make it inherently more vulnerable. These challenges posed by rapid change increase their exposure to risk shocks during the transition process, thereby weakening the overall sustainability performance of AFSCs. Therefore, identifying risks, understanding risk propagation, and mastering resilience capabilities in the AFSCs are vital for promoting the sustainable agrifood system transition.

Risks in AFSCs are complex and can generally be categorized into two types: functional and disruptive risks [110]. The former is related to possible dysfunctions in the day-to-day functioning of AFSCs, such as fluctuations in processing and demand, and lead times in the production of agricultural commodities, while the latter is related to low-frequency, high-impact actions [66]. Current studies in understanding AFSCs' risks from various perspectives, including risk identification [18, 35, 113, 114] and risk assessments [36, 87]. However, these risks are not isolate, interdependence of AFSC's practitioners lead to disruptions or shock caused by any one party's risk will propagate along with supply chains (SCs) effects other practitioners [8]. For example, Tang et al. [101] developed a cascading failure model to analyse risk propagation; however, their focus was on the manufacturing sector rather than the agri-food sector. In the agrifood sector, research by Aday and Aday [1] in food supply chains, COVID-19 has an impact on the entire process from field to consumer. COVID-19 has resulted in restrictions on worker movement, changes in consumer demand, closures of food production facilities, policies restricting food trade, and financial pressures on food supply chains [1]. Similar findings are shown in Jaleta's [62] study. To mitigate these risks in AFSCs, some study also pointed out resilience capabilities to mitigate these risks in AFSCs. For example, Ali et al. [8], Tukamuhabwa [104], and Zhong et al. [117] proposed that various resilience capabilities, such as, flexibility, redundancy, innovation and visibility, to mitigate risks in AFSCs. Coopmans et al. [31] further indicated that diversity may contribute to reducing risk exposure and limiting cascading impacts across AFSCs.

Despite the growing body of literature on agrifood supply chain risk management, existing studies predominantly focus on risk identification or isolated mitigation strategies. Less attention has been paid to the complex interactions between multiple risks, their propagation mechanisms across AFSCs, and the implications for integrated decision-making in agrifood transitions. Moreover, while global sustainability challenges affect agrifood systems worldwide, risk realisation, usually referred to as 'sudden shocks' affects specific food systems, which are geographically localised and/or specialised in specific products and services. van Wassenaer et al. [112] argue, the challenge lies in identifying context-specific challenges and policy options through a resilience lens and translating these into measurable indicators. At the same time, "risk" and "resilience" remain polysemic concepts shaped by differing ontological and epistemological perspectives, which complicates the development of unified assessment frameworks. Against this background, this study develops a conceptual framework of AFSC and analyzes risk factors and their interrelationships within AFSC. The conceptual framework can help decision makers better understand risk factors and risk propagation and guide the implementation of effective risk mitigation measures and the formulation of emergency plans.

Three research questions are developed: (1) What kinds of risk factors exist in AFSCs under the agrifood transition? (2) What are the risk propagation pathways in AFSCs under the agrifood transition? and (3) What resilience capabilities and strategies can be used to mitigate risks under the agrifood transition? The organisation of this study is as follows: Sect. 2 presents a literature review; a conceptual framework and its discussion are down in Sect. 3. Finally, Section e provides a conclusion of this study.

2 Literature Review

In this section, we review identification of risks, risk propagation and resilience capabilities in agrifood transition supply chains, leading to identification of research gaps.

2.1 Risk Factors in AFSCs

Risk refers to an uncertain situation that may lead to adverse consequences for a potential outcome [89]. It is an inherent aspect of modern supply chain operations [93]. Particularly, supply chains composed of numerous participants across multiple tiers are significantly exposed to risk [20]. Agricultural food supply chains are undoubtedly a typical example of high-risk supply chains. In addition to the large number of involved parties, the operational complexity of agricultural food supply chains, such as seasonal production, long and fixed lead times, variations in product quantity and quality standards, trade and buffer inventory constraints, and traceability requirements, makes them highly susceptible to severe disruptions [106].

In supply chain management, risk (or disruption) can be described as any event that interrupts the three main processes of a supply chain and shifts the distribution of possible outcomes [49, 54]. Numerous inherent risk sources exist within agricultural food supply chains [62, 115] and scholars have identified various risks in agricultural product supply chains. Chopra and Sodhi [27] proposed three broad categories of risks and nine specific risk types. Subsequently, Nyamah et al. [84] argued that agrifood supply chains (AFSCs) are primarily influenced by five types of risks: demand risks, policy and regulatory risks, biological and environmental risks, climate and weather risks, and political risks. They also highlighted additional risk factors, including extreme droughts, floods, and contamination caused by poor hygiene conditions and diseases.

More recently, Krstić et al. [68] categorized risks in AFSCs into seven types: natural risks, product-related risks, logistics risks, management risks, market risks, institutional risks, and safety risks. At the same time, Zhao et al. [116] identified 16 prevalent risk factors in agricultural product supply chains, including weak agricultural infrastructure, pest and disease risks, extreme weather conditions, supply–demand imbalances, and price volatility. Following previous studies (e.g., [84, 116]), we categorize AFSCs risks into eight types: supply risk, demand risk, biological and environmental risk, political and regulatory risk, logistics and infrastructure risk, and management and operational risk, among others. This classification is particularly appropriate because it not only encompasses the intrinsic characteristics of agricultural products but also reflects broader structural features of the supply chain. Table 1 summarizes the main types of risks and their influencing factors identified by various practitioners and scholars in AFSCs.

Table 1. Risks factors in agrifood transition supply chains

AFSCs practitioners	Risk types	Risk elements	Author(s) (year)
Famers	Supply risks; Demand risks; Biological and environmental related risks; Political and regulation risks; Logistical and infrastructural risks;	Poor planning; Volatility in fertilizer cost; Uncertainty in yield; Extreme weather; Labour shortage; Volatility in price; Pests and diseases; Biodiversity;	[84, 116]
Processors	Supply risks; Biological and environmental related risks; Political and regulation risks; Political and regulation risks; Logistical and infrastructural risks;	Contamination in production and processing; Food quality and safety issues; poor storage conditions; Machine failures; Lagging technological innovation;	[41, 82]
Wholesalers	Logistical and infrastructural risks; Supply risks; Financial risks; Management and operational risks;	Extreme weather; Labour shortage; Food damage or contamination; Delay in payment or non-payment; Poor storage infrastructure; Imbalance in offer and demand	[4, 6]
Retailers	Supply risks; Financial Risks; Biological and environmental risks; Logistical and infrastructural risks;	Food damage or contamination; Inadequate food quality control; Ineffective system integration; Food contamination; Inefficient shopping experience	[19, 42]
Consumers	Supply risks; Demand risks; Policy and regulation risks;	Customer preference changes; Volatility in market price; Information asymmetry between consumers and retailers; Health risks	[6, 82].

2.2 Risk Propagation in AFSCs

Supply chains (SCs) play an important role in agriculture and are the backbone of global trade [39]. In this context, interactions within supply chain ecosystems are extremely complex, driven by interrelationships and feedback between the supply chain, nature, society, and the economy [59]. Supply chains are becoming increasingly complex Chand et al. [121] due to market uncertainty, shorter product life cycles, rapid iteration of new technologies, heightened global geopolitical risks, and growing personalized demand, leading to simultaneous or sequential disruptions [72]. In other words, risks in a supply chain do not exist in isolation and may affect the performance of other members in the chain as well as the entire supply chain [23, 77]. Risk propagation in SCs has been defined as the process by which the impact of disruptions spreads beyond its initial location (Scheibe and Blackhurst, 2017), including ripple effects Ivanov et al. [61], cascading failures Basole et al. [16] and risk diffusion [15].

In AFSCs, risks propagate along the supply chain and affect the performance of individual members as well as the system as a whole [34]. To investigate risk prpogation in AFSCs, a growing body of literature has examined how different types of risks are interconnected and transmitted across the supply chain. For example, Zhao et al. [116] explored the interrelationships among AFSC risks using a hierarchical framework, showing that risks originating from external environments, such as extreme weather events and political or economic instability, may trigger internal risks, including market price fluctuations, pest and disease outbreaks, and supply and demand imbalances. Similarly, Malik et al. [75] examined the impacts of climate change and extreme weather events on Australian AFSCs and found that, beyond productivity losses, such disruptions also generate social and health consequences, including job and income losses as well as reduced nutritional availability and diet quality. Their results further indicate that fruits, vegetables, and livestock are among the most severely affected sectors, with knock-on effects extending to non-food sectors such as transportation services. Consistent with studies focusing on cascading impacts caused by high-impact, low-probability events, Zhao et al. [117] showed that regional conflicts, such as the Russia–Ukraine war, led to production losses of 16.85% in the food processing sector. In addition, Srivastava et al. [101] argued that changes or inadequacies in government regulations may have serious consequences for fresh food retail supply chains, potentially resulting in low inventory turnover, unmet revenue growth targets, and reduced operating margins.

From the perspective of risk propagation, most existing studies focused on risk identification, description, assessment and risk management Deng et al. [37] and primarily examine the forward and backward transmission of risks along upstream and downstream supply chain. However, risks do not remain isolated at individual nodes but propagate through supply chains as a result of interdependencies among actors, forming a networked risk propagation process [38]. Deng et al. [38] presented and proposed several risk propagation pathways, including risk transfer, risk mutation, risk overlap, risk contagion and risk retrain. Simultaneously, Colon and Hochrainer-Stigler [29] demonstrate how risks spread and interact through both top-down and bottom-up pathways. Existing studies [7, 12, 30, 113] although describe risk propagation in AFSCs, empirical research on the mechanisms risk propagation in AFSCs remains scarce.

2.3 Supply Chains Resilience Capabilities in AFSC

Risk propagation in supply chains (SCs) does not automatically translate into actual losses or recovery to the pre-disruption state [39, 61]. Its ultimate effects depend on the decision-making of supply chain practitioners during the risk propagation process [33, 59]. When facing diverse and evolving risks, SC practitioners need to understand the characteristics of risk propagation and continuously adjust their decision-making related to operations, coordination, and resource reallocation to cope with the uncertainty arising from risk propagation and disruptions e.g., [33, 39, 59, 103]. From this perspective, supply chain resilience (SCRes) can be understood as a capability that emerges through decision making. Accordingly, decision making plays an essential role in supply chains, and the effects of these decisions are directly manifested in supply chain resilience.

Rice et al. [92] proposed one of the earliest definitions of SCRes, describing it as the capability to cope with disruptions and recover operational performance. Subsequently, Ponomarov and Holcomb [90] proposed a more comprehensive definition of SCRes by incorporating multiple phases, including preparation, response, and recovery from disruptions, ultimately achieving the expected level of performance. Later definitions of SCRes are largely consistent with that of Ponomarov and Holcomb [90], while incorporating additional elements. For example, Kamalahmadi and Parast [66] conceptualized SCRes as consisting of four phases, anticipation, resistance, recovery, and response, and emphasized that adaptability, flexibility, and agility play crucial roles in responding to disruptions. Ribeiro and Barbosa-Póvoa [91] defined SCRes as the capability of a supply chain to prepare for, respond to, and recover from disruptions at an acceptable cost and within an acceptable time frame. However, the significance of SCRes extends far beyond supply chain risk control [21]. And it primarily supports firms and supply chain practitioners in developing resilience-related capabilities [52]. These SCRes capabilities represent internal attributes that enable supply chains to anticipate, absorb, and recover from risks and disruptions [86].

In AFSCs, Stone and Rahimifard [102] proposed a comprehensive definition of AFSC resilience, which was later discussed by [52]. They defined AFSC resilience as "the collective ability of agri-food supply chain stakeholders to ensure acceptable, sufficient, and stable food supplies, at the required times and locations, via accurate anticipation of disruptions and the use of strategies which delay impact, aid recovery, and allow cumulative learning post-disruption" [86]. Based on this definition, several studies have identified a range of resilience capabilities in AFSCs aimed at mitigating and preventing the effects of risk propagation. For example, Kamalahmadi and Parast [66] identified eight resilience capabilities that can mitigate risks, including flexibility, redundancy, trust, information sharing, visibility, velocity, leadership, and innovation. Zhao et al. [117] identified more than fifteen capabilities to improve AFSC resilience; in addition to those proposed by Kamalahmadi and Parast [66], they highlighted dynamic competitiveness, supply chain collaboration, and leadership as further contributors to resilience. Other studies, Leat and Revoredo-Giha [70], Ali et al. [6], and Marcucci et al. [79] have also summarized resilience capabilities in AFSCs. Moreover, according to the four stages of supply chain resilience, anticipation, resistance, recovery, and response, Stone and Rahimifard [102] categorized AFSC resilience capabilities into those applicable before, during, and after a shock. For instance, redundancy is relevant in both the pre-shock and post-shock stages [102], while knowledge sharing applies primarily to the pre-shock and post-shock stages [102]. The resilience capabilities used to mitigate disruptions and risks in AFSCs are summarized in Table 2.

Table 2. SCRes Capabilities and Strategies in AFSCs

Phases	SCRes capabilities	SCRes strategies	Author(s)(year)
Pre-shock	Flexibility	Have more customer accounts, flexible supply base, flexible transportation, dynamic assortment planning,	[28, 56, 66, 86, 90]
	Redundancy	Add inventory, postpone strategy, back-up capability, risk management insurance, diversification of marketing channels,	
	Trust	Building relationships with buyers and suppliers, joint purchases	
	Information sharing	Information and communication technology adoption, regularly interaction,	
	Diversification	Multiple suppliers, diversified sourcing regions and products	
	Transparency and visibility	Information sharing, traceability	
	Financial preparedness	Liquidity reserves, insurance	
During- shock	Visibility	Demand forecasting, joint decision-making,	[10, 65, 93, 96, 111]
	Velocity	Decision-making velocity, digitalization, agile production and operations, fast logistics and transportation options	
	Leadership	Top management team's support, government supports, employee wellbeing,	

(*continued*)

Table 2. *(continued)*

Phases	SCRes capabilities	SCRes strategies	Author(s)(year)
	Agility	Speed of decision-making and execution	
	Flexibility	Switching suppliers, multiple sourcing; process outsourcing	[55, 69, 81, 119]
	Redundancy	Safety stock, strategic reserves, backup production lines, backup distribution centres	[80, 81, 102, 118]
After-shock	Knowledge management	Skill and efficiency development, rewards system, building shared understanding, decentralized knowledge network, regularly trainings	[56, 57, 90, 102, 105, 109, 110]
	Innovation	Security system improvement, product innovation,	
	Adaptability	Adjusting strategies, structures, and practices	

As summarized in Table 2, existing research identifies multiple resilience capabilities in AFSCs. However, these capabilities are not merely the result of static structural configurations. For example, Owida et al. [85] proposed a decision-making framework that integrates resilience and sustainability in agri-food and food production systems, demonstrating how flexibility and robustness are achieved through strategic, tactical, and operational decisions under disruption contexts such as the COVID-19 pandemic. Their findings indicate that resilience capabilities emerge from dynamic decision processes that coordinate risk mitigation, resource allocation, and cross-functional collaboration. This perspective highlights the critical role of decision making in shaping resilience outcomes and provides an important foundation for linking risk propagation processes with resilience building in agrifood supply chains, particularly in the context of agrifood transition.

3 Conceptual Framework Development and Discussion

Existing studies have demonstrated risks and resilience capabilities in AFSCs from diverse perspectives. However, there remains a lack of a systematic analytical framework that specifically addresses risk propagation in AFSCs. To address this gap, Fig. 1 defines the core concepts related to risks and risk propagation in AFSCs and, on this basis, develops a conceptual framework. This framework effectively identifies the interrelationships among risks in AFSCs. It is structured into four conceptual sections, including risk identification, risk propagation, and resilience capabilities.

This conceptual framework begins with AFSCs, which consist of five types of practitioners: farmers, processors, wholesalers, retailers, and consumers. Meanwhile, the framework also considers other stakeholders related to AFSCs, such as governments and food hubs. Each practitioner and stakeholder establishes direct or indirect connections within AFSCs. For example, farmers can directly sell products to wholesalers, retailers, and consumers. In terms of risk categories, AFSCs are exposed to a variety of risks that may affect each practitioner and stakeholder. In this layer, the framework identifies 11 common risks generated by AFSC practitioners and further classifies specific risks according to the five types of practitioners. For instance, all AFSC practitioners are often exposed to climate risks, biological and environmental risks, and demand risks. Among them, farmers are particularly sensitive to climate risks, supply risks, political risks, as well as biological and environmental risks. This section provides decision makers with clear guidance for identifying and understanding risk sources faced by AFSC practitioners.

Owing to increasing interconnectedness among practitioners and stakeholders in AFSCs, risks can spread and interact across AFSCs. Accordingly, the risk propagation section illustrates risk propagation pathways across AFSCs. In this layer, risks can be transmitted forward or backward along the supply chain through upstream and downstream relationships. For example, climate risks can lead to a reduction in agricultural output, which in turn causes shortages of raw materials and price fluctuations at the processing stage and ultimately affects supply in the terminal market and consumer welfare. In addition, risks may exhibit complex interactions among different practitioners and stakeholders through mechanisms such as risk transfer, risk contagion, risk mutation, risk overlap, and risk restraint [38]. External risks, such as policy adjustments, macroeconomic fluctuations, or global crises, often affect supply chain operations in a top-down manner [54]. Meanwhile, risks within the supply chain, such as decision-making errors, information asymmetry, or insufficient coordination, may spread and amplify system vulnerability in a bottom-up manner [11, 48]. Ultimately, these risks spread and dynamically evolve within the supply chain network, posing severe challenges to the stability and sustainability of agrifood system transformation. Such risk propagation can lead to a range of social, health, environmental, and economic impacts on individuals and institutions.

To reduce and mitigate AFSC risks and their effects, the final section of the framework focuses on resilience capabilities, drawing on insights from previous studies [90]. This section classifies resilience capabilities into three stages: pre-shock, during-shock, and post-shock, enabling decision makers to implement targeted management measures at different risk stages. Specifically, in the pre-shock phase, AFSCs should enhance their resilience capabilities to identify and anticipate potential risks through redundancy, diversification, and transparency [28, 86, 104]. During the shock phase, flexibility, visibility, and coordination enable AFSCs to respond to and mitigate risk effects [61, 109]. In the post-shock phase, recovery, learning, and adaptability support functional recovery and may even facilitate a transition toward a more robust operational state [110]. These resilience capabilities do not operate in isolation but are linked across different stages through key managerial decisions to contain the spread of risks and support the ongoing agrifood transition.

At the theoretical level, the main contribution of this conceptual framework lies in its comprehensive integration of risk identification, risk propagation pathways, and resilience capabilities. The framework addresses limitations in existing risk management research, particularly with regard to risk propagation, and provides a new perspective for decision makers in AFSCs. Moreover, this study systematically links risk categories and factors with risk propagation and examines how risks can be mitigated through managerial decision-making. It fills an important gap in the literature concerning risk propagation in AFSCs under agrifood system transition.

At the practical level, this study offers a decision-oriented conceptual framework for AFSCs management. First, the framework enables decision makers to identify risk sources and factors across AFSCs [78, 103]. Second, it helps AFSC practitioners understand risk propagation pathways, allowing them to recognize how risks interact across actors and to identify potential cascading effects [61]. This shared understanding facilitates more coordinated and collaborative decision-making among AFSCs practitioners. Moreover, the framework provides guidance on how to mitigate risks by configuring appropriate resilience capabilities across different stages. These resilience capabilities can enhance the ability of AFSCs to anticipate, adapt to, and recover quickly from risks [111]. Through effective management, AFSCs can not only achieve rapid recovery during crises such as natural disasters, pandemics, or resource shortages, but also transition toward more sustainable operational states after crises, thereby strengthening overall resilience.

In conclusion, this study extends prior work by incorporating risk propagation pathways and their potential impacts. Moreover, the proposed conceptual framework addresses three key questions: what risks exist in AFSCs, how risks propagate within AFSCs, and what types of resilience capabilities can be used to mitigate these risks. The framework provides a foundation for analyzing and managing risks and their interrelationships and offers an important reference for decision makers engaged in AFSCs risk management in the agrifood transition. However, no empirical study was conducted to examine risks, risk propagation and resilience capabilities. In the future, this study can be empirically conducted through interviews and thematic analysis based on this conceptual framework.

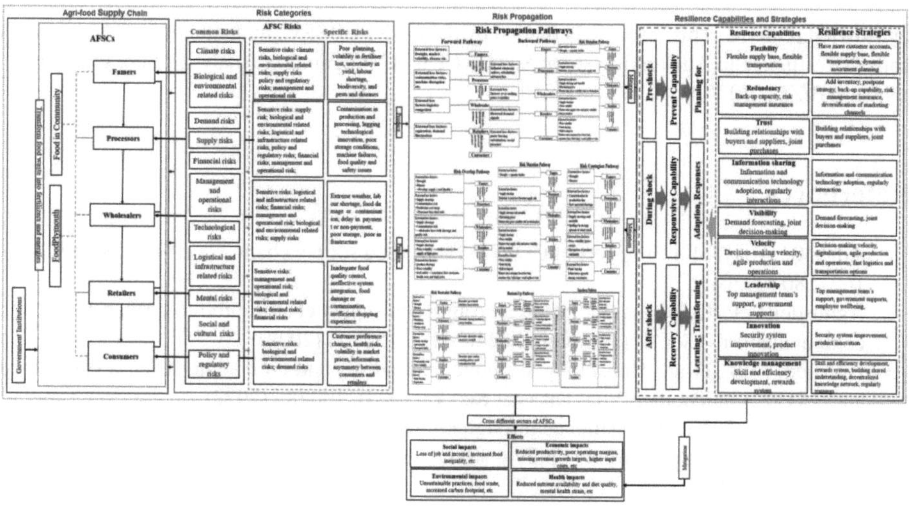

Fig. 1. Risk and risk propagation conceptual framework in AFSCs.

4 Conclusion

This study investigates risks, risk propagation, and resilience capabilities in AFSCs within the context of the agrifood transition. Drawing on a comprehensive literature review, it develops a conceptual framework for decision-making that integrates and explains AFSC risks, risk propagation pathways, as well as resilience capabilities and strategies in sustainable AFSCs.

The literature review shows that practitioners in AFSCs face diverse types of risks, and that these risks differ significantly across AFSC actors. For example, farmers are highly sensitive to climate risks; processors are more affected by supply risks and biochemical risks, while wholesalers and retailers are particularly sensitive to operational and management risks. However, risks do not exist in isolation within a single node. Instead, various risks are often transmitted and evolve among different entities along AFSCs. Specifically, risks can propagate and interact within AFSCs. Based on the characteristics of risk propagation pathways in AFSCs, this study summarizes resilience capabilities from three perspectives: pre-shock, during-shock, and post-shock.

Although this research develops a conceptual framework for AFSCs based on a literature review, several limitations remain. The main limitation is that the framework has not been empirically tested under different decision-making scenarios. Specifically, this study does not examine risks, risk propagation pathways, or resilience capabilities using empirical data. Building on the proposed conceptual framework, future research will conduct empirical studies by collecting data from AFSC practitioners within the context of the agrifood transition. Such studies will examine risk propagation pathways and decision-making dynamics using decision-making tools, thereby enabling a more rigorous assessment of how managerial decisions influence risk propagation and mitigation when AFSC participants are exposed to disruptions.

Acknowledgement. This research is funded by the UKRI/BBSRC, under the RASCAL research project (Resilience in Agrifood System: Supply Chain Configuration Analytics Lab, grant award number BB/Z516703/1).

References

1. Aday, S., Aday, M.S.: Impacts of COVID-19 on food supply chain. Food Qual. Safety. **4**(4), 167–180 (2020)
2. Agnusdei, G.P., Coluccia, B.: Sustainable agrifood supply chains: bibliometric, network and content analyses. Sci. Total Environ. **824**(2022), 153704 (2022)
3. Ahamed, M.S., et al.: Unpacking systemic, cascading, and compound risks: a case based analysis of Asia Pacific. Prog. Disaster Sci. **18**(2023), 100285 (2023)
4. Ali, I., et al.: Achieving resilience through knowledge management practices and risk management culture in agri-food supply chains. Supply Chain Manag. Int. J. **28**(2), 284–299 (2021)
5. Ali, I., et al.: Risk and resilience in agri-food supply chain SMEs in the pandemic era: a cross-country study. Int. J. Log. Res. Appl. **26**(11), 1–19 (2022)
6. Ali, I., et al.: Supply chain resilience to climate change inflicted extreme events in agri-food industry: The role of social capital and network complexity. Int. J. Prod. Econ. **264**(1), 108968 (2023)
7. Ali, I., Gölgeci, I.: Managing climate risks through social capital in agrifood supply chains. Supply Chain Manag. Int. J. **26**, 1–16 (2020)
8. Ali, I., Govindan, K.: Extenuating operational risks through digital transformation of agri-food supply chains. Prod. Plan. Control. **34**(12), 1–13 (2021)
9. Amamou, A., et al.: Exploring novel sustainability metrics for the agri-food supply chain. In: Proceedings of the 10th International Food Operations & Processing Simulation Workshop 21st International Multidisciplinary Modeling & Simulation Multiconference, Tenerife, Spain, pp. 18–20 (2024)
10. Ambulkar, S., et al.: Firm's resilience to supply chain disruptions: scale development and empirical examination. J. Oper. Manag. **33-34**(1), 111–122 (2015)
11. Andika, A., et al.: Transitioning towards zero waste in the agri-food supply chain: a review of sustainable circular agri-food supply chain. Sustain. Futures. **10**(2025), 100917 (2025)
12. Ascoli, D., et al.: Fire-smart solutions for sustainable wildfire risk prevention: bottom-up initiatives meet top-down policies under EU green deal. Int. J. Disaster Risk Reduct. **92**(2003), 103715 (2023)
13. Balezentis, T., et al.: Probabilistic model for assessing the effects of the disruptive events on the viability of the agri-food supply chains: the case of Lithuania. IEEE Trans. Eng. Manag. **71**(2024), 9679–9690 (2024)
14. Baroni, L., et al.: Evaluating the environmental impact of various dietary patterns combined with different food production systems. Eur. J. Clin. Nutr. **61**(2), 279–286 (2007)
15. Basole, R.C., et al.: Computational analysis and visualization of global supply network risks. IEEE Trans. Industr. Inform. **12**(3), 1206–1213 (2016)
16. Basole, R.C., Bellamy, M.A.: Supply network structure, visibility, and risk diffusion: a computational approach. Decis. Sci. **45**(4), 753–789 (2014)
17. Bastian, J., Zentes, J.: Supply chain transparency as a key prerequisite for sustainable agri-food supply chain management. Int. Rev. Retail Distrib. Consum. Res. **23**(5), 553–570 (2013)
18. Behzadi, G., et al.: Agribusiness supply chain risk management: a review of quantitative decision models. Omega. **79**(2018), 21–42 (2018)

19. Borodin, V., et al.: Handling uncertainty in agricultural supply chain management: a state of the art. Eur. J. Oper. Res. **254**(2), 348–359 (2016)
20. Braithwaite, A., Hall, D.: Risky business? Critical decisions in supply chain management. Supply Chain Pract. **1**(2), 44–57 (2001)
21. Brandon-Jones, E., et al.: A contingent resource-based perspective of supply chain resilience and robustness. J. Supply Chain Manag. **50**(3), 55–73 (2014)
22. Brusset, X., Teller, C.: Supply chain capabilities, risks, and resilience. Int. J. Prod. Econ. **184**(2017), 59–68 (2017)
23. Burch, D., et al.: Introduction to symposium on the changing role of supermarkets in global supply chains: from seedling to supermarket: agri-food supply chains in transition. Agric. Hum. Values. **30**(2), 215–224 (2012)
24. Cao, S., et al.: Collaborative risk management in decentralised multi-tier global food supply chains: an exploratory study. Int. J. Logist. Manag. **32**(3), 1050–1067 (2021)
25. Charatsari, C., et al.: Technological innovation and agrifood systems resilience: the potential and perils of three different strategies. Front. Sustain. Food Syst. **6**(2022), 872706 (2022)
26. Chaudhuri, A., et al.: Risk propagation and its impact on performance in food processing supply chain. J. Model. Manag. **11**(2), 660–693 (2016)
27. Chopra, S., Sodhi, M.S.: Supply-chain breakdown. MIT Sloan Manag. Rev. **46**(1), 53–61 (2004)
28. Christopher, M., Peck, H.: Building the resilient supply chain. Int. J. Logist. Manag. **15**(2), 1–14 (2004)
29. Colon, C., Hochrainer-Stigler, S.: Systemic risks in supply chains: a need for system-level governance. Supply Chain Manag. Int. J. **28**(4), 682–694 (2022)
30. Coluccia, B., et al.: Effects of COVID-19 on the Italian agri-food supply and value chains. Food Control. **123**(2021), 107839 (2021)
31. Coopmans, I., et al.: COVID-19 impacts on Flemish food supply chains and lessons for agri-food system resilience. Agric. Syst. **190**(2021), 103136 (2021)
32. Corallo, A., et al.: Leveraging on technology and sustainability to innovate the supply chain: a proposal of agri-food value chain model. Supply Chain Manag. Int. J. **29**(3), 661–683 (2023)
33. Craighead, C.W., et al.: The severity of supply chain disruptions: design characteristics and mitigation capabilities. Decis. Sci. **38**(1), 131–156 (2007)
34. Cricelli, L., et al.: Technological innovation in agri-food supply chains. Br. Food J. **126**(5), 1852–1869 (2022)
35. Dani, S., Deep, A.: Fragile food supply chains: reacting to risks. Int. J. Log. Res. Appl. **13**(5), 395–410 (2010)
36. Paillin, D.B., et al.: Agro-food supply chain risk assessment: A review based on technique and approach. J. Appl. Eng. Technol. Sci. (JAETS). **5**(2), 892–908 (2024)
37. Deng, X., et al.: Risk evaluation of the grain supply chain in China. Int. J. Log. Res. Appl. **27**(1), 83–102 (2021)
38. Deng, X., et al.: Risk propagation mechanisms and risk management strategies for a sustainable perishable products supply chain. Comput. Ind. Eng. **135**(2019), 1175–1187 (2019)
39. Dolgui, A., et al.: Ripple effect in the supply chain: an analysis and recent literature. Int. J. Prod. Res. **56**(1–2), 414–430 (2018)
40. Ericksen, P.J.: Conceptualizing food systems for global environmental change research. Glob. Environ. Chang. **18**(1), 234–245 (2008)
41. Esteso, A., et al.: Conceptual framework for designing agri-food supply chains under uncertainty by mathematical programming models. Int. J. Prod. Res. **56**(13), 4418–4446 (2018)

42. Esteso, A., et al.: Impact of product perishability on agri-food supply chains design. Appl. Math. Model. **96**(2021), 20–38 (2021)
43. Fan, S., et al.: How to prevent and cope with coincidence of risks to the global food system. Annu. Rev. Environ. Resour. **46**(1), 602–617 (2021)
44. FAO: Traceability: A Management Tool for Businesses and Governments. FAO (2016)
45. Fischer, C., Miglietta, P.P.: The links between human diets and health and climate outcomes in the world's macro-regions during the last 50 years. Int. J. Environ. Res. Public Health. **17**(4), 1219 (2020)
46. Fróna, D., et al.: The challenge of feeding the world. Sustainability. **11**(20), 5816 (2019)
47. Gallego-García, S., et al.: Sustainability in the agri-food supply chain: a combined digital twin and simulation approach for farmers. Procedia Comput. Sci. **217**(2023), 1280–1295 (2023)
48. GLOPAN (Glob. Panel Agric. Food Syst. Nutr.): Future Food Systems: For People, Our Planet, and Prosperity (Foresight 2.0). GLOPAN, London (2020)
49. Harland, C., et al.: Risk in supply networks. J. Purch. Supply Manag. **9**(2), 51–62 (2003)
50. Hendricks, K.B., Singhal, V.R.: Association between supply chain glitches and operating performance. Manag. Sci. **51**(5), 695–711 (2005)
51. Hendricks, K.B., Singhal, V.R.: Supply chain disruptions and corporate performance. In: Supply Chain Disruptions: Theory and Practice of Managing Risk, pp. 1–19. Springer London, London (2011)
52. Hidayati, D.R., et al.: Sustainable agrifood value chain—transformation in developing countries. Sustainability. **13**(22), 12358 (2021)
53. HLPE: Food losses and waste in the context of sustainable food systems. http://www.fao.org/3/a-i3901e.pdf
54. Ho, W., et al.: Supply chain risk management: a literature review. Int. J. Prod. Res. **53**(16), 5031–5069 (2015)
55. Hobbs, J.E.: Food supply chain resilience and the COVID-19 pandemic: what have we learned? Can. J. Agric. Econ./Revue canadienne d'agroeconomie. **69**(2), 189–196 (2021)
56. Hohenstein, N.-O., et al.: Research on the phenomenon of supply chain resilience. Int. J. Phys. Distrib. Logist. Manag. **45**(1/2), 90–117 (2015)
57. Holling, C.S.: Resilience and stability of ecological systems. Annu. Rev. Ecol. Syst. **4**(1), 1–23 (1973)
58. Imbiri, S., et al.: Risk propagation and resilience in the agribusiness supply chain: a systematic literature review. J. Agribus. Dev. Emerg. Econ. **14**(4), 712–732 (2023)
59. Ivanov, D.: Predicting the impacts of epidemic outbreaks on global supply chains: a simulation-based analysis on the coronavirus outbreak (COVID-19/SARS-CoV-2) case. Transp. Res. Part E Logist. Transp. Rev. **136**(1), 101922 (2020)
60. Ivanov, D., et al.: The Ripple effect in supply chains: trade-off "efficiency-flexibility-resilience" in disruption management. Int. J. Prod. Res. **52**(7), 2154–2172 (2014)
61. Ivanov, D., Dolgui, A.: A digital supply chain twin for managing the disruption risks and resilience in the era of industry 4.0. Prod. Plan. Control. **32**(9), 775–788 (2021)
62. Jaffee, S., et al.: Rapid agricultural supply chain risk assessment: a conceptual framework. Agricult. Rural Dev. Discuss. Paper. **47**(1), 1–64 (2010)
63. Jaleta, M.E.: Agricultural supply chain analysis during supply chain disruptions: case of Teff commodity supply chain in Ethiopia in the era of COVID-19. Sustain. Agricult. Res. **10**(3), 63 (2021)
64. Jüttner, U., Maklan, S.: Supply chain resilience in the global financial crisis: an empirical study. Supply Chain Manag. Int. J. **16**(4), 246–259 (2011)
65. Kamalahmadi, M., Parast, M.M.: A review of the literature on the principles of enterprise and supply chain resilience: major findings and directions for future research. Int. J. Prod. Econ. **171**(1), 116–133 (2016)

66. Kaur, A., Prakash, G.: Adaptive inventory strategies using deep reinforcement learning for dynamic agri-food supply chains. arXiv. (2025)
67. Kinra, A., et al.: Ripple effect quantification by supplier risk exposure assessment. Int. J. Prod. Res. **58**(18), 1–20 (2019)
68. Krstić, M., et al.: Evaluation of the agri-food supply chain risks: the circular economy context. Br. Food J. **126**(1), 113–133 (2024)
69. Kumar, P., Kumar Singh, R.: Strategic framework for developing resilience in agri-food supply chains during COVID 19 pandemic. Int. J. Logist. Res. Appl. **25**(11), 1–24 (2021)
70. Lamine, C.: Sustainability and resilience in agrifood systems: reconnecting agriculture. Food Environ. Sociol. Ruralis. **55**(1), 41–61 (2014)
71. Leat, P., Revoredo-Giha, C.: Risk and resilience in agri-food supply chains: the case of the ASDA PorkLink supply chain in Scotland. Supply Chain Manag. Int. J. **18**(2), 219–231 (2013)
72. Li, L., et al.: Big data and big disaster: a mechanism of supply chain risk management in global logistics industry. Int. J. Oper. Prod. Manag. **43**(2), 274–307 (2022)
73. Loeber, A.M.: Food Practices in Transition. Routledge (2013)
74. Malassis, L.: Nourrir les Hommes. Dominos-Flammarion, Paris (1994)
75. Malik, A., et al.: Resilient capabilities to tackle supply chain risks: managing integration complexities in construction projects. Buildings. **12**(9), 1322 (2022)
76. Mangla, S.K., et al.: Enablers to implement sustainable initiatives in agri-food supply chains. Int. J. Prod. Econ. **203**(2018), 379–393 (2018)
77. Manuj, I., Mentzer, J.T.: Global supply chain risk management strategies. Int. J. Phys. Distrib. Logist. Manag. **38**(3), 192–223 (2008)
78. Marcucci, G., et al.: Analysis of ripple effect and its impact on supply chain resilience: a general framework and a case study on agri-food supply chain during the COVID-19 pandemic. Oper. Manag. Res. **17**(1), 175–200 (2023)
79. Michel-Villarreal, R., et al.: Resilience and digitalization in short food supply chains: a case study approach. Sustainability. **13**(11), 5913 (2021)
80. Mishra, R., et al.: Impact of disruptions in agri-food supply chain due to COVID-19 pandemic: contextualised resilience framework to achieve operational excellence. Int. J. Logist. Manag. **33**(3), 926–954 (2021)
81. Mkumbukiy, A., et al.: Agrifood systems' resilience for sustainable food security amid geopolitical tensions: a systematic literature review. Front. Sustain. Food Syst. **9**, 1546851 (2025)
82. Moazzam, M., et al.: Measuring agri-food supply chain performance and risk through a new analytical framework: a case study of New Zealand dairy. Prod. Plan. Control. **29**(15), 1258–1274 (2018)
83. Nakandala, D., et al.: Development of a hybrid fresh food supply chain risk assessment model. Int. J. Prod. Res. **55**(14), 4180–4195 (2016)
84. Nyamah, E.Y., et al.: Agri-food supply chain performance: an empirical impact of risk. Manag. Decis. **55**(5), 872–891 (2017)
85. Olafare, K.: The Role of Industry 4.0 Technologies in Promoting Agrifood Supply Chain Sustainability in Africa: Using Nigeria as a Case Study. Nottingham Trent University, Nottingham Business School, Nottingham (2024)
86. Owida, A., et al.: Decision-making framework for a resilient sustainable production system during COVID-19: An evidence-based research. Comput. Ind. Eng. **164**(2022), 107905 (2022)
87. Pettit, T.J., et al.: Ensuring supply chain resilience: development of a conceptual framework. J. Bus. Logist. **31**(1), 1–21 (2010)

88. Plakantara, S.P.: Athanasia Karakitsiou: transforming agrifood supply chains with digital technologies: a systematic review of safety and quality risk management. Oper. Res. Forum. **6**(3), 113 (2025)
89. Ponomarov, S.Y., Holcomb, M.C.: Understanding the concept of supply chain resilience. Int. J. Logist. Manag. **20**(1), 124–143 (2009)
90. Rao, S., Goldsby, T.J.: Supply chain risks: a review and typology. Int. J. Logist. Manag. **20**(1), 97–123 (2009)
91. Ribeiro, P.J., Barbosa-Povoa, A.: Supply chain resilience: definitions and quantitative modelling approaches – a literature review. Comput. Ind. Eng. **115**(1), 109–122 (2018)
92. Rice, J.B., et al.: Supply Chain Response to Terrorism: Creating Resilient and Secure Supply Chains. MIT Center for Transportation and Logistics Report, August (2003)
93. Roberta Pereira, C., et al.: Achieving supply chain resilience: the role of procurement. Supply Chain Manag. Int. J. **19**(5/6), 626–642 (2014)
94. Samson, D., Gloet, M.: Integrating performance and risk aspects of supply chain design processes. Prod. Plan. Control. **29**(15), 1238–1257 (2018)
95. Sheffi, Y., Rice, J.J.B.: A Supply Chain View of the Resilient Enterprise. MIT Sloan Management Review (2005)
96. Spaargaren, G., et al.: Sustainability transitions in food consumption, retail and production. In: Food Practices in Transition, pp. 1–32. Routledge (2013)
97. Srivastava, S.K., et al.: Propagation of risks and their impact on performance in fresh food retail. Int. J. Logist. Manag. **26**(3), 568–602 (2015)
98. Stoeva, S., et al.: Understanding collaboration in short food supply chains: a focus on collaborative relationships, interaction mechanisms and relational benefits. Agric. Food Econ. **12**(1), 1–34 (2024)
99. Stone, J., Rahimifard, S.: Resilience in agri-food supply chains: a critical analysis of the literature and synthesis of a novel framework. Supply Chain Manag. Int. J. **23**(3), 207–238 (2018)
100. Tang, C.S.: Perspectives in supply chain risk management. Int. J. Prod. Econ. **103**(2), 451–488 (2006)
101. Tang, C.S.: Perspectives in supply chain risk management: a review. SSRN Electron. J. **103**(2), 451–488 (2006)
102. Tang, L., et al.: Robustness of assembly supply chain networks by considering risk propagation and cascading failure. Physica A. Stat. Mech. Appl. **459**(2016), 129–139 (2016)
103. Tendall, D.M., et al.: Food system resilience: defining the concept. Glob. Food Sec. **6**(2015), 17–23 (2015)
104. Tilman, D., Clark, M.: Food, agriculture & the environment: can we feed the world & save the earth? Daedalus. **144**(4), 8–23 (2015)
105. Tukamuhabwa, B.R.: Supply chain orientation and supply chain risk management capabilities: mechanisms for supply chain performance of agro-food processing firms in Uganda. J. Afr. Bus. **24**(4), 649–672 (2023)
106. Tukamuhabwa, B.R., et al.: Supply chain resilience: definition, review and theoretical foundations for further study. Int. J. Prod. Res. **53**(18), 5592–5623 (2015)
107. Van Der Vorst, J.G.: Effective Food Supply Chains: Generating, Modelling and Evaluating Supply Chain Scenarios. Wageningen University and Research (2000)
108. van Wassenaer, L., et al.: Food system resilience: ontology development and impossible trinities. Agric. Food Secur. **10**(1), 38 (2021)
109. Vern, P., et al.: Unlocking the potential: leveraging blockchain technology for agri-food supply chain performance and sustainability. Int. J. Logist. Manag. **36**(2), 474–500 (2024)
110. Walker, B., et al.: Resilience, adaptability and transformability in social-ecological systems. Ecol. Soc. **9**, 2 (2004)

111. Wieland, A., Marcus Wallenburg, C.: The influence of relational competencies on supply chain resilience: a relational view. Int. J. Phys. Distrib. Logist. Manag. **43**(4), 300–320 (2013)
112. Xu, S., et al.: Disruption risks in supply chain management: a literature review based on bibliometric analysis. Int. J. Prod. Res. **58**(11), 3508–3526 (2020)
113. Xue, Y., et al.: Supply chain risks in agri-food systems: a comprehensive review of economic vulnerabilities and mitigation approaches. Front. Sustain. Food Syst. **9**(2025), 1–16 (2025)
114. Yadav, S., et al.: Using internet of things (IoT) in agri-food supply chains: a research framework for social good with network clustering analysis. IEEE Trans. Eng. Manag. **70**(3), 1–10 (2022)
115. Yeboah, N.E., et al.: Agricultural supply chain risk identification- a case finding from Ghana. J. Manag. Strategy. **5**(2), 31–48 (2014)
116. Zhao, G., et al.: Identification, establishment of connection, and clustering of social risks involved in the agri-food supply chains: a cross-country comparative study. Ann. Oper. Res. **338**(2), 1241–1282 (2024)
117. Zhao, G., et al.: Links between risk source identification and resilience capability building in agri-food supply chains: a comprehensive analysis. IEEE Trans. Eng. Manag. **71**(2022), 1–18 (2022)
118. Zhao, G., et al.: Risk analysis of the agri-food supply chain: a multi-method approach. Int. J. Prod. Res. **58**(16), 1–26 (2020)
119. Zhong, J., et al.: Supply chain resilience capability factors in agri-food supply chains. Oper. Manag. Res. Adv. Pract. through Res. **17**(3), 850–868 (2024)
120. Zhu, H.Y., et al.: Structural robustness of the international food supply network under external shocks and its determinants. arXiv. (2025)

Security *vs.* Sustainability? A Multi-criteria Decision-Method Framework for Information Security and Sustainability

Yann Goetgheluck[1,2(✉)], Pierre-Emmanuel Arduin[1], and Myriam Merad[2]

[1] DRM UMR CNRS 7088, Université Paris-Dauphine, PSL, Paris, France
{yann.goetgheluck,pierre-emmanuel.arduin}@dauphine.psl.eu

[2] LAMSADE UMR CNRS 7243, Université Paris-Dauphine, PSL, Paris, France
myriam.merad@dauphine.psl.eu

Abstract. This paper proposes a multi-criteria decision-method (MCDM) framework for integrated governance of information security and environmental sustainability. The framework addresses a recurrent tension between security-oriented resilience/compliance and sustainability-oriented efficiency/impact reduction by making trade-offs and potential synergies explicit. The contribution is methodological. We define five *scenarios* as category-level weight profiles (security, sustainability, hybrid) and structure twenty evaluation criteria (7 security, 7 sustainability, 6 hybrid). Because expert elicitation is often hesitant and incomplete, we combine incomplete Hesitant Fuzzy Preference Relations (HFPR), Analytic Hierarchy Process (AHP) consistency control, and Hesitant Fuzzy Bonferroni Mean (HFBM) aggregation to obtain traceable and outlier-robust weights. A first case study with two experts (security and sustainability) yields an intermediate result: a company-specific *preference orientation* closest to scenario S2 (security with sustainability constraints), not a final scenario-performance ranking. The framework therefore provides a reproducible baseline for integrated risk governance and ESG alignment, while full matrix population, cross-method ranking comparison, and robustness analyses (sensitivity, missingness, uncertainty propagation) are defined as next steps.

Keywords: Multi-Criteria Decision-Method · Information Security · Sustainability · Hesitant Fuzzy Preference Relation · Analytic Hierarchy Process · Hesitant Fuzzy Bonferroni Mean

1 Introduction

Organizations increasingly face a governance tension between cybersecurity and environmental sustainability. Security programs prioritize resilience, continuity, and compliance, while sustainability programs prioritize efficiency and impact

A. P. Cabral Seixas Costa et al. (Eds.): ICDSST 2026, LNBIP 582, pp. 82–99, 2026.
https://doi.org/10.1007/978-3-032-25146-6_6

reduction [1–3]. This tension is amplified by interdependent infrastructures and cascading risks [4].

These objectives are often managed separately, which reinforces the perception of an unavoidable trade-off. Yet integration is possible: work on *highly sustainable organizations* suggests that resilience improves when multiple risks are jointly governed through structured decision-support [5,6].

Related domains (water, agriculture, supply chains, humanitarian operations, energy) already combine security and sustainability considerations [7–12]. In Information Security Management System (ISMS) contexts, however, a comparable integrated framework remains limited. To address this gap, this paper proposes a transparent approach for explicit trade-off and synergy analysis between information security and environmental sustainability.

We use Multi-Criteria Decision Method (MCDM) to structure this tradespace [13–17]. The framework compares five *scenarios* (defined here as weight profiles, not concrete decision scenarios) across twenty criteria covering security, sustainability, and hybrid dimensions. At this stage, the contribution is methodological: we specify assumptions, inputs, and verification steps for later organization-specific ranking once full performance data are populated.

The remainder of the paper is organized as follows. Section 1 defines core MCDM concepts, the five scenarios, the twenty criteria, and the weighting logic under hesitant judgments. Section 2 justifies the MCDM method choice. Section 3 presents the pilot weight-computation protocol and empirical positioning. Section 4 discusses interpretation and limitations, and the conclusion synthesizes contributions and future work.

2 Multi-criteria Decision-Method Foundations

2.1 MCDM Concepts and Definitions

MCDM is a structured analytical approach for comparing scenarios when multiple, often conflicting, objectives must be considered simultaneously. It is well suited to this study because decisions in cybersecurity and sustainability create both internal and external effects. For instance, cyber incidents can affect not only the focal organization but also customers and suppliers, while security and infrastructure choices can generate environmental externalities. MCDM therefore provides an appropriate framework to formalize trade-offs and synergies between information security and environmental sustainability through a consistent set of criteria. We adopt the following standard terminology [14,18]:

Scenario: A distinct, complete course of strategy;
$S = \{S_i \mid i = 1, 2, \ldots, m\}$
where m is the number of scenarios under consideration, here $m = 5$.
Criteria: Dimensions used to evaluate and compare options.
$C = \{C_j \mid j = 1, 2, \ldots, n\}$
where n is the number of evaluation criteria, here $n = 20$.

Weight: The relative importance of a criterion according to the decision maker
$W = \{W_j \mid j = 1, 2, \ldots, n\}$
where each $W_j \in [0, 1]$ and $\sum_{j=1}^{n} W_j = 1$.

Scenario profile: In this paper, S_x denotes a scenario represented by a predefined category-level weight profile over three criterion groups (security, sustainability, hybrid). These profiles are *inputs* to the decision analysis, not final decision outcomes. Together, they define the strategic spectrum shown in Fig. 1.

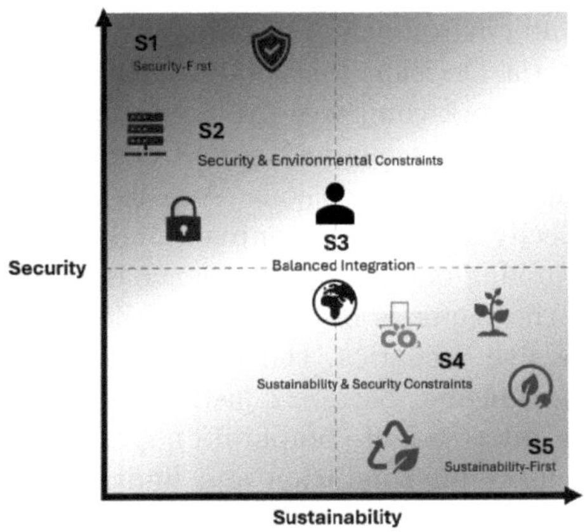

Fig. 1. Strategic positioning of the five governance scenarios along the security-sustainability spectrum.

The standard MCDM problem is formulated as:

$$\text{Decision Matrix } X = \begin{pmatrix} x_{1,1} & x_{1,2} & \cdots & x_{1,20} \\ x_{2,1} & x_{2,2} & \cdots & x_{2,20} \\ \vdots & \vdots & \ddots & \vdots \\ x_{5,1} & x_{5,2} & \cdots & x_{5,20} \end{pmatrix} \tag{1}$$

where x represents the performance score of scenario S_i with respect to criterion C_j. In other words, each cell x is the observed value of one scenario on one criterion (e.g., hours, %, tonnes, or counts depending on C_j).

An MCDM method (e.g., weighted-sum, outranking, compromise ranking) can then rank scenarios S_i based on aggregated performance across all criteria C_j, weighted by their importance W_j.

2.2 Scenarios

The five scenarios represent the main governance postures from security-dominant to sustainability-dominant.

[**S1: Security-First**] Security is the dominant objective; sustainability is not considered (even as a constraint). Example: Maximum redundancy, continuous monitoring, aggressive patching regardless of energy impact.

[**S2: Security with Environmental Constraints**] Security remains dominant, while sustainability acts as a constraint. Example: Energy-efficient firewalls and SIEM that still meet security baselines.

[**S3: Balanced Integration**] Security and sustainability are treated as equally important objectives. Example: Prioritize high-risk vulnerabilities while deferring low-impact, high-energy remediations.

[**S4: Sustainability with Security Constraints**] Sustainability is dominant, while security acts as a constraint. Example: Green IT deployment with baseline controls for compliance.

[**S5: Sustainability-First**] Sustainability is dominant; security is considered after sustainability objectives are met. Example: Defer non-critical security updates/monitoring to reduce energy use.

These five scenarios cover the main governance positions used in this study. More granular profiles are possible, but this set is sufficient for comparison and avoids impractical extremes that fully neglect either security or sustainability (Fig. 1).

2.3 Criteria

We next detail the sources underpinning the three groups of evaluation criteria.

The criteria come from two sources and are organized into three groups (security, sustainability, hybrid). Security criteria are derived from the CMM (Cybersecurity Maturity Model) factors of Güler and Büyüközkan [19]. In this paper, their cybersecurity scope is interpreted as compatible with ISMS governance because it includes organizational alignment, cross-functional coordination, and management commitment, not only technical threat handling [19–21].

Security Criteria (C1–C7) Inspired by CMM Factors. For security, we operationalize seven CMM-related success factors as measurable criteria applicable across the five scenarios.

1. **C1: Annual organizational budget share allocated to ISMS development (= F1.1).** Measures financial priority for information security; interpretation must consider realized outcomes, not spending alone. Higher value is preferred. Unit: % of annual organizational budget.
2. **C2: Critical vulnerabilities older than 30 days and exposed on internet (= F1.2 + Supply Chain (intern & extern) + Exposure Management).** Counts internet-exposed high-severity vulnerabilities unpatched

for more than 30 days, including third-party components. Lower value is preferred. Unit: number of high-severity vulnerabilities older than 30 days and internet-exposed.
3. **C3: Mean Time To Detect and Respond (= F1.2 + F4.3 + process-related).** Measures response speed (Mean Time To Detect/Mean Time To Respond) for technical and process-related incidents[1]; read with SLA[2] compliance and incident severity. Lower value is preferred. Unit: hours.
4. **C4: Governance Effectiveness (= F1.3 + F5).** Assesses governance maturity via control coverage of critical risks, incident-to-known-risk ratio, and policy/control compliance. Higher value is preferred. Unit: composite score (0–100), based on % of critical risks covered by active controls, ratio of incidents attributed to known risks, and % of policies with implemented controls.
5. **C5: Security Awareness and Culture (= F2 + human components).** Assesses training coverage, observable secure behavior, and cultural indicators (e.g., reporting confidence, voluntary participation). Higher value is preferred. Unit: composite score (0–100), based on % of employees completing annual security training, evidence of secure behavior in incidents and access patterns, and cultural measures (reporting confidence, voluntary participation).
6. **C6: System availability (= F3.2 + F3.3 + F4.2 + F4.3 + direct business impact).** Measures uptime of critical services; interpret with service degradation and business-impact incidents. Higher value is preferred. Unit: % uptime.
7. **C7: Mission-critical continuity gaps (= F1.2.4 + F2.2.4 + direct business impact).** Counts mission-critical services without tested continuity mechanisms; read with RTO/RPO[3] results and drill success rates. Lower value is preferred. Unit: number of services.

These seven criteria define the security dimension in a measurable and auditable way for subsequent weighting and scenario analysis.

Sustainability Criteria (C8–C14) Aligned with GRI Standards. For sustainability, criteria are aligned with GRI[4] disclosures to ensure traceability and comparability with common ESG reporting practice [23]:

1. **C8: Annual organizational budget share allocated to sustainability development (= GRI 102-2).** Measures budget priority given to sustainability and should be interpreted with realized environmental outcomes. Higher value is preferred. Unit: % of annual organizational budget.

[1] "Problem Exists Between Keyboard And Chair" (PEKBAC) [22].
[2] Service Level Agreement: a contractual commitment defining maximum acceptable response or resolution times for service disruptions.
[3] Recovery Time Objective (RTO): maximum acceptable downtime; Recovery Point Objective (RPO): maximum acceptable data loss.
[4] Global Reporting Initiative [23].

2. **C9: IT carbon and energy emissions (GRI 103-2).** Tracks annual CO_2[5] emissions and energy use across IT operations; workload normalization is recommended for comparability. Lower value is preferred. Unit: tonnes CO_2e/year and kWh/year.
3. **C10: Renewable energy share (GRI 103-2).** Measures the share of total energy sourced from renewables, supported by traceable sourcing evidence. Higher value is preferred. Unit: %.
4. **C11: Sustainability projects initiated (GRI 103-1).** Counts active environmental initiatives and should be read with project impact and completion metrics. Higher value is preferred. Unit: number of projects/year.
5. **C12: Annual CO_2 reduction (GRI 102-4).** Tracks year-over-year CO_2 reduction against targets, with explicit baseline and boundary documentation. Higher value is preferred. Unit: % change per year.
6. **C13: Energy intensity (GRI 103-4).** Measures electrical energy per unit of output (kWh/Mb processed) using a consistent functional unit for benchmarking. Lower value is preferred. Unit: kWh/Mb.
7. **C14: Water consumption (GRI 103-2).** Quantifies annual water use for cooling, cleaning, and operations, interpreted with reuse/recycling rates. Lower value is preferred. Unit: m^3/year.

These seven criteria support sustainability assessment that is compatible with established ESG reporting practices.

Hybrid Criteria (C15–C20): Win-Win Integration Levers. The six hybrid criteria are original and capture practices designed to improve security and sustainability jointly. Unless stated otherwise, indicators are annualized for comparability.

1. **C15: Low-consumption security infrastructure.** Counts security systems deployed on energy-efficient or renewable-powered infrastructure and is interpreted with measured kWh/CO_2 reductions. Higher value is preferred. Unit: number of systems/year + supporting impact metrics (kWh/year, tonnes CO_2e/year).
2. **C16: Green innovation projects combining security and sustainability.** Measures projects that explicitly integrate both objectives and should be complemented by evidenced security and environmental outcomes. Higher value is preferred. Unit: number of projects/year + supporting impact metrics (hours, kWh/year, tonnes CO_2e/year).
3. **C17: Joint green and secure technology investment.** Tracks annual R&D spending on dual-objective initiatives and pairs input with realized security and environmental gains. Higher value is preferred. Unit: % of R&D budget + supporting impact score (0–100).
4. **C18: Supplier evaluation on dual criteria.** Counts suppliers formally assessed on both security and carbon criteria using explicit performance

[5] Carbon dioxide (CO_2) emissions are a key indicator of environmental impact.

benchmarks. Higher value is preferred. Unit: number of suppliers/year + benchmark compliance score (0–100).

5. **C19: Integrated reporting linking cybersecurity and ESG.** Measures governance artifacts that explicitly connect cybersecurity and environmental metrics, with attention to decision use (e.g., budget or control changes). Higher value is preferred. Unit: number of artifacts/year + decision-use score (0–100).
6. **C20: Joint training on cybersecurity and sustainability.** Measures the share of employees trained in integrated programs, complemented by effectiveness indicators (behavior change and preventive practices). Higher value is preferred. Unit: % of employees/year + effectiveness score (0–100).

For each criterion, direction is explicit in Table 1: ↑ for maximization and ↓ for minimization (e.g., vulnerabilities and emissions).

Weight. To keep the weighting process transparent and operational, each scenario is defined by a three-component profile over our three distinct concepts (security, sustainability, hybrid). Each component represents the share of total weight allocated to that category of criteria within the scenario. The sum of the three components must equal 1 to ensure a valid weight distribution across all criteria. Formally, for each scenario S_x:

$$\alpha_{S_x} = \left(\alpha_{S_x}^{sec}, \alpha_{S_x}^{sus}, \alpha_{S_x}^{hyb}\right), \quad \alpha_{S_x}^{sec} + \alpha_{S_x}^{sus} + \alpha_{S_x}^{hyb} = 1.$$

The illustrative profiles used in this paper are: $S1 = (0.70, 0.15, 0.15)$, $S2 = (0.55, 0.25, 0.20)$, $S3 = (0.375, 0.375, 0.25)$, $S4 = (0.25, 0.55, 0.20)$, and $S5 = (0.15, 0.70, 0.15)$. Within each scenario, the category share is distributed uniformly across criteria in that category (7 security, 7 sustainability, 6 hybrid criteria):

$$W_j^{(S_x)} = \begin{cases} \alpha_{S_x}^{sec}/7, & j \in \{1, \ldots, 7\}, \\ \alpha_{S_x}^{sus}/7, & j \in \{8, \ldots, 14\}, \\ \alpha_{S_x}^{hyb}/6, & j \in \{15, \ldots, 20\}. \end{cases}$$

Detailed criterion-level values for all scenarios are available in the supplementary materials (available from the first author by email upon reasonable request, subject to confidentiality constraints).

Table 1 synthesizes all twenty criteria, their definitions, preference directions, and units of measurement. The full scenario-by-criterion weight table is provided in the Supplementary Material (Section S2).

Explicit Example for Our Work (Scenario S3). With $S3 = (0.375, 0.375, 0.25)$:

$$W_{C1} = W_{C2} = W_{C3} = W_{C4} = W_{C5} = W_{C6} = W_{C7} = 0.375/7 \approx 0.0536,$$

Table 1. Evaluation Criteria for Information Security and Environmental Sustainability

ID	Criterion	Target	Unit
	Security		
C1	Annual organizational budget share allocated to ISMS development	↑	% of annual organizational budget
C2	Critical vulnerabilities older than 30 days and internet-exposed	↓	Number of vulnerabilities
C3	Mean Time To Detect and Respond (MTTD/MTTR) for incidents	↓	Hours
C4	Governance effectiveness	↑	Composite score (0–100)
C5	Security awareness and culture maturity	↑	Composite score (0–100)
C6	System availability of critical services	↑	% uptime
C7	Mission-critical services without tested continuity mechanisms	↓	Number of services
	Sustainability		
C8	Annual organizational budget share allocated to sustainability development	↑	% of annual organizational budget
C9	IT carbon and energy emissions	↓	tonnes CO_2e/year + kWh/year
C10	Renewable energy share in total consumption	↑	%
C11	Sustainability projects initiated	↑	Number of projects/year
C12	Annual CO_2 reduction	↑	% change/year
C13	Energy intensity of infrastructure	↓	kWh/Mb
C14	Water consumption	↓	m^3/year
	Hybrid		
C15	Low-consumption security infrastructure deployments	↑	Number/year (+ kWh/year, tonnes CO_2e/year)
C16	Green innovation projects combining security and sustainability	↑	Number/year (+ impact metrics)
C17	Joint green and secure technology investment	↑	% of R&D budget (+ impact score)
C18	Supplier evaluation on dual criteria (security + carbon)	↑	Number/year (+ benchmark score)
C19	Integrated reporting linking cybersecurity and ESG	↑	Number/year (+ decision-use score)
C20	Joint training on cybersecurity and sustainability	↑	% of employees/year (+ effectiveness score)

$$W_{C8} = W_{C9} = W_{C10} = W_{C11} = W_{C12} = W_{C13} = W_{C14} = 0.375/7 \approx 0.0536,$$
$$W_{C15} = W_{C16} = W_{C17} = W_{C18} = W_{C19} = W_{C20} = 0.25/6 \approx 0.0417.$$

Hence, the explicit weight vector for $S3$ is:

$$W^{(S_3)} = \begin{pmatrix} 0.0536, 0.0536, 0.0536, 0.0536, 0.0536, \\ 0.0536, 0.0536, 0.0536, 0.0536, 0.0536, \\ 0.0536, 0.0536, 0.0536, 0.0536, 0.0417, \\ 0.0417, 0.0417, 0.0417, 0.0417, 0.0417 \end{pmatrix}.$$

This first stage established a structured theoretical baseline: five scenarios, explicit category-level weight profiles, and transparent criterion-level allocations. It defines the analytical perimeter for the next phase. Detailed scenario weight profiles are available in the Supplementary Material (Section S2).

The next phase is empirical and methodological. Before populating the decision matrix or positioning scenarios in a Gartner-style quadrant view [24], we must select a weighting method that is robust to real elicitation constraints.

3 MCDM Choice

The transition from framework design to empirical assessment requires a weighting method consistent with real expert elicitation. In our context, judgments are often hesitant, incomplete, and partially inconsistent.

We therefore selected *incomplete Hesitant Fuzzy Preference Relations (HFPR) with AHP consistency control and Hesitant Fuzzy Bonferroni Mean (HFBM) aggregation.* This combination was chosen after screening 60 MCDM methods (including 20 widely used ones), because it directly fits these constraints: uncertainty in pairwise judgments, missing comparisons, need for explicit consistency checks, and robust multi-expert aggregation.

HFPR–AHP with HFBM was benchmarked against representative alternatives (PROMETHEE, ELECTRE, VIKOR, TOPSIS, and BWM). PROMETHEE, VIKOR, TOPSIS, and ELECTRE were retained as finalists because they are established, methodologically mature, based on distinct ranking logics, and directly applicable once a complete performance matrix is available [9,15,16,25].

- **Hesitation and incompleteness**: HFPR accepts multiple linguistic terms and missing pairwise entries [26].
- **Consistency control**: AHP provides an explicit consistency-ratio check before finalizing weights [9,15,16,26].
- **Robust expert aggregation**: HFBM reduces outlier influence while rewarding agreement [26].
- **Modularity**: The resulting weight vector can be reused by downstream ranking methods (PROMETHEE, VIKOR, TOPSIS, ELECTRE) once x_{ij} is populated [9,15,16,25].

In short, HFPR–AHP with HFBM is adapted to small, heterogeneous expert panels and yields validated, outlier-resistant weights compatible with standard ranking engines.

3.1 Weight Computation Method (Incomplete HFPR–AHP with HFBM)

This subsection summarizes the *protocol* used to compute criterion weights. At this stage, only the procedure is reported (no final empirical weights).

1. **Elicit pairwise judgments.** Experts compare criteria with the linguistic scale in Table 2, producing incomplete HFPRs with possible hesitation.
2. **Enforce reciprocity.** Known entries are normalized to satisfy reciprocal structure and neutral diagonal conditions.
3. **Complete missing entries.** Missing comparisons are inferred via additive consistency rules [26].
4. **Aggregate experts.** Individual HFPRs are merged with HFBM; (p, q) tune agreement reinforcement and outlier damping.
5. **Check consistency.** The collective matrix is converted into an AHP-style priority matrix and validated with CR.
6. **Compute final weights.** Normalized priorities are transformed into local (and, if needed, global) criterion weights.

The linguistic scale used in this protocol is summarized in Table 2. To remain consistent with the order of magnitude of criterion weights used in this study, each term is mapped to a normalized interval (l, m, u) on a low-weight scale. The complete elicitation support table is reproduced in the Supplementary Material (Section S3). Here, l is the *lowest plausible value*, m is the *most plausible (central) value*, and u is the *highest plausible value*. For example, "Good" $=$ $(0.05, 0.06, 0.08)$ means the judgment is centered on 0.06, but can reasonably vary between 0.05 and 0.08.

Table 2. Linguistic scale aligned with criterion-weight magnitudes

Linguistic term	Normalized interval (l, m, u)
Perfect (P)	$(0.08, 0.10, 0.10)$
Very Good (VG)	$(0.06, 0.08, 0.10)$
Good (G)	$(0.05, 0.06, 0.08)$
Medium (M)	$(0.03, 0.05, 0.06)$
Bad (B)	$(0.02, 0.03, 0.05)$
Very Bad (VB)	$(0.00, 0.02, 0.03)$

4 Weight Computation

In this section, we present the methodological protocol and normalization rules used to map relative weights into the five scenario profiles ($S1$–$S5$). At this stage, we report one empirical case study. This empirical section is a pilot feasibility study: results are exploratory and should not be interpreted as statistically generalizable beyond the case organization.

Planned Expert Elicitation Process. We conduct a pilot questionnaire-based expert consultation for one software publishing company, with two experts ($K = 2$): one ISMS/security expert and one sustainability expert. These respondents illustrate the target profiles for this work (e.g., CISO/RSSI, Information Security Engineer, CSR/ESG manager), but this pilot is not intended to be statistically representative. This multi-profile panel is intentional because relevant evidence is distributed across functions (security, sustainability, operations, governance), and no single role systematically owns all required data. Each expert will perform pairwise comparisons of the twenty criteria using the adaptive linguistic scale from Table 2. Experts may provide multiple linguistic terms when uncertain, resulting in HFPRs. Missing comparisons are allowed (target missing rate $\leq 15\%$ per matrix). For baseline aggregation, we set $p = q = 1$ and report one-way sensitivity checks with $(p, q) \in \{(1, 2), (2, 1)\}$. In this pilot, the matrix size is driven by the number of criteria ($n = 20$), while the number of experts ($K = 2$) affects only the aggregation step.

Mathematical Formulation. The incomplete HFPR is represented as a matrix $H = (h_{ij})_{n \times n}$ where h_{ij} is a HFE (Hesitant Fuzzy Element) representing the preference degree of criterion C_i over C_j. The matrix satisfies:

$$h_{ij}^{(s)} + h_{ji}^{(l-s+1)} = 1, \quad h_{ii} = \{0.5\} \tag{2}$$

where $h_{ij}^{(s)}$ denotes the s-th smallest element in h_{ij}, and l is the number of elements in the HFE.

When comparisons are missing, we complete the matrix using the additive consistency property [26]:

$$h_{ij} = h_{ik} + h_{kj} - 0.5 \tag{3}$$

for an intermediate criterion C_k.

Aggregation with HFBM. Individual expert judgments are aggregated using the HFBM operator. For K experts providing HFEs $h_{ij}^{(1)}, h_{ij}^{(2)}, \ldots, h_{ij}^{(K)}$, the aggregated preference is computed as:

$$\text{HFBM}^{p,q}(h_{ij}^{(1)}, \ldots, h_{ij}^{(K)}) = \left\{ \left(\frac{1}{K(K-1)} \sum_{\substack{k,l=1 \\ k \neq l}}^{K} (h_{ij}^{(k)})^p \cdot (h_{ij}^{(l)})^q \right)^{\frac{1}{p+q}} \right\} \tag{4}$$

where parameters $p, q \geq 0$ control the interrelationship between arguments. We set $p = q = 1$ for balanced aggregation.

Priority Weight Extraction. From the aggregated collective HFPR matrix $\bar{H}$, the priority weight vector $w = (w_1, w_2, \ldots, w_n)^T$ is derived using the following procedure:

1. **Converting HFEs to crisp values:** For each HFE h_{ij}, compute its expected value:

$$E(h_{ij}) = \frac{1}{|h_{ij}|} \sum_{\gamma \in h_{ij}} \gamma \tag{5}$$

2. **Normalizing each row:** Construct the priority matrix $P = (p_{ij})$ where:

$$p_{ij} = \frac{E(h_{ij})}{\sum_{k=1}^{n} E(h_{ik})} \tag{6}$$

3. **Computing criterion weights:** The weight of criterion C_j is:

$$w_j = \frac{1}{n} \sum_{i=1}^{n} p_{ij} \tag{7}$$

Consistency Verification. To ensure the quality and reliability of expert judgments, consistency is verified using Saaty's consistency ratio:

$$\mathrm{CR} = \frac{\mathrm{CI}}{\mathrm{RI}} = \frac{(\lambda_{\max} - n)/(n-1)}{\mathrm{RI}} \tag{8}$$

where $\lambda_{\max}$ is the maximum eigenvalue of the priority matrix, and RI is the random index. In this study, CR is computed on the aggregated 20×20 matrix covering C1–C20 after completion and defuzzification. This 20×20 dimension is imposed by the criterion set ($n = 20$), not by the panel size ($K = 2$). Aggregated judgments are accepted if $\mathrm{CR} \leq 0.1$. If inconsistencies arise, the most problematic pairwise comparisons are revisited with the expert panel.

Empirical Result. At this stage, we report one empirical case study from an anonymized software publishing organization (hereafter, the case organization) to preserve confidentiality. Two experts participated (one security expert and one sustainability expert). They provided criterion-importance judgments together with available criterion data. This is a single-company pilot and is therefore used to demonstrate methodological feasibility rather than statistical generalization. Accordingly, the empirical output is interpreted as an intermediate orientation signal to guide subsequent full-scale data collection and cross-case validation. Data were collected through a structured questionnaire based on the linguistic scale.

Questionnaire Structure. The questionnaire was organized into three blocks:

1. **Expert profile and perimeter.** Role, business unit, and criteria coverage (full or partial).
2. **Pairwise judgments on criteria importance.** Experts compared criteria using the linguistic scale in Table 2. Multiple terms were allowed when uncertainty existed (hesitant judgments).
3. **Criterion data sheet.** For each criterion, experts provided the available value, unit, period, and hesitation level (Low/Medium/High).

The full questionnaire templates and filled expert profiles are available in the Supplementary Material (Section S7), subject to confidentiality constraints.

Template Used for Pairwise Judgments (Extract). For each pair (C_i, C_j), experts answered: "How much more important is C_i than C_j for your organization?" using the terms {P, VG, G, M, B, VB}. Missing comparisons were accepted when an expert had no reliable information.

First Empirical Calculations and Scenario Positioning. Using the first observed judgments reported in the Supplementary Material (Section S4), we compute a preliminary criterion-priority vector in four explicit steps.

The hesitation level is coded as **Low** (single term), **Medium** (two terms), or **High** (three or more terms). When a respondent does not have the information, or prefers not to share it, the judgment is coded as missing ("/").

Step 1 – Linguistic-to-Numeric Mapping. Each linguistic term is converted to its central value m from Table 2. These values are then used to compute criterion priorities:

$$P{=}0.10,\ VG{=}0.08,\ G{=}0.06,\ M{=}0.05,\ B{=}0.03,\ VB{=}0.02.$$

Step 2 – Two-Expert HFBM Aggregation. With $K = 2$ and $(p, q) = (1, 1)$, the HFBM formula simplifies to a geometric mean for each criterion:

$$\tilde{w}_j = \sqrt{m_j^{(sec)}\, m_j^{(sus)}}.$$

The complete computations for all criteria are reported below and in the Supplementary Material (Section S5) for cross-checking and reproducibility (Table 3):

Step 3 – Normalization. Let $S = \sum_{j=1}^{20} \tilde{w}_j = 0.8744$. Final preliminary weights are:

$$w_j = \frac{\tilde{w}_j}{S}.$$

Table 3. Complete criterion-level computation from linguistic judgments

Criterion	$m_j^{(sec)}$	$m_j^{(sus)}$	$\tilde{w}_j = \sqrt{m_j^{(sec)} m_j^{(sus)}}$	w_j
C1	0.10	0.08	0.0894	0.1023
C2	0.08	0.06	0.0693	0.0792
C3	0.06	0.06	0.0600	0.0686
C4	0.08	0.05	0.0632	0.0723
C5	0.08	0.05	0.0632	0.0723
C6	0.10	0.08	0.0894	0.1023
C7	0.08	0.05	0.0632	0.0723
C8	0.03	0.03	0.0300	0.0343
C9	0.03	0.03	0.0300	0.0343
C10	0.03	0.03	0.0300	0.0343
C11	0.03	0.03	0.0300	0.0343
C12	0.05	0.02	0.0316	0.0362
C13	0.03	0.03	0.0300	0.0343
C14	0.03	0.02	0.0245	0.0280
C15	0.05	0.03	0.0387	0.0443
C16	0.03	0.03	0.0300	0.0343
C17	0.02	0.02	0.0200	0.0229
C18	0.05	0.02	0.0316	0.0362
C19	0.03	0.03	0.0300	0.0343
C20	0.02	0.02	0.0200	0.0229

The complete normalized vector is therefore (also reproduced in the Supplementary Material, Section S5):

$$w = (0.1023,\ 0.0792,\ 0.0686,\ 0.0723,\ 0.0723,\ 0.1023,\ 0.0723,\ 0.0343,\ 0.0343,\ 0.0343,\ 0.0343,\ 0.0362,\ 0.0343,\ 0.0280,\ 0.0443,\ 0.0343,\ 0.0229,\ 0.0362,\ 0.0343,\ 0.0229).$$

Step 4 – Category Shares and Scenario Comparison. Aggregating w_j by criterion groups yields the empirical profile:

$$\hat{\alpha}^{sec} = \sum_{j=1}^{7} w_j = 0.5693, \quad \hat{\alpha}^{sus} = \sum_{j=8}^{14} w_j = 0.2357, \quad \hat{\alpha}^{hyb} = \sum_{j=15}^{20} w_j = 0.1949.$$

We then compare $\hat{\alpha}$ to the scenario vectors ($S1 \ldots S5$) using Euclidean distance:

$$d(S_x) = \|\hat{\alpha} - \alpha_{S_x}\|_2 .$$

Computed distances are approximately:

$$d(S1) = 0.1627,\ d(S2) = 0.0245,\ d(S3) = 0.2453,\ d(S4) = 0.4481,\ d(S5) = 0.6272.$$

The corresponding company matrices are:

$$W^{(\text{company})} = \begin{pmatrix} 0.1023, 0.0792, 0.0686, 0.0723, 0.0723, \\ 0.1023, 0.0723, 0.0343, 0.0343, 0.0343, \\ 0.0343, 0.0362, 0.0343, 0.0280, 0.0443, \\ 0.0343, 0.0229, 0.0362, 0.0343, 0.0229 \end{pmatrix}.$$

Hence, the anonymized case organization is positioned closest to S2. It is then closer to S1 than to S3, and clearly farther from S4 and S5. The profile indicates a security-led posture with sustainability constraints and a non-negligible hybrid component. Figure 2 provides a two-axis representation inspired by the Gartner-style quadrant and our strategic map.

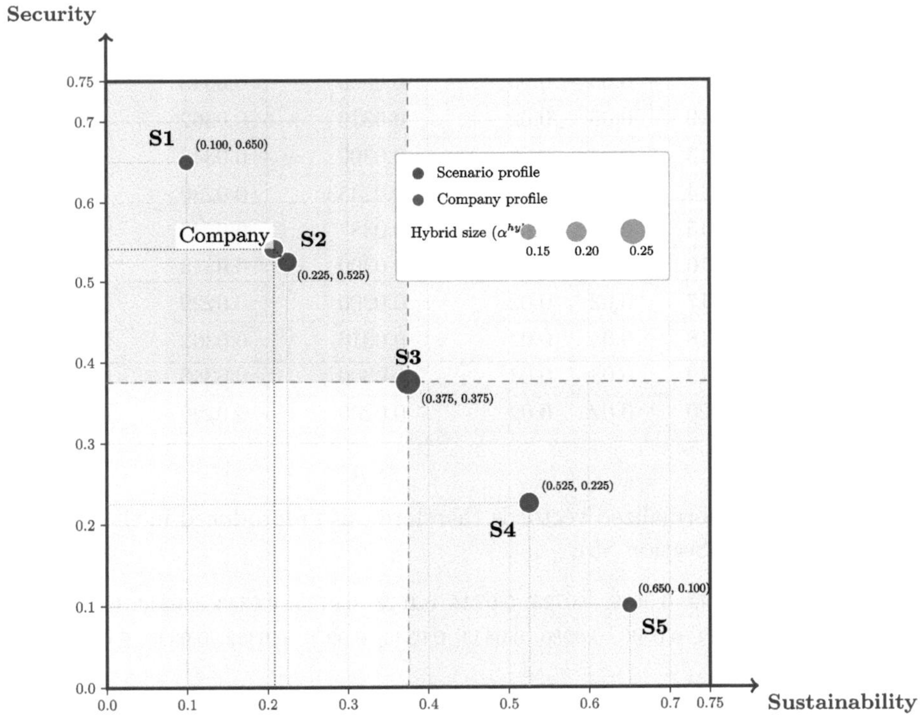

Fig. 2. Empirical positioning of the company on the Security–Sustainability plane.

Numeric Construction of Fig. 2 The plot uses a Cartesian scale from 0 to 0.75 on both axes, which places $S3 = (0.375, 0.375)$ at the geometric center of the plotting area. To make the hybrid component affect both axes symmetrically, we use a diagonal projection:

$$x' = \alpha^{sus} + \lambda(\alpha^{hyb} - h_0), \qquad y' = \alpha^{sec} + \lambda(\alpha^{hyb} - h_0),$$

with $\lambda = 0.5$ and $h_0 = 0.25$ (neutral hybrid level centered on $S3$). This is linear on each axis. Higher hybrid values move points up-right, and lower hybrid values move points down-left. Hybrid is also made explicitly visible by bubble size:

$$r\ (\text{pt}) = 0.5 + 20\,\alpha^{hyb}.$$

Blue bubbles denote scenarios ($S1$–$S5$), while the red bubble denotes the company profile (Table 4).

Table 4. Numerical values used to construct Fig. 2

Profile	α^{sec}	α^{sus}	α^{hyb}	$x = \alpha^{sus}$	$y = \alpha^{sec}$	(x', y') used in Fig. 3	bubble radius (pt)
S1	0.70	0.15	0.15	0.15	0.70	(0.100, 0.650)	3.50
S2	0.55	0.25	0.20	0.25	0.55	(0.225, 0.525)	4.50
S3	0.375	0.375	0.25	0.375	0.375	(0.375, 0.375)	5.50
S4	0.25	0.55	0.20	0.55	0.25	(0.525, 0.225)	4.50
S5	0.15	0.70	0.15	0.70	0.15	(0.650, 0.100)	3.50
Company	0.5693	0.2357	0.1949	0.2357	0.5693	(0.2082, 0.5418)	4.40

5 Discussion

5.1 Empirical Interpretation and Current Perimeter

The first empirical iteration provides a clear signal: the anonymized case organization is closest to S2. This corresponds to a security-led posture where sustainability is integrated as a constraint rather than a dominant objective.

This result reflects *elicited priorities*, not observed scenario performance. It should therefore be interpreted as an intermediate preference orientation. Figure 2 helps visualize both directional positioning (security vs sustainability) and hybrid intensity.

5.2 Limits and Implications

The main limitation is structural: this phase uses criterion-importance data only. The framework identifies *what matters most*, not yet *which scenario performs best* on observed values. A full decision claim requires complete $X = (x_{ij})$, normalization choices, and final rankings.

Threats to validity are fourfold. First, this is a single-company pilot with a very small panel ($K = 2$), which limits statistical representativeness. Second, expert-judgment bias may remain (role perspective, organizational culture, and anchoring effects). Third, missing data are likely non-random, which can distort inferred preferences even with completion rules. Fourth, sectoral generalizability is limited at this stage and requires replication across additional organizations and industries. Despite these limits, the framework remains operationally useful because it makes assumptions explicit, preserves traceability from judgments to weights, and identifies data-governance priorities for the next phase.

Conclusion and Future Work

This paper proposes a transparent MCDM framework to analyze security–sustainability tensions through five scenarios and twenty criteria, including six hybrid integration levers. At this stage, the empirical output is intentionally intermediate: a company-specific preference orientation (closest to S2), not a final performance ranking.

Methodologically, HFPR–AHP–HFBM was selected for hesitant and incomplete elicitation in small expert panels. HFPR handles hesitation and missing comparisons, AHP provides consistency control, and HFBM limits outlier dominance during aggregation.

Three implications follow: decision quality depends on both method choice and data-governance maturity; early-stage weights provide actionable transparency; and hybrid criteria should be treated as first-class governance levers.

Future work will (1) complete criterion data and metadata, (2) populate X and run full rankings with cross-method concordance checks, (3) assess robustness to sensitivity, missingness, and uncertainty propagation, and (4) extend validation to additional organizations and sectors.

Supplementary Materials Availability. Additional methodological and empirical materials (detailed criterion-level scenario weights, questionnaire templates, and intermediate computation tables) are available from the corresponding author upon reasonable request, subject to confidentiality constraints. Further inquiries can be directed to the corresponding author.

References

1. Chowdhury, P., Paul, S.K.: Applications of MCDM methods in research on corporatesustainability. Manage. Envir. Qual. Int. J. **31**(2), 385–405 (2020)
2. Julisch, K., Hall, M.: Security and Control in the Cloud. Inf. Secur. J. A Global Persp. **19**(6), 299–309 (2010)
3. Chen, A.J., Boudreau, M.C., Watson, R.T.: Information systems and ecological sustainability. J. Syst. Inf. Technol. **10**(3), 186–201 (2008)
4. Armenia, S., Tsaples, G., Carlini, C.: Critical Events and critical infrastructures: a system dynamics approach (2018)
5. Merad, M.: Organisations hautement durables. Tec & doc (2013)
6. Franco, E., Armenia, S., Christou, O., Tsaples, G., Uribe-Toril, J., Papathanasiou, J.: A system dynamics model to promote awareness of the green transition challenges in higher education institutions. (2025)
7. Wang, X., Liu, Z., Wu, J.: Inter-basin water transfer and water security: a landscape sustainability science perspective. J. Envir. Manage. **390**, 126326 (2025)
8. Sharma, Y.K., Yadav, A.K., Mangla, S.K., Patil, P.P.: Ranking the success factors to improve safety and security in sustainable food supply chain management using fuzzy AHP. Materials Today Proc. **5**(5), 12187–12196 (2018)

9. Papathanasiou, J., Ploskas, N., Bournaris, T., Manos, B.: A decision support system for multiple criteria alternative ranking using TOPSIS and VIKOR: a case study on social sustainability in agriculture (2016)
10. Qazi, A., Quigley, J., Dickson, A., Ekici, S.O., Gaudenzi, B.: Detectability based prioritization of interdependent supply chain risks. (2016)
11. Drakaki, M., Gören, H.G., Tzionas, P.: An Intelligent Multi-agent System Using Fuzzy Analytic Hierarchy Process and Axiomatic Design as a Decision Support Method for Refugee Settlement Siting. Springer International Publishing, Cham (2018)
12. Abdul, D., Wenqi, J., Tanveer, A., Sharaf, I.M.: Analyzing the sustainability factors to prioritize sustainable energy source: a novel integrated MCDM model. Energy **337**, 138637 (2025)
13. Wind, Y., Saaty, T.L.: Marketing applications of the analytic hierarchy process. Manage. Sci. **26**(7), 641–658 (1980)
14. Taherdoost, H., Madanchian, M.: Multi-Criteria Decision Making (MCDM) Methods and Concepts. Encyclopedia **3**(1), 77–87 (2023)
15. Burak, S., Samanlioglu, F., Ülker, D.: Evaluation of irrigation methods in söke plain with hf-ahp-promethee ii hybrid mcdm method. Agric. Water Manag. **271**, 107810 (2022)
16. Li, L., Jia, C., Li, X.: A novel intuitionistic fuzzy vikor method to mcdm based on intuitionistic fuzzy covering rough set. Expert Syst. Appl. **293**, 128713 (2025)
17. Li, J., Niu, L.L., Chen, Q., Wang, Z.X., Li, W.: Group decision making method with hesitant fuzzy preference relations based on additive consistency and consensus. Complex Intell. Syst. **8**(3), 2203–2225 (2022)
18. Merad, M.: Aide à la décision et expertise en gestion des risques. Tec & doc (2010)
19. Güler, M., Büyüközkan, G.: Cybersecurity maturity assessment using an incomplete hesitant fuzzy AHP method and Bonferroni means operator. Exp. Syst. Appl. **282**, 127268 (2025)
20. International Organization for Standardization: ISO/IEC 27001:2022: Information security management systems. Standard (2022)
21. Al-Dhahri, S., Al-Sarti, M., Abdul, A.: Information Security Management System. Int. J. Comput. Appl. **158**(7), 29–33 (2017)
22. Arduin, P.E.: Insider threats. John Wiley and Sons (3 2018)
23. Global Reporting Initiative (GRI): GRI - GRI Standards English Language (2026). Accessed on February 21, 2026
24. Gartner, I.: Magic quadrant (2026). Accessed on February 21, 2026
25. Nusev, G., Boškoski, P., Bohanec, M., Boshkoska, B.M.: A DSS model for selection of computer on module based on PROMETHEE and DEX methods. (2018)
26. Beg, I., Rashid, T.: TOPSIS for Hesitant Fuzzy Linguistic Term Sets. Int. J. Intell. Syst. **28**(12), 1162–1171 (2013)

Decision Support Systems and Models for Energy and Green Transition

Decision Support Systems for Evaluating Technological Innovation Impacts in Smart Cities: The Case of Hydrogen-Based Urban Mobility

Limmatola Anna(✉) and Botte Marilisa

Department of Architecture, Federico II University of Naples, Naples, Italy
{anna.limmatola,marilisa.botte}@unina.it

Abstract. The accelerating pace of technological innovation in urban environments calls for robust, systematic approaches to assess its multidimensional impacts on cities. In this context, Decision Support Systems (DSS) play a key role in supporting policymakers, Chief Information Officers (CIOs), and urban planners in evaluating and guiding strategic technological choices. This paper proposes adopting the Smart City framework as a conceptual modeling approach to represent the city as an integrated system, enabling the structured assessment of technological innovations across environmental, economic, social, and infrastructural dimensions within DSS-based analyses.

To illustrate a concrete application, the paper examines the introduction of hydrogen-based transportation systems as an innovative energy solution for urban mobility. The analysis investigates the potential impacts of hydrogen adoption on urban energy systems, transportation networks, environmental sustainability, and governance processes.

The findings highlight the strategic role of DSS in enabling informed, transparent, and resilient decision-making in smart cities, particularly in the evaluation and governance of complex and emerging technological innovations.

Keywords: Decision Support Systems · Smart Cities · Hydrogen Mobility

1 Introduction

The current phase of technological transition is precipitating a systemic reconfiguration of the anthropogenic environment, fueled by a convergence of innovations that operate transversally across multiple dimensional and functional scales. This process redefines not only the technical performance of individual urban components but the very essence of living space. This metamorphosis does not constitute a mere incremental update; rather, it represents an ontological shift in the territory, where physical and digital components merge into a single, hyper-connected entity. At the infrastructural level, this transformation is driven by the massive integration of Urban Sensing systems and sensor networks based on the Internet of Things (IoT). These devices, ubiquitously distributed

A. P. Cabral Seixas Costa et al. (Eds.): ICDSST 2026, LNBIP 582, pp. 103–118, 2026.
https://doi.org/10.1007/978-3-032-25146-6_7

throughout the urban fabric, transform physical space into a "sentient" digital organism capable of generating continuous streams of raw data. When properly processed, this data becomes strategic information for the real-time monitoring of territorial dynamics. The ability to instantaneously measure environmental variables, heat flows, and human mobility enables a transition from reactive planning to proactive and predictive management, where data serves as the catalyst linking policy decisions to the actual needs of the citizenry. Within this context of pervasive digitalization, the energy sector is undergoing a critical and irreversible transition toward decentralization. The transcendence of the centralized energy production model, typical of the 20th century, is occurring through the implementation of innovative frameworks such as Renewable Energy Communities (RECs). These structures not only promote local self-sufficiency but also introduce a new democratic dimension to resource management, where the territory is no longer a passive consumer but an active node of production and exchange. Simultaneously, the built environment—responsible for a significant share of global emissions—is evolving through the systematic adoption of Building Information Modeling (BIM) [1]. This information modeling methodology accompanies the building throughout its entire life cycle, allowing for the high-precision prediction of thermal, energetic, and structural behavior. Digital simulation optimizes building envelopes, drastically reducing the carbon footprint and ensuring that new additions to the built environment contribute to the overall resilience of the urban system. This systemic evolution is reflected with particular vigor in the field of mobility, where technological innovation acts as the primary driver for redesigning public space [2]. The advent of Autonomous Vehicles (AVs) and Cooperative Intelligent Transport Systems (C-ITS) necessitates a profound reflection that extends beyond mere automation; it requires a radical redesign of interchange nodes, road cross-sections, and urban parking regimes. If vehicles can autonomously optimize their routes and park in remote areas, vast amounts of urban land previously dedicated to parking can be reclaimed, opening unprecedented scenarios for urban afforestation or the creation of new commons [3]. Crucial to this transition is the shift toward new energy vectors, prominently including large-scale electrification and the use of green hydrogen. These new energy paradigms for transport do not merely represent a necessary fuel switch to meet international climate agreements; they constitute a direct intervention in public health and the equilibrium of urban ecosystems. The systematic reduction of nitrogen oxides (NO_X), particulate matter (PM), and acoustic pollution tangibly improves the quality of life in densely populated areas [4], mitigating negative externalities related to respiratory diseases and environmental stress. This overview of innovations outlines a consistent paradigm shift across major urban domains. In the specific field of transport, we are witnessing a definitive transition from the static and rigid management of infrastructure toward dynamic, flexible, and interconnected flow models. Roadways are no longer mere strips of asphalt for transit but have become smart digital corridors capable of altering their function based on real-time demand. Regarding spatial and urban planning, the availability of advanced analytical technologies enables a more fluid and flexible vision of space. Traditional planning, based on rigid zoning and long-term forecasting, is yielding to adaptive urban configurations capable of rapid modulation in response to shifting usage needs, new service standards, and the challenges imposed by climate change. Urban space thus becomes a dynamic palimpsest, where technology facilitates equitable and sustainable habitability. Furthermore, on the energy front, technological transition results in the progressive autonomy of urban districts. Energy production, once relegated to marginal industrial zones, is now deeply integrated into the existing fabric—on rooftops, building facades, and transport infrastructure—radically altering historical hierarchies between the core and periphery. However, it is essential to recognize that the impact of these innovations is not confined to isolated “silos”; rather, they generate complex, intertwined externalities that alter urban morphology as a whole. This interdependence requires a rigorous, holistic vision, interpreting the city as an interconnected and fragile ecosystem. The velocity and pervasiveness of these technological shifts preclude a passive or purely regulatory approach. Instead, they mandate the adoption of preventive, scientifically robust, and multidimensional evaluation methodologies. It is therefore imperative to integrate critical analysis processes and next-generation Decision Support Systems (DSS) capable of processing the vast quantities of data produced by new technologies to measure, forecast, and govern territorial impacts. In this context, the implementation of the Urban Digital Twin emerges as an essential instrument: a dynamic virtual replica of the territory which, fueled by real-time data streams, enables the simulation of complex scenarios and the monitoring of infrastructural performance throughout the entire life cycle. Examples of such implementations can be found in the city of Helsinki [5] and the city of Vienna [6]. Only through a meticulous evaluation that combines algorithmic precision with a strategic vision of urban development can we ensure that technological innovation becomes a real added value for the community, promoting territorial models that are resilient, healthy, and future-oriented.

1.1 Scope and Structure

To demonstrate the effectiveness of Decision Support Systems in estimating the urban impacts of sustainability-oriented technological innovations, this study adopts hydrogen-based railways as a representative use case [7, 8]. Rather than addressing the full spectrum of technological transformations affecting cities, the analysis deliberately focuses on a specific mobility-related innovation to illustrate how a DSS can support the structured evaluation of its effects on urban systems. Within this context, the transport sector is considered as a meaningful domain through which the implications of technological change on the city can be observed and assessed, particularly in relation to mobility–territory interactions and urban sustainability. Hydrogen railways are therefore examined not as an end in themselves, but as an exemplary case to validate the proposed DSS-based approach and to demonstrate its ability to capture the environmental, infrastructural, and spatial impacts of innovative and sustainable transport technologies on cities. Therefore, this document contributes a comprehensive and systematic review of hydrogen applications in the railway sector, with particular emphasis on the evaluation frameworks adopted to support the implementation of this emerging technology. While hydrogen-powered rail systems are increasingly recognized as a promising solution for decarbonizing non-electrified lines, existing studies typically assess their feasibility through sector-specific and fragmented perspectives. Most of the available literature focuses on isolated dimensions, such as transport performance, environmental impact, techno-economic feasibility, or energy efficiency, without fully accounting for the broader systemic implications of hydrogen integration. This reductionist approach limits the ability to capture indirect effects, cross-sectoral interactions, and long-term strategic consequences associated with large-scale deployment. In contrast, this work proposes an integrated and holistic evaluation approach grounded in the Smart City framework, which considers the railway system as part of a complex and interconnected urban ecosystem. By adopting this broader perspective, the proposed framework enables the assessment not only of direct operational and environmental benefits but also of secondary and enabling effects, such as interactions with urban energy infrastructures, renewable energy integration, land use planning, energy system flexibility, and urban sustainability objectives. This comprehensive perspective is essential to capture the full value proposition of hydrogen technologies and to avoid suboptimal decisions based on partial or short-term evaluations.

Furthermore, the proposed approach provides decision-makers, planners, and policymakers with a more robust and multidimensional decision-support tool, capable of informing long-term infrastructure planning, investment prioritization, and policy design. By explicitly incorporating systemic interdependencies and long-term urban development goals, this work advances current evaluation practices and contributes to bridging the gap between technological feasibility assessments and strategic urban planning. Ultimately, this contribution supports a more informed, resilient, and sustainable transition toward hydrogen-based railway systems within the broader context of smart and climate-neutral cities.

The remainder of the paper is organized as follows. Section 2 introduces the Smart City framework and discusses its suitability as the most appropriate conceptual approach for addressing the objectives of this study. Section 3 analyzes the impacts of hydrogen-based railway systems on urban contexts, with specific attention to their environmental, infrastructural, and spatial implications. Section 4 presents selected case studies that empirically support and validate the impact assessment discussed in the previous section. Section 5 constitutes the core of the paper, focusing on the role of Decision Support Systems, their usefulness in evaluating technological innovations, and the main DSS approaches adopted in this domain. Finally, Sect. 6 summarizes the main findings and provides concluding remarks.

2 The Smart City as a Systemic Modeling Framework

The complexity of the technological innovations described thus far necessitates an organic conceptual framework capable of synthesizing fragmented technical interventions into a unified vision of territorial governance; this paradigm finds its most complete expression in the Smart City model [9]. Within this context, the Smart City is not uncritically interpreted as a mere summation of digital devices or intelligent infrastructures. Rather, it is defined as a multidimensional methodological framework designed to model the interdependencies between urban subsystems (Fig. 1). Adopting this model allows for the transcendence of sector-specific territorial views, providing an integration platform where technological innovation ceases to be an exogenous element and becomes an endogenous variable within the planning process. Consequently, the city is mapped through interconnected pillars that reflect the trajectories of scientific progress and their spatial repercussions.

Within this structure, Smart Energy and Smart Mobility act in synergy to redefine the urban metabolism: the former through the intelligent and bidirectional management

Fig. 1. The Smart City ecosystem: A network of integrated services.

of networks, and the latter by transforming transport into a fluid, service-oriented system [10]. The impacts of such integration are manifested in a drastic reduction of environmental externalities and an optimization of resources that profoundly alter the physical and infrastructural landscape. In parallel, the Smart Health pillar introduces monitoring technologies and telemedicine, which, by integrating into the urban fabric, ensure a timely response to citizen needs. This integration directly impacts social resilience and reduces the operational pressure on physical hospital facilities. This domain is indissolubly linked to Smart Safety, where the deployment of advanced sensors and predictive analytics for public security and emergency management transforms the perception of urban space, rendering it an intrinsically more protected and monitored environment. The paradigm is further consolidated by the Smart Living and Smart Building domains, where technological innovation aims to elevate the quality of daily life through automation and environmental comfort. The impact of Smart Living extends beyond individual well-being; it fosters greater inclusivity and accessibility to urban services, encouraging the integration of vulnerable population segments. Smart modeling thus becomes an indispensable tool for translating technological pervasiveness into a development strategy that is no longer based on intuition or obsolete historical data, but on a dynamic understanding of impacts. Only through this rigorous structuring can it be guaranteed that innovation in the fields of energy, health, safety, and transport converges toward the singular goal of collective well-being. This transforms the city into an active laboratory for social and structural sustainability, capable of adapting in real-time to the challenges of modernity [11].

In light of the above, the Smart City framework emerges as an appropriate and effective approach for assessing the impact of technologies on urban systems. Its holistic perspective captures the city as an integrated socio-technical system, allowing technological, environmental, economic, and governance dimensions to be analyzed coherently. This makes the framework particularly well-suited to support DSS-based evaluations of complex innovations and their systemic effects on cities.

3 Green Hydrogen and Rail Mobility: A Paradigm for the Sustainable Transition of the Smart City

Building on the broader range of technological transformations affecting urban systems discussed above, this contribution concentrates on the impacts of transport-related innovations on the city, with particular attention to the relationship between mobility and territory. Among the various technological changes influencing urban environments, the analysis focuses on how mobility systems shape urban sustainability and spatial dynamics. In this perspective, special emphasis is placed on environmentally sustainable solutions, specifically hydrogen-based railways, and on their potential effects on cities.

Indeed, hydrogen, specifically in its "green" form obtained via electrolysis powered by renewable sources, is not merely an alternative fuel [12], but a fundamental enabler for the entire Smart City ecosystem, as can be observed in Fig. 2.

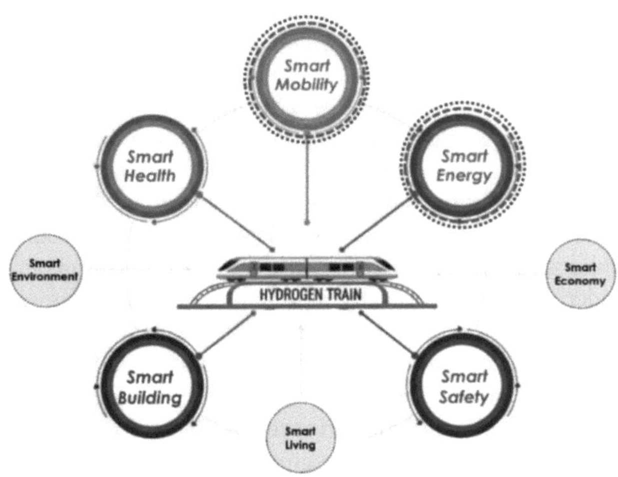

Fig. 2. Hydrogen railways as a strategic breakthrough for the Smart City.

In terms of Smart Energy, it serves as a critical long-term storage system, stabilizing grids and managing the production surpluses of energy communities. The adoption of this vector impacts the smart city pillars transversally: it enhances Smart Health by eliminating carbon dioxide and localized pollutant emissions; strengthens Smart Safety through digitally monitored gas management protocols; and promotes Smart Living by drastically reducing acoustic pollution—a decisive factor for psychophysical well-being in urban and suburban contexts. A focal point of this analysis is the impact of hydrogen-powered railways (H2-rail), a technology representing the synthesis of infrastructural innovation, materials engineering, and environmental sustainability [13]. The introduction of rolling stock equipped with Hydrogen Fuel Cell (HFC) systems represents a disruptive intervention for non-electrified lines, which still constitute a significant portion of global and regional rail networks.

This technology radically transforms the conception of mobility, acting as a primary catalyst for the Smart Mobility pillar. Unlike internal combustion engines, fuel cell systems operate through an electrochemical process that combines onboard stored hydrogen with atmospheric oxygen to generate electricity. This ensures operational ranges and refueling times comparable to traditional means, but with significantly higher energy efficiency and a total absence of greenhouse gas (GHGs) emissions during operation. From a Smart Energy perspective, the hydrogen railway transcends its role as a mere end-user to become the pivot of a complex energy architecture. The constant and programmable demand for hydrogen from rail services acts as an "anchor load" for the establishment of territorial Hydrogen Valleys [4]. In these ecosystems, rail refueling stations evolve into multimodal energy hubs capable of supplying local bus fleets, heavy-duty logistics, and private vehicles, thereby optimizing infrastructural costs and fostering economies of scale in electrolytic hydrogen production. This model of energy integration facilitates regional grid stabilization by using hydrogen as a storage medium for surplus energy generated by local wind or solar farms, closing the loop of a circular and energy-autonomous economy. Regarding spatial and urban planning, hydrogen rail offers novel solutions for historically disadvantaged routes characterized by orographic constraints, restricted tunnel gauges, or high-value landscapes where the installation of catenaries and high-voltage pylons would be technically unfeasible or aesthetically detrimental. By avoiding the visual and structural impact of "heavy" electrification, this technology preserves the integrity of built heritage and natural landscapes while enabling the enhancement of transport services. This capacity for "territorial mending" translates into a profound impact on Smart Living: by guaranteeing modern and frequent connections to inland and peripheral areas, the hydrogen railway actively combats depopulation and bridges the digital and physical divide. Territorial inclusivity thus becomes a tangible reality, ensuring the right to sustainable mobility regardless of urban density. As shown by [14], however, it is essential to evaluate the user experience with extreme care throughout the transition process. Nevertheless, such criticalities do not emerge in the transition to hydrogen traction; the latter proves to be "transparent" to the user, who benefits from high standards of efficiency and comfort without being subjected to the performance trade-offs typical of other energy-saving technologies. In the domains of Smart Health and Smart Safety, the transition from diesel to hydrogen yields immediate sanitary benefits. The sole byproduct of hydrogen oxidation within the fuel cell is water vapor, fundamentally eliminating the emission of particulate matter, nitrogen oxides and sulfur dioxide in inhabited centers. This eradication of toxic emissions, combined with the near-total absence of noise pollution and vibrations typical of thermal engines, transforms railway corridors into high-quality environmental zones. The environmental sustainability of this energy source represents the core of this technological shift [15]. When hydrogen is produced via electrolysis (green hydrogen), rail transport achieves complete Well-to-Wheel (WtW) decarbonization [16]. Life Cycle Assessment (LCA) methodologies confirm that, despite the manufacturing complexity of cells and tanks, the GHG emission balance is drastically lower than any hydrocarbon-based solution.

In conclusion, the hydrogen railway does not merely modernize a transport service; it reshapes the systemic interdependencies between energy, health, environment, and society, serving as a fundamental component of an intelligent territory that harmonizes frontier innovation with ecological equilibrium. To substantiate this perspective, the following section presents a set of selected case studies that illustrate concrete applications of hydrogen rail systems and highlight their impacts on urban contexts.

3.1 The Proposed Decision Support System (DSS) Framework

A comprehensive Decision Support System (DSS) framework for the assessment and deployment of hydrogen-powered railway systems should be designed as an integrated, scenario-based, and multi-layer analytical platform capable of rigorously comparing hydrogen traction with conventional diesel and electrified railway alternatives under present and future energy transition conditions. The framework must adopt a life-cycle and system-level perspective, incorporating the full value chain, including energy production, storage, distribution, traction technologies, and infrastructure requirements [17]. The analytical structure should be based on a coherent and hierarchically organized set of Key Performance Indicators (KPIs), grouped into major evaluation domains: economic, environmental, technical-operational, infrastructural, and socio-economic. Economic KPIs should include capital expenditures (CAPEX), operational expenditures (OPEX), maintenance costs, total cost of ownership (TCO), lifecycle cost (LCC), and cost per train-kilometer. Environmental KPIs must capture greenhouse gas emissions on a well-to-wheel basis, local air pollutant emissions (NOx, particulate matter), energy consumption, primary energy intensity, and potential for integration with renewable energy sources. Technical and operational KPIs should address system efficiency, vehicle performance, operational range, refueling or recharging time, reliability, availability, maintainability, and technological maturity. Infrastructure-related KPIs must consider infrastructure investment requirements, network compatibility, scalability, and implementation complexity. Finally, socio-economic KPIs should include energy security, reduction of fossil fuel dependency, job creation potential, industrial development opportunities, public acceptance, and alignment with long-term decarbonization and policy objectives.

To ensure methodological rigor and relevance for policy support, the DSS must incorporate advanced Multi-Criteria Decision-Making (MCDM) methodologies capable of handling heterogeneous, multidimensional, and potentially conflicting evaluation criteria [18, 19]. Techniques such as the Analytic Hierarchy Process (AHP) and Analytic Network Process (ANP) can be employed to structure the decision problem hierarchically and derive consistent weighting factors, while outranking methods such as ELECTRE and PROMETHEE can be applied to perform robust pairwise comparisons and identify dominance relationships among alternatives. Furthermore, the integration of Multi-Actor Multi-Criteria Analysis (MAMCA) is essential to explicitly account for the perspectives, priorities, and preferences of multiple stakeholder groups, including public authorities, railway operators, infrastructure managers, energy providers, industry stakeholders, and local communities [20]. Stakeholder-specific weighting vectors should be elicited through structured expert elicitation, participatory workshops, or Delphi-based consensus processes, ensuring transparency and legitimacy in the representation of value judgments. The DSS should incorporate normalization procedures to ensure comparability across heterogeneous KPIs and aggregation algorithms capable of producing both stakeholder-specific and global performance indices. In addition, uncertainty and variability in key input parameters—such as hydrogen production costs, electricity prices, carbon pricing, technology learning curves, and policy incentives—must be explicitly addressed through sensitivity analysis, scenario analysis, and, where appropriate, probabilistic methods such as Monte Carlo simulation. The final output of the DSS should consist of a transparent and reproducible ranking of hydrogen, diesel, and electric railway scenarios, complemented by robustness indicators, stability intervals, and sensitivity maps that quantify the influence of individual criteria and stakeholder preferences on the final ranking. Such a framework provides not only a static comparative assessment but also a dynamic and adaptive decision-support tool capable of evolving with technological progress and policy developments.

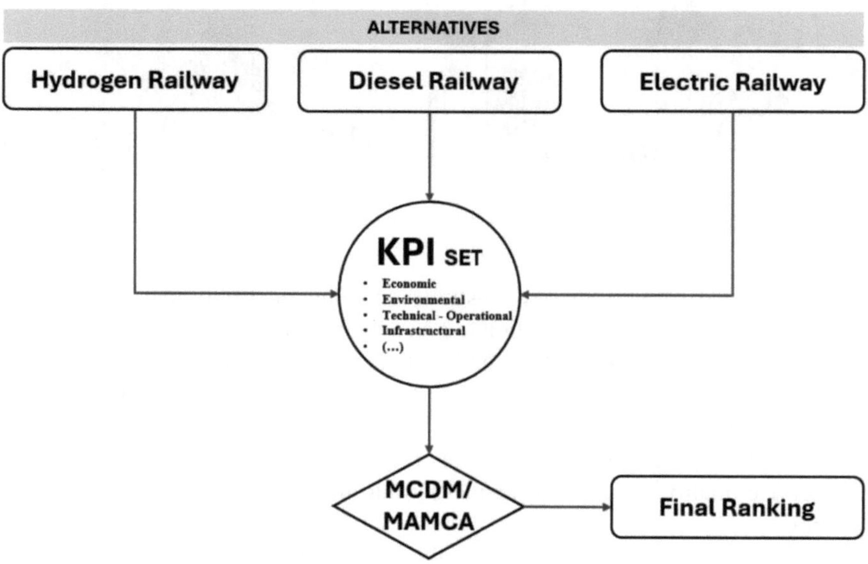

Fig. 3. DSS architecture.

By integrating techno-economic modeling, environmental assessment, stakeholder preference analysis, and robust multi-criteria aggregation, the proposed DSS enables evidence-based policymaking and supports strategic investment decisions aimed at identifying the most sustainable, cost-effective, and socially acceptable railway decarbonization pathway (Fig. 3).

4 Implementation Cases: From Theoretical Framework to Territorial Reality

Following the definition of the Smart City paradigm and the role of the hydrogen railway as an energy vector, it is essential to translate the theoretical framework into real-world operational contexts. This is achieved through the analysis of three case studies that demonstrate the tangible repercussions of the performance, environmental, and social impacts previously discussed. The first global reference case is represented by the Alstom Coradia iLint, in service since 2019 (Fig. 4). The iLint stands as the world's first hydrogen fuel cell-powered train to have successfully completed its commercial operation phase in Germany [21].

Fig. 4. The Alstom Coradia iLint train operating in Lower Saxony.

The experience in Lower Saxony constitutes a fundamental empirical validation, confirming that the technological transition to green hydrogen can be entirely "transparent" for the users. Indeed, the use of the iLint highlighted a drastic reduction in noise pollution and vibrations compared to the diesel units previously in use, while simultaneously ensuring high standards of thermal comfort and commercial speed. This demonstrates that the adoption of zero-emission systems requires no compromise in terms of perceived service quality, integrating seamlessly into the daily mobility dynamics of the population.

In parallel, the British experience with the HydroFLEX project introduces a crucial variable in the management of infrastructural transformation: technological retrofitting.

By converting existing electric trains into hydrogen-powered bimodal units, the University of Birmingham demonstrated how hydrogen acts as an enabling factor for the accelerated decarbonization of historical railway lines (Fig. 5). This approach allows for the circumvention of physical constraints and the prohibitive costs associated with traditional electrification, particularly in the presence of ancient tunnels or bridges [22]. Impact analysis showed that the adoption of hydrogen preserves the resilience of existing transport networks, providing a strategic solution to govern the energy transition without the need for invasive infrastructural interventions, thereby ensuring service continuity and the safeguarding of railway architectural heritage.

Finally, the case of the Jonica Railway in Italy—with completion scheduled for 2027—embodies the translation of the Smart City concept into a territorial "Hydrogen

Fig. 5. The HydroFLEX, UK's experience.

Valley" [23]. In this area, characterized by a non-electrified railway backbone of strategic importance for Southern Italy, hydrogen is not intended merely as a technological replacement for rolling stock, but as the trigger for a circular energy ecosystem (Fig. 6).

Fig. 6. The Jonica railway in Italy.

The reconfiguration of the Jonica Railway will allow the line to serve as a spine for the distribution of clean energy: railway refueling stations evolve into integrated energy hubs capable of powering not only trains but also local public transport bus fleets and last-mile logistics vehicles.

This case study highlights how transport innovation can generate multidimensional added value, promoting territorial cohesion and local economic development through a holistic vision of urban transformation, turning an environmental challenge into an opportunity for profound infrastructural renewal.

5 Territorial Modeling and Data Governance

The speed and pervasiveness of the transformations induced by innovative energy vectors, such as green hydrogen, rule out a passive management approach and instead call for structured, ex-ante evaluation processes. To this end, this study outlines a set of methodological approaches and analytical tools—intended as illustrative examples rather than exhaustive or prescriptive solutions—that can support the estimation of the impacts of hydrogen-based technologies on urban systems. These approaches span multiple analytical scales, ranging from basic spatial visualization to advanced multidimensional evaluation frameworks [24, 25], and collectively form the methodological backbone of DSS-based impact assessment. At the most immediate level, spatial analysis and visualization tools based on Geographic Information Systems (GIS), such as ArcGIS and QGIS [26, 27], enable an initial understanding of the territorial distribution and spatial compatibility of hydrogen-related infrastructure. Through the overlay of physical, socio-economic, and environmental layers, GIS-based Spatial Decision Support Systems (SDSS) support the assessment of urban suitability for Hydrogen Valleys and the localization of new hydrogen-powered railway nodes. Within this phase, approaches such as Transit-Oriented Development (TOD) are employed to evaluate how new rail infrastructures may influence land-use patterns, functional density, and the quality of the built environment in surrounding areas [28]. Moving from the territorial scale to the transport engineering domain, network assignment methodologies play a central role in estimating how hydrogen railways alter mobility dynamics. Using macroscopic transport modeling tools such as TransCAD and PTV Visum [29, 30], these methods allow for counterfactual "what-if" analyses that simulate changes in traffic flows, travel times, and accessibility. By incorporating behavioral assumptions derived from Discrete Choice Theory, such models provide insight into user responses to infrastructural innovation and help anticipate whether hydrogen rail investments may contribute to territorial rebalancing or introduce new network pressures [31]. The integration of these transformations into the physical and infrastructural fabric of the city is addressed through Building Information Modeling (BIM). BIM environments, supported by software such as Autodesk Revit, enable the detailed digital representation of hydrogen railway stations and associated facilities, ensuring coherence between structural, technological, and energy components [32]. Extending beyond asset-level modeling, the evolution toward the Urban Digital Twin represents the synthesis of these heterogeneous tools and methodologies. The Digital Twin can be understood as a comprehensive and dynamic digital entity of the city, capable of integrating GIS, transport models, BIM data, and real-time information to simulate interactions between infrastructure, environment, and society over time. Notable examples, such as the Urban Digital Twin of Los Angeles, demonstrate how traffic dynamics and air quality models can be jointly evaluated to balance mobility efficiency and environmental performance [33].

Finally, the multidimensional impacts of hydrogen railways are consolidated through evaluation frameworks such as Cost–Benefit Analysis (CBA) and Multi-Criteria Decision Analysis (MCDA), embedded within the DSS architecture [34, 35]. Using software tools such as Definitive and Super Decisions, these methods decompose complex decision problems into structured hierarchies of objectives and Key Performance Indicators, including emissions reduction, accessibility, and territorial cohesion. The adoption of Multi-Actor Multi-Criteria Analysis (MAMCA) further enhances transparency and inclusiveness by explicitly incorporating stakeholders and citizen perspectives, often supported by Public Participation GIS (PPGIS) tools [36–38]. Together, these methodologies—ranging from spatial visualization to advanced multicriteria evaluation—demonstrate how DSS can operationalize the assessment of hydrogen-related innovations in urban contexts. Table 1 synthesizes this logical framework, linking each phase of the planning and evaluation process with the corresponding analytical approaches and selected software tools, and illustrating how the Urban Digital Twin ultimately enables the integrated, systemic vision of the city as stated above.

Table 1. Integrated methodological framework for the analysis and governance of multidimensional territorial impacts.

Process phase	Methodology	Examples of tools
Spatial planning	Spatial decision support systems	ArcGIS, QGIS
Transport systems engineering	Macroscopic network assignment and discrete choice theory	TransCAD, PTV Visum
Infrastructure design	Building information modeling	Autodesk Revit, Civil 3D
Evaluation process	Cost-benefit analysis, multi-criteria decision analysis	@RISK, Visual PROMETHEE
Stakeholder engagement	Multi-actor multi-criteria analysis	PPGIS, Geo Intelligence Data

6 Conclusions and Future Perspectives

This study has examined the role of Decision Support Systems in addressing the growing complexity of technological innovation within urban environments, emphasizing the need for integrated and methodologically robust evaluation frameworks. By conceptualizing the city as a coherent and interconnected system through the Smart City framework, the analysis has shown how DSS can support a comprehensive understanding of the systemic effects generated by emerging technologies across multiple urban dimensions. The analysis focused on green hydrogen rail traction has demonstrated that this vector not only constitutes an alternative to fossil fuels but serves as a strategic enabling factor for the decarbonization of non-electrified networks and the stabilization of urban energy grids through long-term storage systems. Beyond the direct environmental benefits, such as the elimination of localized pollutant emissions and the drastic reduction of acoustic impact, the hydrogen solution has emerged as a decisive driver for territorial cohesion, capable of ensuring high standards of comfort and performance without the criticalities associated with energy-saving policies that often penalize the end-user experience by reducing onboard comfort. In this scenario, the implementation of next-generation Decision Support Systems, integrated with Urban Digital Twin technology, confirms itself as an indispensable future perspective for territorial data governance. The digital twin, acting as a dynamic virtual replica fueled by real-time data streams, enables the transcendence of subjective intuition in favor of evidence-based planning, allowing for the simulation of complex scenarios and the monitoring of infrastructural performance throughout the entire life cycle. Looking forward, research must evolve railway infrastructures into multimodal energy hubs, so-called Hydrogen Valleys, capable of radiating sustainability to local public transport and urban logistics, while simultaneously integrating artificial intelligence models for predictive consumption optimization. Parallel to this, it will be essential to refine digital public participation tools to ensure that the technological transition is not only technically efficient but also equitable and socially accepted. Ultimately, the synthesis between algorithmic precision, innovation in energy vectors, and a strategic vision of urban development outlines a model of an intelligent territory where technology ceases to be an exogenous element and becomes an endogenous variable of the planning process, ensuring that every technical advancement translates into tangible value for public health, spatial justice, and the ecological equilibrium of the contemporary landscape. In this context, the findings highlight how DSS enable decision-makers to systematically evaluate trade-offs, interdependencies, and long-term consequences associated with complex technological choices. By integrating heterogeneous data, scenario analysis, and multidimensional impact assessment, DSSs enhance transparency, accountability, and resilience in urban policy processes. Overall, the work demonstrates the critical importance of Decision Support Systems as essential instruments for assessing the impacts of technological innovations on cities and for guiding informed and sustainable urban transformation. Only through this holistic and multidimensional approach will it be possible to govern the velocity of ongoing change, transforming environmental challenges into opportunities for authentic renewal for future generations.

References

1. Cepa, J.J., Pavon, R.M., Alberti, M.G., Ciccone, A., Asprone, D.: A review on the implementation of the BIM methodology in the operation maintenance and transport infrastructure. Appl. Sci. **13**(5), 3176 (2023)
2. Cascetta, E., Botte, M., Limmatola, A.: Le sette rivoluzioni dei trasporti e la struttura delle città: fra storia e futuri possibili. Ingegneria Ferroviaria. **80**, 401–439 (2025)
3. Silvestri, F., De Fabiis, F., Coppola, P.: Consumers' expectations and attitudes to-wards owning, sharing, and riding autonomous vehicles. Case Stud. Transp. Policy. **15**, 101112 (2024)
4. Ganji, M. et al.: Comprehensive study on hydrogen production for sustainable transportation planning: strategic, techno-economic, and environmental impacts. Hydrogen. **6**(2), 24 (2025)
5. Hämäläinen, M.: Urban development with dynamic digital twins in Helsinki city. IET Smart Cities. **3**(4), 201–210 (2021)
6. Deistler, J. et al.: Smart City Strategy Vienna. Our Way to Becoming a Model Climate City. Vienna Municipal Administration (2022)
7. D'Acierno, L., De Matteis, L., Stefanelli, R.: Adoption of hydrogen in railway traction: opportunities and limits in the case of non-electrified railway lines. Transp. Res. Procedia. **90**, 130–137 (2025)
8. Gallo, M.: Optimal location of refuelling stations for hydrogen-powered freight vehicles: models and methods. Transp. Res. Procedia. **90**, 5–12 (2025)

9. Kunzmann, K.R.: Smart cities: a new paradigm of urban development. Crios. **4**(1), 9–20 (2014)
10. Orlowski, A., Romanowska, P.: Smart cities concept: smart mobility indicator. Cybern. Syst. **50**(2), 118–131 (2019)
11. Salman, M.Y., Hasar, H.: Review on environmental aspects in smart city concept: water, waste, air pollution and transportation smart applications using IoT techniques. Sustain. Cities Soc. **94**, 104567 (2023)
12. Ministero dell'Ambiente della Sicurezza Energetica (MASE): Strategia Naziona-le Idrogeno, MASE. (2024). https://www.mase.gov.it/portale/documents/d/guest/strategia-nazionale-idrogeno-pdf. Last access: 13 Jan 2026
13. Böhm, M., Del Rey, A.F., Pagenkopf, J., Varela, M., Herwartz-Polster, S., Calde-rón, B.N.: Review and comparison of worldwide hydrogen activities in the rail sector with special focus on on-board storage and refueling technologies. Int. J. Hydrog. Energy. **47**(89), 38003–38017 (2022)
14. D'Acierno, L., Botte, M.: Passengers' satisfaction in the case of energy-saving strategies: a rail system application. In: IEEE International Conference on Environment and Electrical Engineering and IEEE Industrial and Commercial Power Systems Europe (EEEIC/I & CPS Europe), pp. 1–5. IEEE (2018)
15. Gentilucci, R.: Idrogeno verde al posto del diesel nei trasporti ferroviari: il caso italia-no. Agenda Digitale Eu (2025)
16. Osman, A.I. et al.: Life cycle assessment of hydrogen production, storage, and utilization toward sustainability. Wiley Interdiscip. Rev. Energy Environ. **13**(3), 526 (2024)
17. Hoffrichter, A., Miller, A.R., Hillmansen, S., Roberts, C.: Well-to-wheel analysis for electric, diesel and hydrogen traction for railways. Transp. Res. Part D. Transp. Environ. **17**(1), 28–34 (2012)
18. Çelikdemir, S., Çelikdemir, M.Y., Özdemir, M.T.: A multi-criteria analytical hierarchy process framework for the design of Türkiye's transcontinental hydrogen pipe-line corridor. Int. J. Hydrog. Energy. **152**, 150208 (2025)
19. Albadwi, A., Elzein, A., Bashir, A.A.A., Selçuklu, S.B., Kaya, M.F.: Multi-criteria decision-making in hydrogen production, storage, and transportation: an over-view. Renew. Sust. Energ. Rev. **222**, 115947 (2025)
20. Alvarez, P.A., Ishizaka, A., Martinez, L.: Multiple-criteria decision-making sorting methods: a survey. Expert Syst. Appl. **183**, 115368 (2021)
21. Palmer, C.: Hydrogen-powered trains start to roll. Engineering. **11**, 9–11 (2022)
22. Calvert, C., Allan, J., Amor, P., Hillmansen, S., Roberts, C., Weston, P.: Concept development and testing of the UK's first hydrogen-hybrid train (HydroFLEX). Railway Eng. Sci. **29**(3), 248–257 (2021)
23. Gandini, L.M., Ricci, S., Verrascina, F.: Servizi ferroviari con treni a idrogeno: vincoli normativi e valutazione di un caso di studio Hydrogen-powered trains operation: normative constraints and assessment on a cases study. Ingegneria Ferroviaria. **697**, 10 (2023)
24. McFadden, D.: Economic choices. Am. Econ. Rev. **91**(3), 351–378 (2001)
25. Wegener, M.: Overview of land use transport models. In: Handbook of Transport Geography and Spatial Systems, pp. 127–146. Emerald Group Publishing Limited (2004)
26. Kim, H.Y., Lee, H.K.: Enhanced validity and reliability of spatial decision sup-port systems (SDSS) for sustainable transportation decision-making. Appl. Geogr. **51**, 65–71 (2014)
27. Zhu, X.: GIS and spatial decision support. In: Geographic Information Systems, pp. 1–33. Nova Science Publishers, New York (2013)
28. Cervero, R.: Transit-Oriented Development in the United States: Experiences, Challenges, and Prospects. Transportation Research Board (2004)

29. Abdul, K.S.: Tecnologia e software per la gestione del flusso di traffico. In: The International Conference on Artificial Intelligence and Applied Mathematics in Engineering, pp. 713–722. Springer International Publishing, Cham (2022)
30. Bjorketun, U.: Traffic assignment in a GIS environment. In: WIT Transactions on the Built Environment, vol. 26. WIT Press (2025)
31. Cascetta, E.: Transportation Systems Analysis: Models and Applications. Springer, New York (2009)
32. Botte, M., Zampi, A., Oreto, C., D'Acierno, L.: The use of road microsimulation software within BIM environments: a preliminary assessment. J. Adv. Transp. **2021**, 1–11 (2021)
33. Hand, A.: Urban Mobility in a Digital Age: a Transportation Technology Strategy for Los Angeles. The City of Los Angeles Office of the Mayor and Department of Transportation (2016)
34. Greco, S., Figueira, J., Ehrgott, M.: Multiple Criteria Decision Analysis, vol. 37. Springer, New York (2016)
35. Malczewski, J.: GIS-based multicriteria decision analysis: a survey of the literature. Int. J. Geogr. Inf. Sci. **20**(7), 703–726 (2006)
36. Macharis, C., Turcksin, L., Lebeau, K.: Multi actor multi criteria analysis (MAMCA) as a tool to support sustainable decisions: state of use. Decis. Support. Syst. **54**(1), 610–620 (2012)
37. Czepkiewicz, M. et al.: Public participation GIS for sustainable urban mobility planning: methods, applications and challenges. Rozwój Regionalny i Polityka Regionalna. **35**, 9–35 (2016)
38. Raymond, C. et al.: An Impact Evaluation Framework to Support Planning and Evaluation of Nature-Based Solutions Projects. Report Prepared by the EKLIPSE Expert Working Group on Nature-Based Solutions to Promote Climate Resilience in Urban Areas. Centre for Ecology and Hydrology (2017)

A Decision Support Framework for Urban Green Transition: Spatiotemporal Evolution and Determinants of Green Total Factor Productivity in China

Ping Huang[1,2], Xiaohui Chen[3](✉), and Yufan He[4]

[1] School of Digital Economy and Finance, Guizhou University of Commerce, Guiyang 550000, China
[2] Guizhou Provincial Academician Workstation for Remote Sensing Image Processing and Plateau Disaster Prevention and Control, Guiyang 550000, China
[3] Faculty of Economics and Business Administration, Yibin University, Yibin 644000, China
m_cxh@163.com
[4] School of Business Administration, Guizhou University of Commerce, Guiyang 550000, China

Abstract. China's commitment to its dual-carbon goal underscores its dedication to environmentally sustainable development and exhibits notable achievements in green development. The profound impact of the digital economy on fostering green practices cannot be overstated. This study employs a three-stage Data Envelopment Analysis (DEA) method to assess the extent to which the digital environment in Chinese prefecture-level cities has influenced green total factor productivity (GTFP). The findings reveal several significant insights: (1) GTFP demonstrates a trend similar to the leftward side of an inverted U-shaped curve, indicative of the positive influence of the digital environment on enhancing green productivity; (2) According to the comparison results before and after the adjustment, when excluding the impact of digital environment, the GTFP significantly decreases; (3) From the perspective of spatial patterns, the digital environment has clearly played a positive role in promoting the concentration, distribution, and diffusion of GTFP; (4) From the perspective of temporal evolution, the digital environment has led to a significant "north-south divide" in the improvement of green total factor productivity, with a noticeable Heihe-Tengchong Line. It also offers important policy implications for giving full play to the economic effects of the digital environment and promoting high-quality development of cities.

Keywords: Digitalization · Green total factor productivity · Manufacturing digitalization · Digital financial inclusion

1 Introduction

In 2021, the National Bureau of Statistics unveiled its "Statistical Classification of the Digital Economy and its Core Industries (2021)," designating the digital efficiency enhancement industry as one of the "core industries of the digital economy." Since then,

A. P. Cabral Seixas Costa et al. (Eds.): ICDSST 2026, LNBIP 582, pp. 119–132, 2026.
https://doi.org/10.1007/978-3-032-25146-6_8

this sector has experienced a rapid expansion. Projections outlined in the 14th Five-Year Plan anticipate that by 2025, the value-added share of the digital efficiency enhancement industry, among the "core industries of the digital economy," will reach 10% of the GDP. The ascent of the digital efficiency enhancement industry has been characterized by its pervasive influence across diverse economic and production domains. Its integration has catalyzed the digitalization of manufacturing and inclusive finance. This transformative momentum extends to all facets of production and daily life, steering them towards environmentally conscious, low-carbon, and resource-efficient industries.

The increasing integration of digital technologies into manufacturing has sparked a significant shift towards digitalization across all facets of production. Firstly, the digital economy offers opportunities to optimize the utilization of natural resources. By streamlining production processes and minimizing redundant operational steps, digitalization enhances natural resource efficiency and mitigates losses associated with resource dependency, thus fostering a transition towards greener industrial practices [1]. Secondly, incorporating data as a new production factor within industrial systems improves the efficiency of information flow both within and between enterprises. This integration enhances the core competitiveness of the manufacturing industry, propelling it towards greater productivity and innovation [2, 3]. Thirdly, embedding digital technology into the utilization of natural resources stimulates innovation and improves resource collection and utilization efficiency. This integration influences the broader transformation towards greener industrial practices by leveraging the potential of digital technology to optimize resource usage [4]. The digital environment significantly impacts green credit, low-carbon environmental protection, and low-carbon life, exerting profound effects on the financial services sector [5, 6]. The October 2023 State Council Implementing Opinions on Promoting the High-quality Development of Inclusive Finance underscores the imperative of cultivating a robust digital inclusive financial ecosystem. This innovative financial sector aims to cater to the diverse, personalized, and multi-level financial needs of individuals while offering inclusive financial services in areas such as green production, green credit, low-carbon environmental protection, and low-carbon living [7, 8].

Existing literature lacks a comprehensive examination of the impact of the digital environment on green total factor productivity (GTFP), both before and after the adjustment. By beginning at the micro level and selecting prefecture-level cities as the research focus, this paper offers a valuable addition to the GTFP literature, providing insights from the perspective of the digital economic environment. This paper employs a three-stage DEA model, which eliminates external factors and random interference, thereby resulting in GTFP estimates that better reflect reality.

2 Research Hypotheses

The adjustment of the digital environment on GTFP involves the adjustment of input factors and output factors.

First, concerning input factors, the digital environment conserves input resources for green development, enhancing resource efficiency and the utilization of green credit. Irrespective of labor, capital, or energy factors, within the digital context, input resources are utilized more efficiently. In the digitalization of the manufacturing sector, diverse types of input factors are preserved and effectively leveraged. Firstly, through online monitoring and data analysis of energy consumption and equipment operation in enterprises, companies optimize energy management, enhance energy efficiency, diminish utilization of natural resources, and support the industry's green transformation [9]. Secondly, the digital economy presents opportunities for more efficient utilization of natural resources. By upgrading production with its process flows and directly reducing intermediate and redundant operations, the digital economy enhances the utilization efficiency of natural resources, boosts production capacity and value creation, and maintains optimal utilization efficiency of natural resources, thus enabling traditional manufacturing to achieve high-efficiency production [10–12]. Thirdly, it facilitates optimal resource allocation, enhances the efficiency of economic operation, reduces material losses, and fosters the manufacturing industry's green transformation. The digital information network possesses green attributes. Its low-cost, high-speed information transmission optimizes resource allocation within and beyond industries, facilitating the integration and sharing of various resource elements. Digital inclusive finance addresses the financing needs for green development. By offering more green financial resources to environmentally friendly and resource-conserving enterprises, it provides robust backing for the economy's green and sustainable development. Digital inclusive finance employs big data to promptly acquire relevant information on green enterprises' environmental protection processes and equipment, enabling financial institutions to significantly mitigate associated financial risks. Financial institutions, armed with such information, are more inclined to allocate funds to enterprises [13].

Second, in terms of output factors, the digital environment enhances GTFP, reduces output costs, and improves output quality. Firstly, within the digital environment of the manufacturing industry, effective green outputs are amplified. On the one hand, the reduced marginal cost of acquiring factors for green innovation leads to an increase in green innovation output [14]. On the other hand, it diminishes negative outputs. Leveraging digital tools across various manufacturing segments enhances intelligence levels, fosters innovation within the core segments of industrial production, and curtails the adverse environmental impacts associated with traditional industrial models [15]. Secondly, under digital inclusive finance, there's a drive toward the adoption of high-quality life consumption, such as green living and consumption. Data precision in user matching, coupled with increased income, stimulates consumption. With the growing prevalence of environmental consciousness and awareness, consumption trends favor green products and services, thus promoting the development of green enterprises from the demand side. This, in turn, fuels the growth of platforms and products for green consumption and facilitates the advancement of green economic development [16–18]. Based on the analysis above, this study proposes Hypothesis 1: The digital environment adjusts both the inputs and outputs of GTFP.

3 Research Design

3.1 Model

3.1.1 Three-Stage DEA

The three-stage DEA model calculates the efficiency of various decision-making units (DMUs) by excluding the influence of environmental factors and random errors, and analyzes them within a standardized environment. The steps are as follows:

First, we utilize the traditional DEA model to calculate as Eq. (1). This study adopts the method by Liu, treating each identical DMUs across different years as a distinct, independent entity. Original input-output data are fed into the Banker-Charnes-Cooper (BCC) model to derive the input slack value before adjustment, the GTFP in the first stage, and the pure technical efficiency (PTE) and scale efficiency (SE), where GTFP = PTE * SE.

$$\begin{aligned} & min\theta - \varepsilon\left(\hat{e^{\tau}} S^{-} + e^{\tau} S^{+}\right) \\ & s.t. \begin{cases} \sum_{j=1}^{n} X_j\lambda_j + S^{-} = \theta X_0 \\ \sum_{j=1}^{n} Y_j\lambda_j - S^{+} = Y_0 \\ \lambda_j \geq 0, S^{-}, S^{+} \geq 0 \end{cases} \end{aligned} \tag{1}$$

Where j = 1, 2, …, n represents the DMUs, X and Y represent the input and output indicators, respectively. λ_j are the weighting coefficients of the decision unit, θ represents the effective values, S^{-} and S^{+} are slack variables.

Second, a stochastic frontier analysis (SFA) model with panel data is employed. This involves determining the impact of random errors, management efficiency, and environmental factors on the input slack variables of the DMU based on the difference between the original input data and the optimal input values obtained in the first stage, in conjunction with the data of environmental variables as Eq. (2):

$$S_{ni} = f(Z_i; \beta_n) + \nu_{ni} + \mu_{ni} (n = 1, 2, \ldots N; i = 1, 2 \ldots, I) \tag{2}$$

S_{ni}represents the input slack variable, Z_i represents the environmental variable, β_n represents the coefficient of the environmental variable, and $\nu_{ni} + \mu_{ni}$ represents the mixed error, including statistical noise and managerial inefficiency.

Third, the adjusted DEA model employs the adjusted input data from the second stage and reintegrates them into the BCC model alongside the original output data. This step yields the final true efficiency scores for each DMU.

3.1.2 Kernel Density Estimation

Kernel density estimation is a non-parametric estimation method. Its essence lies in using an estimated and smoothed density curve to describe the distribution of random variables, which can reveal the dynamic evolutionary trends of the distribution of the research object. The calculation formula is as Eq. (3).

$$f(x) = \frac{1}{Nh}\sum_{i=1}^{N} K\left(\frac{x_i - \overline{x}}{h}\right), K(x) = \frac{1}{\sqrt{2\pi}} \exp\left(\frac{-x^2}{2}\right) \tag{3}$$

In the formula: x_i represents the level of high-quality agricultural development in each province, $\overline{x}$ is the mean value of high-quality agricultural development, N is the number of observed values in the evaluated region, and h is the bandwidth and K(x) is the Gaussian kernel density function.

3.1.3 Standard Deviational Ellipse

The Standard Deviational Ellipse (SDE) is a spatial statistical method that reflects spatial distribution. SDE can visualize the center of gravity (depicted by the center of the ellipse), distribution range, intensity, shape, and direction of economic indicators. Suppose there are *n* sub-regions within a region, and d_i represents the economic density of region *i*, then the formula for calculating the center is as Eq. (4).

$$M(\overline{X}, \overline{Y}) = \left| \frac{\sum_{i=1}^{n} d_i x_i}{\sum_{i=1}^{n} d_i} \frac{\sum_{i=1}^{n} d_i y_i}{\sum_{i=1}^{n} d_i} \right| \tag{4}$$

Formulas for calculating the major and minor axes is as Eq. (5). In formulas (5), x_i and y_i represent the coordinates of sub-region i, and $(\overline{X}, \overline{Y})$ are the coordinates of the center of the standard deviational ellipse.

$$SDE_x = \sqrt{\frac{\sum_{i=1}^{n} (x_i - \overline{X})^2}{n}}, SDE_y = \sqrt{\frac{\sum_{i=1}^{n} (y_i - \overline{Y})^2}{n}} \tag{5}$$

3.2 Variables

3.2.1 Green Total Factor Productivity (GTFP)

The input and output indicators selected for measuring GTFP are shown in Table 1. Input variables include labor, capital, and energy inputs, and specific indicators is as Table 1. Desired output variables consist of real GDP, total consumption per capita, and urban green areas; non-desired outputs comprise industrial waste (wastewater, sulfur dioxide, and dust).

3.2.2 Environmental Variables

(1) Manufacturing digitalization index (MDI)

China's National Bureau of Statistics (NBS) categorizes the manufacturing sector into 29 secondary industries, from agricultural and food processing to automobile manufacturing. Each of these secondary industries has distinct requirements for implementing digitalization. To capture this diversity, we compiled specific keywords tailored to the needs of each secondary industry. With these keywords, we extracted data on the number of enterprises incorporating these terms within their business scope, along with their registered capital, from the State Administration for Market Supervision and Administration (SAMSA). This approach yielded 58 data points for each city (=29*2). For instances where data points were absent, we substituted them with random numbers between 0 and

Table 1. Input-output table for GTFP

First-level indicators	Second-level indicators	Tertiary indicators
Input indicators	Labor input	Year-end number of employees
		Number of industrial enterprises above designated size
	Capital input	Fixed asset investment
		Urban construction land area
		Expenditure on scientific research
	Energy input	Total water supply
		Total gas supply
		Overall electricity consumption
Output indicators	Non-expected output	Wastewater discharge
		Sulfur dioxide emission
		Dust
	Expected output	Real GDP
		Per capita total consumption
		Urban greening area

0.0001. Subsequently, factor analysis was employed to construct the Urban Manufacturing Digitalization Index. Moreover, we calculated rMDI as another proxy variable for the level of manufacturing digitalization by taking the natural logarithm of (1 + MDI).

(2) Digital financial inclusion

This study relies on the Peking University Digital Inclusive Finance Comprehensive Index, to assess the developmental status of cities in China regarding digital financial inclusion. This index carries significant authority and influence, covering various aspects such as financial service coverage, service quality, and technological innovation within the financial sector.

3.2.3 Data Sources

Considering the availability and comprehensiveness of data, this study selected a sample of 280 prefecture-level cities across 30 provinces, autonomous regions, and municipalities nationwide, excluding Tibet, Hong Kong, Macao, and Taiwan. The research period spans from 2008 to 2020. Data utilized in this study primarily originate from the China Urban Statistical Yearbook, the China Environmental Statistical Yearbook, the China Energy Statistical Yearbook, as well as provincial statistical yearbooks, and the National Bureau of Statistics.

4 Results

4.1 GTFP Change Results Before and After Adjustment

4.1.1 Phase I Traditional DEA Results

Table 2 reports the mean values for GTFP, PTE, and SE at 0.739, 0.806, and 0.916, respectively. For cities of different sizes, provincial capital cities exhibit slightly lower GTFP compared to non-capital cities. This discrepancy primarily arises from the faster growth of PTE in provincial capital cities, counterbalanced by slower growth in SE. Notably, SE in provincial capital cities lags behind that of non-capital cities until 2019 and 2020. This discrepancy can be attributed to the conducive environment in capital cities for enhancing technical efficiency, leveraging resource endowments, and accumulating technology advantages. However, the formation of a stable scale effect in capital cities takes a longer time.

Table 2. GTFP in phase I

Year	All cities			Provincial capital city			Non-provincial capital city		
	TFP	PTE	SE	TFP	PTE	SE	TFP	PTE	SE
2008	0.612	0.693	0.892	0.543	0.624	0.881	0.682	0.763	0.903
2009	0.615	0.702	0.882	0.558	0.654	0.862	0.671	0.750	0.901
2010	0.633	0.718	0.888	0.585	0.683	0.866	0.680	0.752	0.910
2011	0.667	0.742	0.902	0.629	0.720	0.877	0.704	0.764	0.926
2012	0.687	0.766	0.901	0.662	0.762	0.877	0.712	0.771	0.926
2013	0.689	0.767	0.902	0.678	0.778	0.875	0.700	0.756	0.928
2014	0.721	0.801	0.898	0.720	0.820	0.870	0.723	0.783	0.926
2015	0.750	0.822	0.910	0.743	0.834	0.884	0.757	0.809	0.936
2016	0.818	0.882	0.930	0.829	0.907	0.917	0.807	0.856	0.944
2017	0.809	0.862	0.939	0.830	0.885	0.938	0.787	0.839	0.940
2018	0.842	0.892	0.943	0.863	0.915	0.938	0.821	0.868	0.947
2019	0.884	0.921	0.959	0.918	0.951	0.965	0.850	0.891	0.954
2020	0.881	0.916	0.960	0.909	0.936	0.971	0.852	0.896	0.950
Mean	0.739	0.806	0.916	0.728	0.805	0.902	0.750	0.807	0.930

4.1.2 Second-Stage SFA Regression Results

In the second stage, an SFA regression with panel data is formulated, employing the slack variables of labor input, capital input, and energy input as dependent variables. Table 3 displays the regression results obtained using Frontier 4.1.The digitalization of the manufacturing industry and digital financial inclusion exhibit a significant and negative correlation with the slack of input variables related to labor, suggesting that cities with stronger manufacturing industry and digital financial inclusion can economize on human resources and thus foster green economic and social development. Regarding energy variables, manufacturing digitalization and digital financial inclusion are negatively correlated with slack in water and gas inputs, indicating that increased digitalization in manufacturing and digital financial inclusion lead to more efficient water and gas usage, thereby promoting green development. On the other hand, manufacturing digitalization and digital financial inclusion exhibit a significant and positive correlation with slack in electricity inputs, suggesting that the reliance on computers and intelligent equipment in digital manufacturing increases electricity consumption, potentially hindering green development.

4.1.3 Adjusted DEA Analysis

Applying the DEA model again, we derived the adjusted GTFP as presented in Table 4. The GTFP decreases from 0.739 to 0.350 compared to the pre-adjustment period, a decrease of more than 52.6%. This suggests that the efficiency level is not high. Thus, Hypothesis 1 is confirmed, i.e., GTFP is enhanced in the digitized environment. However, when the influence of factors relating to the digital environment and other variables is removed, GTFP decreases significantly. Before and after adjusting to the digital environment, the influence on GTFP is primarily due to the low level of SE. The recent rapid development of China's manufacturing industry digitization and digital financial inclusion, driven by various digital economy policies and significant market demand, has led to economic expansion, improved management, and control capabilities. This, in turn, has contributed to the formation of economies of scale, thereby enhancing GTFP. Consequently, it can be inferred that the digital environment promotes the improvement of China's GTFP by increasing SE.

When considering city heterogeneity, the capital cities demonstrate higher TFP compared to non-capital cities. PTE tends to be higher in non-capital cities, while SE is higher in capital cities. This pattern may be attributed to the focus of non-capital cities on specific sectors or products for greening practices, leading to improved resource utilization efficiency but limited scale of green development. Consequently, their PTE surpasses that of larger enterprises.

4.2 Changes in the Spatial Patterns of GTFP Before and After Adjustment

SDE calculations were performed for GTFP in the first and third stages, resulting in Table 5. The geographical distribution and evolution of GTFP reflect significant differences in the following. Firstly, the SDE in the third stage is larger than in the first stage, especially in 2008 and 2012, showing a more dispersed distribution of GTFP. Without digital support, innovation activities primarily relied on traditional manufacturing and technology transfer methods. The lower mobility of innovation resources led to greater regional disparities in innovation capabilities. Secondly, without the support of digital inclusive finance, companies and individuals may have faced greater barriers to accessing funds and technological resources. The slower pace of manufacturing digitalization also meant that GTFP's innovation capabilities relied more on specific regions, resulting

Table 3. Second-stage regression results

Variables	Labor input 1	Labor input 2	Capital input 1	Capital input 2	Capital input 3	Energy input 1	Energy input 2	Energy input 3
coefficients	2.257***	260.699***	−1108403.000***	8.818***	−68252.550***	1353.893***	−22163.830***	−305548.600***
environment variable 1	−0.026***	−2.560***	950.645***	−0.144***	100.786***	−20.568***	−34.155***	708.656***
environment variable 2	−0.176***	−7.421***	−2584.349***	−0.414***	100.903***	−55.262***	−21.710***	537.029***

Note: ***Indicates statistical significance at 0.01.The sign of the regression coefficients of environmental variables in Table 3 indicates their impact on input slack. Positive coefficients suggest that environmental factors unfavorably influence the reduction of input slack. In other words, as input levels increase, the problem of wastage intensifies, and vice versa. The one-sided errors of the LR test for all input slack variables are significant at the 1% level, and their values are close to 1, indicating that the model setup is accurate and the SFA model is applicable. Additionally, all parameter estimates pass the 1% t-test, affirming that the selected indicators meet the study's requirements. The independent variables consist of the two environmental factors identified in the preceding section. Regressions are not established separately for each year to ensure consistency in the direction of impacts. Prior to analysis, environmental variables were standardized.

Table 4. GTFP of prefecture-level cities in the third stage

Year	All cities			Provincial capital city			Non-provincial capital city		
	TFP	PTE	SE	TFP	PTE	SE	TFP	PTE	SE
2008	0.321	0.966	0.330	0.449	0.942	0.466	0.192	0.990	0.195
2009	0.350	0.969	0.359	0.494	0.947	0.508	0.206	0.990	0.210
2010	0.382	0.972	0.390	0.528	0.954	0.542	0.235	0.991	0.239
2011	0.412	0.975	0.420	0.564	0.959	0.578	0.260	0.991	0.263
2012	0.441	0.977	0.449	0.599	0.964	0.612	0.282	0.991	0.286
2013	0.464	0.978	0.473	0.628	0.966	0.641	0.301	0.990	0.305
2014	0.490	0.982	0.496	0.667	0.974	0.675	0.313	0.989	0.318
2015	0.511	0.984	0.516	0.696	0.977	0.704	0.325	0.991	0.329
2016	0.548	0.990	0.551	0.751	0.987	0.753	0.346	0.993	0.349
2017	0.579	0.990	0.581	0.779	0.988	0.780	0.378	0.992	0.382
2018	0.608	0.991	0.611	0.812	0.989	0.814	0.404	0.993	0.408
2019	0.628	0.994	0.628	0.834	0.994	0.832	0.422	0.995	0.424
2020	0.618	0.993	0.619	0.822	0.992	0.821	0.415	0.995	0.417
Mean	0.350	0.990	0.355	0.663	0.972	0.671	0.314	0.991	0.317

in greater imbalances in innovation development. Thirdly, in the absence of a digital environment, the changes in GTFP's flattening and azimuth angles were less dramatic than in the first stage, indicating that under traditional conditions, the geographical distribution of green technological innovation changed more slowly, with lower rates of innovation resource transfer and diffusion. In conclusion, digital inclusive finance and manufacturing digitalization have clearly played a positive role in promoting the concentration, distribution, and diffusion of GTFP. In the digital environment, innovation activities have been more efficiently concentrated in advantageous regions and gradually spread to more areas.

4.3 Temporal Evolution Analysis of GTFP Before and After Adjustment

4.3.1 Comparison Between the Northern and Southern Regions

The GTFP performance disparities between the northern and southern regions before and after adjustments to the digital environment underscore distinct priorities in digital utilization, environmental management, and resource efficiency. The significant reduction in GTFP in the northern region suggests a heavy dependence on digital technologies, emphasizing their important role in enhancing efficiency. Conversely, while leveraging digital tools, the southern region appears to prioritize environmental management and resource efficiency, thus mitigating reliance on a single technological advancement.

Table 5. Results of Standard Deviational Ellipse Calculations

Year	Stage	Perimeter /km	Major Axis /km	Minor Axis /km	Flattening /km	Azimuth
2008	Phase I	6138.923	1187.081	741.176	445.905	23.460
2012		6143.828	1172.789	760.895	411.895	22.555
2016		6105.070	1184.372	732.213	452.160	23.918
2020		6000.814	1141.149	748.562	392.587	24.524
2008	Phase III	5760.698	1131.507	672.977	458.530	16.370
2012		5715.935	1103.647	692.184	411.464	18.486
2016		5675.898	1096.512	686.583	409.930	19.289
2020		5534.304	1045.440	698.972	346.468	20.742

Note: (1) Considering space limitations, the changes in the economic center of gravity are displayed at four-year intervals, with results from 2008, 2012, 2016, and 2020; (2) The standard deviational ellipse calculation is based on 68% of the elements being concentrated within it; (3) Flattening is the difference between the major and minor axes.

4.3.2 Comparison Along the Heihe-Tengchong Line

The "Heihe-Tengchong Line" highlights the uneven distribution of China's population across its territory. The results of pre- and post-adjustment comparisons highlight that by removing the effects of digitization, resultant GTFP focuses more on resource-efficient production processes and minimized environmental impact. While digitization notably boosts GTFP, removing these factors provides clearer insights into the potential and hurdles for regions to attain green growth. Despite the western region's stronger PTE performance, its overall GTFP remains relatively low, therefore underscoring the opportunity for GTFP enhancement and green development. It's evident that when environmental factors are factored in, cities on the western and eastern sides exhibit distinct characteristics. The influence of the "Hu Huanyong line" is particularly pronounced in the digital environment.

5 Discussion

This study studies the adjustment of GTFP within a digital environment, revealing a significant contribution of the digital environment to GTFP enhancement. Unlike prior research on digitization and green practices [19], this study specifically considers digitization as an environmental factor and assesses its impact on GTFP adjustments. This approach aligns more closely with real-world dynamics and departs from the conventional regression analyses employed in earlier studies, yet yields conclusions consistent with those of Wang et al. [20]. Moreover, our study reveals that none of the digitization characteristic variables exhibit significant correlations with pure technical efficiency, indicating that the digitization environment has yet to facilitate the pathway towards greening through technical advancements—a result akin to that of Cao et al. [21] and Fang et al. [22]. This underscores the absence of a direct channel through which digitization enhances the efficiency of green technologies, a novel insight from our research.

In the analysis of regional heterogeneity, we consider the regional issues that have been highlighted in recent years and analyze the heterogeneity in terms of the North-South difference and the Hu Huanyong line. Notably, our study reveals that the delineation of the GTFP "Hu Huanyong line" becomes more pronounced in the context of the digital environment, which is a relatively novel insight. Compared to prior research by Yuet et al. [23] and Akberdina et al. [24], based on the trends in GTFP and scale efficiency on either side of the Hu Huanyong Line, cities on the eastern side exhibit superior performance in GTFP and SE, whereas those on the western side demonstrate robust PTE. Eastern cities appear adept at leveraging economies of scale efficiently, while their counterparts in the west may excel in specific technologies or industry sectors, such as wind and solar energy—an observation consistent with the findings of Jia et al. [25], Xiao et al. [26].

6 Conclusions

This study employs the three-stage DEA method to assess the extent to which the digital environment impacted the GTFP of Chinese prefecture-level cities from 2008 to 2020. The method adopted is rigorous and effective, and it yields the following comparative findings before and after adjustment: (1) Over the period from 2008 to 2020, GTFP exhibits an upward trajectory, positioning it on the ascending side of an inverted U-shaped curve—a testament to the positive influence of the digital environment on GTFP enhancement; (2) When excluding the effects of digital environmental factors and other variables, a significant decline in GTFP is observed. Before and after adjusting digital environment, the primary determinant of GTFP is identified as the low level of SE; (3) From the perspective of spatial patterns, the center of gravity of GTFP has shifted, and the digital environment has clearly played a positive role in promoting the concentration, distribution, and diffusion of GTFP; (4) From the perspective of temporal evolution, the digital environment has significantly enhanced green total factor productivity, showing a clear "north-south divide" and a noticeable Hu Huanyong Line.

Acknowledgement. This research was funded by Research on the New Ecosystem of China-ASEAN Cross-Border E-Commerce under the Integration of Talent, Technology, and Industry (ZXLHSA-2025-3); Funded by the Humanities and Social Sciences Research Project of Colleges and Universities in Guizhou Province (2025RW021). Title: Research on the Paths and Modes of Digital Transformation of Traditional Industries in Guizhou; 'Research on the optimization of regulatory policies for systemic risks of commercial banks after the inclusion of data resources in tables' (24BJY082).

References

1. Wang, J., Liu, Y., Wang, W., Wu, H.: How does digital transformation drive green total factor productivity? Evidence from Chinese listed enterprises. J. Clean. Prod. **406**, 136954 (2023). https://doi.org/10.1016/j.jclepro.2023.136954

2. Xie, J., Chen, L., Xu, S., et al.: Digital twin-driven design for elevator fairings via multi-objective optimization. Int. J. Adv. Manuf. Technol. **131**, 1413–1426 (2024). https://doi.org/10.1007/s00170-024-13049-1
3. Leng, J., Zhang, H., Yan, D., et al.: Digital twin-driven manufacturing cyber-physical system for parallel controlling of smart workshop. J. Ambient. Intell. Humaniz. Comput. **10**, 1155–1166 (2019). https://doi.org/10.1007/s12652-018-0881-5
4. Brunzini, A., Ciccarelli, M., Sartini, M., Papetti, A., Germani, M.: A comparative study for the assessment of marker-less mixed reality applications for the operator training. Int. J. Comput. Integr. Manuf. **37**(12), 1559–1581 (2024). https://doi.org/10.1080/0951192X.2024.2314793
5. Huang, P., Chen, X.: The impact of data factor-driven industry on the green total factor productivity: evidence from the China. Sci. Rep. **14**, 25377 (2024). https://doi.org/10.1038/s41598-024-77189-w
6. Rakic, S., Pero, M., Sianesi, A., Marjanovic, U.: Digital servitization and firm performance: technology intensity approach. Inzinerine Ekonomika. **33**(4), 398–413 (2022). https://doi.org/10.5755/j01.ee.33.4.29649
7. Zhou, C., Zhou, J., Chen, G., Xu, C.: Digital finance and greener emissions: evidence from China. Appl. Econ. **56**(49), 5830–5844 (2024). https://doi.org/10.1080/00036846.2023.2266599
8. Zhang, G., Chen, Y., Wang, G., et al.: Spatial-temporal evolution and influencing factors of digital financial inclusion: county-level evidence from China. Chin. Geogr. Sci. **33**, 221–232 (2023). https://doi.org/10.1007/s11769-023-1333-5
9. Tang, M., Liu, Y., Hu, F., Wu, B.: Effect of digital transformation on enterprises' green innovation: empirical evidence from listed companies in China. Energy Econ. **128**, 107135 (2023). https://doi.org/10.1016/j.eneco.2023.107135
10. Liu, Y., Cheng, J., Dai, J.: Harnessing digital transformation for green innovation in energy transition: a study on R&D investments and spatial spillover in China. Econ. Chang. Restruct. **57**(1), 1–13 (2024). https://doi.org/10.1007/s10644-024-09610-3
11. Han, Y., Zhao, F., Zhao, B.: Navigating a sustainable transition: green digital finance in manufacturing. Econ. Chang. Restruct. **57**(1), 1–16 (2024). https://doi.org/10.1007/s10644-024-09607-y
12. Ning, J., Yin, Q., Yan, A.: How does the digital economy promote green technology innovation by manufacturing enterprises? Evidence from China. Front. Environ. Sci. **10**, 967588 (2022). https://doi.org/10.3389/fenvs.2022.967588
13. He, Y.: Chinese Fintech goes global: political challenges and business strategies. Asia Policy. **19**(1), 35–50 (2024) https://www.jstor.org/stable/27307994
14. Ma, D., Zhu, Y., Yang, Y.: How Green finance affects export production quality: fresh evidence from China. Energy Econ. **131**, 107381 (2024). https://doi.org/10.1016/j.eneco.2024.107381
15. Long, Y., Liu, L., Yang, B.: The effects of enterprise digital transformation on low-carbon urban development: empirical evidence from China. Technol. Forecast. Soc. Chang. **201**, 123259 (2024). https://doi.org/10.1016/j.techfore.2024.123259
16. Fu, H., Zheng, C., Yang, P.: Digital inclusive finance and green transformation of manufacturing enterprises: empirical analysis based on the dual perspectives of demand and supply. Technol. Forecast. Soc. Chang. **200**, 123152 (2024). https://doi.org/10.1016/j.techfore.2023.123152
17. Sun, Y.: Digital transformation and corporates' green technology innovation performance–the mediating role of knowledge sharing. Financ. Res. Lett. **62**, 105105 (2024). https://doi.org/10.1016/j.frl.2024.105105
18. Al Khub, A., Saeudy, M., Gerged, A.M.: Digital financial inclusion in emerging economies: evidence from Jordan. J. Risk Financ. Manag. **17**(2), 66 (2024). https://doi.org/10.3390/jrfm17020066

19. Elfaki, K.E., Ahmed, E.M.: Digital technology adoption and globalization innovation implications on Asian Pacific green sustainable economic growth. J. Open Innov.: Technol. Mark. Complex. **10**(1), 100221 (2024). https://doi.org/10.1016/j.joitmc.2024.100221
20. Wang, W., Wang, J., Wu, H.: The impact of energy-consuming rights trading on green total factor productivity in the context of digital economy: evidence from listed firms in China. Energy Econ. **131**, 107342 (2024). https://doi.org/10.1016/j.eneco.2024.107342
21. Cao, G., She, J., Cao, C., Cao, Q.: Environmental protection tax and green innovation: the mediating role of digitalization and ESG. Sustainability. **16**(2), 577 (2024). https://doi.org/10.3390/su16020577
22. Fang, C., Yuan, Y., Chen, J., Gao, D., Peng, J.: Examination of green productivity in China's mining industry: an in-depth exploration of the role and impact of digital economy. Sustainability. **16**(1), 463 (2024). https://doi.org/10.3390/su16010463
23. Yu, C., Yu, J., Gao, D.: Smart cities and greener futures: evidence from a quasi-natural experiment in China's smart city construction. Sustainability. **16**(2), 929 (2024). https://doi.org/10.3390/su16020929
24. Akberdina, V., Kuzmin, E., Kyriakopoulos, G.L., Kumar, V.: Sustainability of digital transformation for the environment. Front. Environ. Sci. **12**, 1371047 (2024). https://doi.org/10.3389/fenvs.2024.1371047
25. Jia, K., Li, L.: The moderate level of digital transformation: from the perspective of green total factor productivity. Math. Biosci. Eng. **21**(2), 2254–2281 (2024). https://doi.org/10.3934/mbe.2024099
26. Xiao, Y., Lin, M., Wang, L.: Impact of green digital finance on sustainable development: evidence from China's pilot zones. Financ. Innov. **10**(1), 10 (2024). https://doi.org/10.1186/s40854-023-00552-9

Leveraging DRSA and SRAM for an Optimised Green Car Adoption Model

Yu-ling Lin[1], Salem Chakhar[2,3](✉), and Rui Yang[2]

[1] Goldsmiths, University of London, London, UK
Yuling.Lin@gold.ac.uk
[2] University of Portsmouth, Portsmouth, UK
{salem.chakhar,rui.yang}@port.ac.uk
[3] Centre of Operational Research and Decision Analysis, University of Portsmouth, Portsmouth, UK

Abstract. This paper presents a two-phase analysis approach to develop a comprehensive green car adoption model that incorporates government, consumer, and manufacturer perspectives. In the first phase, individual preference models are created for each stakeholder. Consumer and manufacturer adoption models are derived directly from total sales and total profit, respectively, while the government model is constructed using the Dominance-based Rough Set Approach (DRSA) to consider various influencing factors. In the second phase, these stakeholder-specific models are aggregated using a rank aggregation method to generate a unified green car adoption model. The resulting model reflects collective preferences, providing a framework for informed green car policy-making.

Keywords: Decision Support · Dominance-based rough set approach · Decision rules · Rank aggregation · Green cars adoption

1 Introduction

An effective green car adoption model must balance the Triple Bottom Line (TBL) dimensions: environmental (car greenness), economic (manufacturer profit), and social (consumer utility) [7], as sustainability arises when all three dimensions are integrated into decision-making. Governments, as market regulators, must consider these environmental, social, and economic impacts when crafting policies. However, many policies often prioritise environmental concerns while neglecting social and economic factors, leading to conflicts between green objectives and economic growth [15,18].

To address this gap, we propose a two-phase methodology that incorporates the perspectives of government, manufacturers, and consumers into a comprehensive green policy framework. In Phase 1, consumer and manufacturer adoption models are derived from total sales and total profit, respectively, while the government model is developed using the Dominance-based Rough Set Approach

A. P. Cabral Seixas Costa et al. (Eds.): ICDSST 2026, LNBIP 582, pp. 133–148, 2026.
https://doi.org/10.1007/978-3-032-25146-6_9

(DRSA) [2,4,8,19] to account for various influencing factors. Phase 2 aggregates the models from Phase 1 using the Sequential Rank Aggregation Method (SRAM) [14].

The DRSA is a well-established ordinal classification and knowledge-extraction method that has been successfully applied in many real-world settings [3,5,6,11,13,16]. It was chosen here because it supports qualitative, non-compensatory and rule-based preference structures, which align more closely with governmental decision-making than compensatory approaches such as AHP [17], TOPSIS [12] or GAMETE [20]. Its key advantage is the generation of transparent if-then rules that reduce cognitive load and provide a clear basis for justifying and communicating policy choices, ensuring the level of explainability required in public-sector contexts.

The paper is structured as follows: Sect. 2 reviews DRSA and SRAM; Sect. 3 presents the analytical approach; Sect. 4 applies the method and discusses results; Sect. 5 briefly reports on the strengths and limitations of the current study and outlines avenues for future work; Sect. 6 concludes the paper.

2 Background

2.1 Dominance-Based Rough Set Approach for Ranking Problems

This section outlines the ranking version of DRSA [8,19,21]. In rough set theory, information on *decision objects* is organised in a four-tuple *information table* $\mathbf{S} = \langle A, Q, V, f \rangle$, where A is a finite set of objects, Q a finite set of condition attributes, V_q the domain of attribute q, $V = \cup_{q \in Q} V_q$ and $f : A \times Q \to V$ the information function, with $f(x, q)$ giving the evaluation of object x on attribute $q \in Q$. Attributes are grouped into Q_N, Q_O, and Q_R, representing cardinal, ordinal, and regular attributes. The domains of Q_N and Q_O are ordered by preference, and such attributes are treated as *criteria*.

Pairwise Comparisons Table. Policy makers provide learning examples in two forms: (i) pairwise comparisons of objects from A, or (ii) rankings of objects, which can be transformed into pairwise comparisons. The two preference relations are comprehensive outranking S and non-outranking S^c. For instance, $y \prec x \prec z$ implies ySx, ySz, xSz, zS^cy, zS^cx, and xS^cy. Let $B \subseteq A \times A$ denote the set of pairs assigned to S or S^c. These comparisons form the *Pairwise Comparisons Table* (PCT), where each row is a pair (x, y), the last column indicates S or S^c, and other columns record differences or ordered evaluations. The PCT is then used by DRSA to approximate preference relations.

Dominance Relation. Granules of knowledge are dominance cones in attribute space. A pair (x, y) dominates (w, z), denoted $(x, y)\varDelta(w, z)$, if:

$$\begin{aligned} &(f(x,q) - f(y,q)) \succeq (f(w,q) - f(z,q)), \forall q \in Q_N \\ &f(x,q) \succeq f(w,q) \wedge f(y,q) \preceq f(z,q), \qquad \forall q \in Q_O \\ &f(x,q) = f(w,q) \wedge f(y,q) = f(z,q), \qquad \forall q \in Q_R. \end{aligned} \tag{1}$$

Here $\succeq$ and $\preceq$ denote weak preference and its inverse. For decreasing criteria, the symbols are inverted. This relation is reflexive and transitive, i.e., a partial order.

Rough Approximation. For any $(x, y) \in A \times A$, the *dominating set* $\Delta^+(x, y)$ contains pairs dominating (x, y), while the *dominated set* $\Delta^-(x, y)$ contains pairs dominated by (x, y). Lower and upper approximations of S are: $L(S) = (x, y) \in B : \Delta^+(x, y) \subseteq S$ and $U(S) = (x, y) \in B : \Delta^-(x, y) \cap S \neq \emptyset$. Analogously for S^c. Lower approximations contain pairs certainly in S (or S^c), upper approximations those possibly in S (or S^c). Boundaries are $Bn(S) = U(S) - L(S)$ and $Bn(S^c) = U(S^c) - L(S^c)$.

Decision Rules. Approximations yield 'if–then' rules describing pairs in the PCT. The condition part specifies attribute differences or values, and the decision part assigns S or S^c. Rules can be: (i) certain (from lower approximations), (ii) possible (from upper), or (iii) approximate (from boundaries). Let $Q_N = q_1, \ldots, q_e$, $Q_O = q_{e+1}, \ldots, q_p$, and $Q_R = q_{p+1}, \ldots, q_z$. Then the general structure of certain rules from the lower approximation of S is as follows:

If $q_1(x) - q_1(y) \succeq r_1$ and $\cdots$ and $q_e(x) - q_e(y) \succeq r_e$ and
$q_{e+1}(x) \succeq r_{e+1}$ and $q_{e+1}(y) \preceq s_{e+1}$ and $\cdots$ and $q_p(x) \succeq r_p$ and $q_p(y) \preceq s_p$ and
$q_{p+1}(x) = r_{p+1}$ and $q_{p+1}(y) = s_{p+1}$ and $\cdots$ and $q_z(x) = r_z$ and $q_z(y) = s_z$,
then xSy.

Certain rules from the lower approximation of S^c have an analogous syntax, in which every condition of the form $\succeq$ is replaced by $\preceq$, every $\preceq$ is replaced by $\succeq$, and the decision part becomes xS^cy.

Construction and Exploitation of Preference Graphs. The induced certain or possible rules can be used to construct a preference graph $G = (A, E)$ where A is the set of objects in the information table and E is the set of arcs corresponding to S and S^c preference relations between objects. This means that an S-arc (S^c-arc) between objects x and y exists only if the pair of objects $(x, y) \in A \times A$ is covered by at least one decision rule with an assignment to relation S (S^c).

The resulting graph can be exploited by different scoring techniques in order to obtain a final ranking on the decision objects in A. Different classical scoring techniques can be used for this purpose, such as the Net Flow Score, the Repeated Net Flow Score and their variants. The final ranking is simply deduced by ordering the decision objects based on their scores.

2.2 Sequential Rank Aggregation Method

This section provides an overview of the SRAM [14]. Assuming first that consumer-, manufacturer- and government-oriented rankings are strict orders (i.e., without ties). Let X^l $(l = 1, 2, 3)$ denote these initial rankings. We also denote by X_i^l the ranks of green car Z_i according to initial ranking l. The objective of the ranking aggregation operation is to construct a new ranking Y by combining the initial rankings X^l $(l = 1, 2, 3)$. The aggregated ranking should summarise as well as possible the three initial rankings. One possible solution to this issue is to select an aggregated ranking Y with a minimal distance to the initial rankings. Different distance measures can be used for this purpose, but the two most common are the Spearman footrule distance and Kendall's tau distance. In this paper, however, we employ the squared Euclidean distance. Although it is not a metric in the strict sense, as it does not satisfy the triangle inequality, it is widely used in optimisation settings where only relative distances need to be compared. The squared Euclidean distance is as simple to compute as the Spearman footrule distance, gives progressively greater weight to objects that are farther apart, and is computationally more efficient than Kendall's tau distance.

Let us first denote by Y_i the aggregated rank of green car Z_i. Then, we construct a distance matrix D_{ij}, $(i, j = 1, \ldots, n)$. The squared Euclidean Distance between X^l $(l = 1, 2, 3)$ and Y is then obtained by squaring the standard Euclidean distance as follows:

$$d^2(X^l, Y) = (X_1^l - Y_1)^2 + \ldots + (X_n^l - Y_n)^2 \quad (2)$$

$$d^2(X^l, Y) = \sum_{i=1}^{n} (X_i^l - Y_i)^2 \quad (3)$$

The sum of squared Euclidean distances between the aggregated ranking Y and the initial rankings X^1 and X^2 is then computed as follows:

$$d^2(X^1, Y) + d^2(X^2, Y) = \sum_{l=1}^{3} \sum_{i=1}^{n} (X_i^l - Y_i)^2 \quad (4)$$

The computation of the aggregated ranking Y is modelled as an assignment problem and solved by integer programming as follows. First, for each initial ranking X^l $(l = 1, 2)$, we construct an $n \times n$ distance matrix D^l. An element D_{ij}^l of D^l measures the distance between the rank of object Z_i in initial ranking X^l and the aggregated rank j of object Z_i in the aggregated ranking Y. We also denote by C_{ij}^l $(i, j = 1, \ldots, n;\ l = 1, 2)$ the dummy variables of the problem.

Then, the 0-1 integer programme that should be solved is defined as follows:

$$\min \quad Z = \sum_{l=1}^{3}\sum_{i=1}^{n}\sum_{j=1}^{n} D_{ij}^{l} \cdot C_{ij}^{l}$$

s.t.

$$\sum_{i=1}^{n} C_{ij}^{l} = 1, \quad j = 1 \ldots n; \quad l = 1,2,3 \tag{5}$$

$$\sum_{j=1}^{n} C_{ij}^{l} = 1, \quad i = 1 \ldots n; \quad l = 1,2,3$$

$$C_{ij}^{l} \in \{0,1\}, \quad i,j = 1 \ldots n; \quad l = 1,2,3$$

The first set of constraints ensures that each object is assigned a single rank, while the second set of constraints ensures that each rank is assigned once to a single object. The last set of constraints specifies that all variables should be binary.

The final ranking is obtained based on the dummy variables C_{ij}^{l} $(i, j = 1 \ldots n)$. The dummy variable $C_{ij}^{l} = 1$ means that object x_j is ranked i in the aggregated ranking Y.

The current version of the SRAM model does not apply any weighting to the three stakeholder perspectives; instead, it aggregates them using an equal-weight structure consistent with the balanced orientation of the triple bottom line. Furthermore, although a weighted extension of SRAM exists, the stakeholders involved in this study explicitly requested the use of equal weights to avoid privileging any single viewpoint. Future work could explore differential weighting to reflect specific policy priorities or strategic emphases.

3 Analysis Approach

The proposed analysis approach is structured into two phases (see Fig. 1). In the first phase, Stakeholder Preference Modelling, individual preference models are constructed for green car adoption from the perspectives of government, consumers, and manufacturers, resulting in three stakeholder-specific rankings. In the second phase, Comprehensive Green Car Ranking Construction, these stakeholder-specific rankings are aggregated using a ranking aggregation method.

3.1 Phase 1: Assessment of Existing Green Policy-Making Models

The objective of the first phase is to assess the existing government, consumer, and manufacturer green car adoption models. These models will take the form of three different rankings of green cars. The rankings for consumers' and manufacturers' green car adoption models are defined based on total sales and total profit, respectively. The arguments supporting this choice are discussed hereafter. The computation of the ranking relative to the government's green car adoption model is more complex since it requires the consideration of several factors.

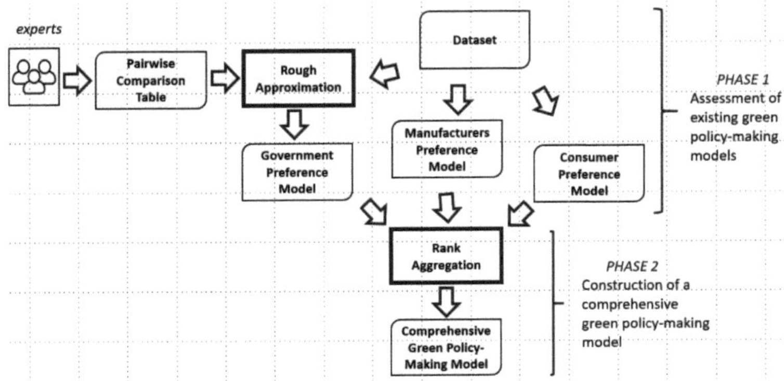

Fig. 1. Principles of analysis approach.

Government Adoption Model. The computation of the ranking for government-oriented analysis relies on the ranking version of the Dominance-based Rough Set Approach (DRSA) [8,21], which is a typical preference learning and knowledge discovery method. It was chosen for its ability to handle the ordinal nature of the considered factors and overall greenness levels. Another key argument in favour of using DRSA is its enhanced explainability, which is achieved through the production of if-then decision rules that are simple and easily understandable by policymakers. These rules represent an explicit relationship between factors and overall greenness levels, thus providing better interpretability than most alternative machine learning methods.

The main inputs of the ranking version of DRSA are a learning set containing the scores of a subset of green cars with respect to the considered factors (see Table 1) and a Pairwise Comparison Table (PCT) containing user-defined pairwise comparisons between all pairs in the learning set. The entries of the PCT represent pairwise comparisons that assign the green cars into two preference relations, S and S^c, where xSy means that x is preferred over y (i.e., the greenness level of car x is at least as good as the greenness level of car y), while S^c is the complement of S. The approximation of the PCT leads to a collection of if-then decision rules that can be applied to construct a preference graph. The obtained preference graph is then exploited by a scoring/ranking method, typically the Net Flow Score, to derive the final government-oriented ranking of green cars.

Consumers Adoption Model . As previously mentioned, the consumer-oriented ranking was defined based on total sales. The use of Total Sales to model consumers' behaviour is supported by several existing studies. For instance, the authors in [10] argue that consumer purchasing behaviour data, such as sales volume, are often used as an indicator to explain and describe consumer choice.

Total Sales offers a transparent and policy-relevant proxy; however, additional factors such as brand loyalty, charging infrastructure availability, and long-term maintenance costs may also influence consumer decisions and could be incorporated in future research.

Manufacturers Adoption Model. Profit is commonly accepted as the manufacturer's goal in general, and particularly in the automotive industry [1,9]. Therefore, in the manufacturer green car adoption model, this article takes the manufacturer's perspective on total profit (economic performance).

In the dataset used in the considered case study, the profit was estimated according to car brand, model, and price. There are two profit ranges in our dataset: 20% to 30% or 30% to 50%. The adoption model for manufacturers is obtained by ordering cars according to average profit between the minimum and maximum in decreasing order.

Profit serves as an appropriate and widely used indicator, yet manufacturers' decisions may also be affected by factors such as brand strategy, production constraints, and long-term market positioning, which future work may examine.

3.2 Phase 2: Construction of a Comprehensive Green Policy-Making Model

The objective of the second phase is to construct a comprehensive green policy-making model that is based on the manufacturer-oriented, consumer-oriented, and government-oriented rankings obtained in the first phase. A possible solution to construct the comprehensive green policy-making model is to use a rank aggregation method. Let τ_1, τ_2, and τ_3 be the car ratings defined according to government, manufacturer, and consumer preference models, respectively. The objective of phase 2 is to construct a new rating ν by aggregating the initial ratings τ_1, τ_2, and τ_3. The aggregated rating should summarise, as well as possible, the initial ratings. In other words, the aggregated rating should be as 'close' as possible to all individual ordered lists simultaneously. One possible solution to this problem is to select an aggregated rating ν with a minimal distance to the initial ratings. A natural requirement and intuitive objective function can then be defined as follows:

$$\phi(\nu) = \sum_{j=1}^{3} d(\nu, \tau_j), \tag{6}$$

where ν is a proposed aggregated rating. The best aggregated rating ν^* can then be defined as the one that would minimise the total distance in Eq. (6):

$$\phi(\nu^*) = \arg\min_{\nu \in V} \sum_{j=1}^{3} d(\nu, \tau_j) \tag{7}$$

where V denotes the set of potential candidate ratings. Various distance measures can be employed for this purpose, with the two most commonly used being the Spearman footrule distance and Kendall's tau distance. An alternative is the Squared Euclidean distance. Although the Squared Euclidean distance is not a metric in the strict sense, it is often used in optimisation problems where only relative distances need to be compared. Consequently, the Sequential Rank Aggregation Method (SRAM) [14], which is based on the Squared Euclidean distance, is used to construct the comprehensive green policy-making model.

4 Application

4.1 Data Collection

The car dataset used in this case study consists of 114 entries collected over several years. The data were obtained from Taiwan's Directorate General of Highways and the Ministry of Transportation and Communications. Each car is described with respect to a set of factors listed in Table 1. The first two columns of the table provide the name and description of each factor. The remaining columns specify the nature of the data, the measurement unit, and the preference direction, which may be gain-type, cost-type, or none for nominal attributes.

Table 1. List of attributes.

Name	Description	Data type	Unit	Preference
Brand	Car brand	Symbolic		None
Model	Car model	Symbolic		None
Price	Car market price	Continuous	10K TWD	Cost
SweptVolume	Volume of fluid through which a piston or plunger	Continuous	CC	Cost
	moves when it makes a stroke in an engine			
FE	Fuel efficiency	Continuous	KM/FE	Gain
Horsepower	Car horsepower	Continuous		Cost
Torque	Car torque	Continuous		Cost
FuelTax	Fuel Tax	Continuous	TWD	Cost
CarTax	Car Tax	Continuous	TWD	Cost
TotalSales	Total sales	Integer		Gain

4.2 Pairwise Comparison Table Construction

The PCT construction was challenging due to the large number of green cars. Asking experts to compare all 114 cars pairwise would have been impractical. To resolve this, we used the following procedure. In a preliminary study, policymakers classified the cars into 10 ordered categories based on Fuel Efficiency (FE) and other confidential parameters. Each car was assigned to a category C_1 to C_{10}, with C_{10} being the most preferred. Policymakers then compared pairs within each category, resulting in pairwise comparisons summarised in preference graphs (Fig. 2), where nodes represent cars and arcs denote the preference relation S. These graphs can be represented as PCTs. For example, the PCTs associated with decision classes can be represented in Tables 2 for C_5 and C_7.

As mentioned earlier, the ordinal scale was defined by the policymakers based on the fuel efficiency factor and other technical parameters, which were not shared with the authors due to confidentiality. However, the overall structure of the categorisation process and the decision logic used by policymakers are fully documented to support transparency and enable partial reproducibility.

Pairwise comparisons induced intra-category preference relations. Inter-category relations were deduced by assuming cars in group C_t $(1 \leq t \leq 10)$ are ranked better than those in groups $C_1, \ldots, C_{t-1}$. Preference relations were then deduced as follows:

- $xSy, \forall x \in C_{t+1}^{-}, y \in C_t^{+}$,
- $xS^cy, \forall x \in C_t^{+}, y \in C_{t+1}^{-}$.

In Fig. 2, the sets of the most and least preferred cars, C_t^+ and C_t^-, are denoted by nodes with bold boundaries and double boundaries, respectively. Consider, for instance, the inter-category preference relations between categories C_7 and C_6. Based on Fig. 2, we get: $C_7^- = \{9, 108\}$ and $C_6^+ = \{59, 79\}$. Based on the statement above, the following inter-category preference relations should be considered: $(9, 59) \in S$, $(9, 79) \in S$, $(108, 59) \in S$, and $(108, 79) \in S$.

The intra- and inter-category pairwise comparisons are then merged into a single PCT. DRSA is then used to approximate the preference relations S and S^c in the PCT. The obtained rough approximations of the preference relations S and S^c are used to induce generalised descriptions of pairs from the PCT in terms of if-then decision rules. This process resulted in 189 "certain at least" decision rules and 157 "certain at most" decision rules. The obtained decision rules were then applied to the considered dataset to construct the preference graph. The preference graph resulting from the application of decision rules was exploited by the Net Flow Score ranking method to obtain the final ranking.

4.3 Analysis of Results

Correlation Analysis. Table 3 and Fig. 3 present Pearson correlation coefficients between the three stakeholder models and their relationship to the aggregated ranking, computed as a composite measure balancing Government, Consumer, and Manufacturer preferences. The results show low to moderate correlations: government and consumer models exhibit a weak positive correlation (0.314), while the government and manufacturer (-0.483) and consumer and manufacturer (-0.417) models are negatively correlated, indicating substantial divergence in stakeholder preferences. These correlations are statistically significant ($p < 0.001$), highlighting key differences in stakeholder priorities. Despite these, the aggregated ranking correlates positively with both the government (0.495) and consumer (0.675) models, though weakly with the manufacturer model (0.115), suggesting it synthesises diverse perspectives, especially those prioritising environmental concerns.

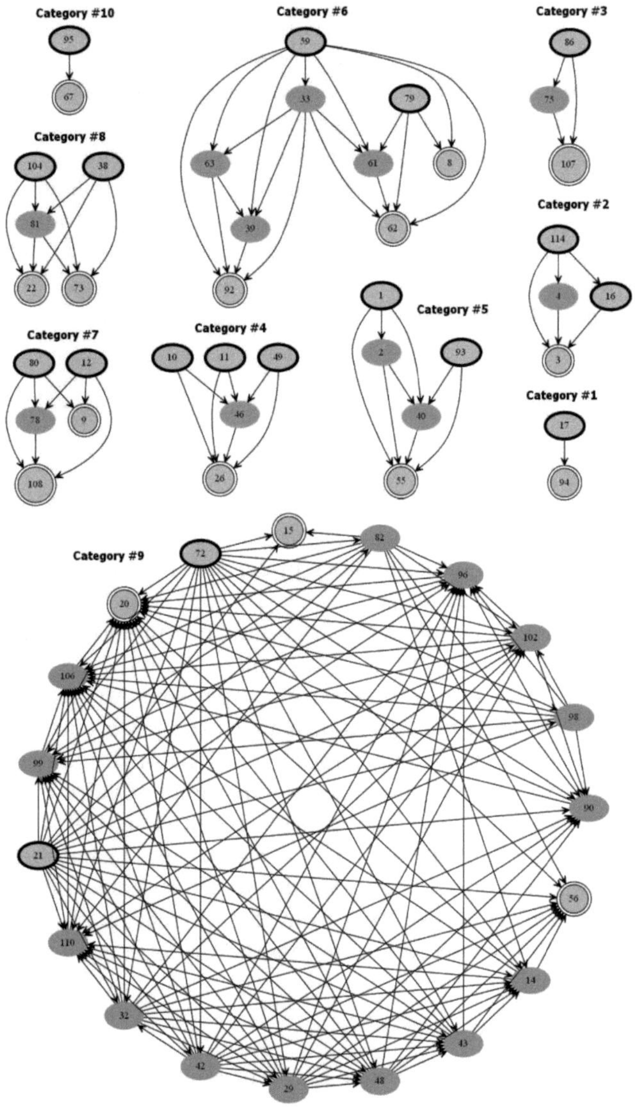

Fig. 2. Preference graphs. Nodes represent green cars, and arcs denote the preference relation S. Nodes with bold boundaries indicate the most preferred cars, while double boundaries indicate the least preferred. The remaining nodes are intermediate. (Color figure online)

While the table provides exact correlation values, the heatmap in Fig. 3 offers visual clarity on symmetry and clustering patterns that are less apparent in numeric form. One notable insight is the asymmetry in how strongly different stakeholders align with the aggregated model. The Consumer model stands out as the most influential in shaping the Aggregated Ranking, not only due to its

Table 2. PCT for Categories C_7 (a), C_6 (b), and C_5 (c)

(a)

	9	12	78	80	108
9					
12	S		S		S
78					S
80	S		S		S
108					

(b)

	8	33	39	59	61	62	63	79	92
8									
33			S		S	S	S		
39									S
59	S	S	S			S	S		S
61									
62			S						
63		S							S
79	S			S		S			
92									

(c)

	1	2	40	55	93
1		S	S	S	
2			S	S	
40				S	
55					
93			S	S	

Table 3. Correlation analysis between individual and aggregated models.

Pearson's	Green car adoption model			
	Government	Consumer	Manufacturer	Aggregated
Government	1	0.314***	-0.483***	0.495***
Consumer		1	-0.417***	0.675***
Manufacturer			1	0.115
Aggregated				1

high correlation (0.675) but also because of its dominant warm hue, in contrast to the cooler tone of the Manufacturer's correlation. The Manufacturer model, isolated by cooler colours both horizontally and vertically, appears marginalised in the consensus-building process, suggesting a potential misalignment between industry priorities and collective decision-making outcomes. Visually, the Government and Consumer models cluster more closely, both positively correlated with each other and with the Aggregated Ranking, forming a kind of policy-public axis. This contrasts with the Manufacturer model, which stands apart. The colour intensity gradient also helps intuitively distinguish the magnitude of stakeholder agreement or conflict, offering an at-a-glance understanding of stakeholder dynamics without needing to read exact numbers.

Variation in Aggregated Rankings Across Car Brands. The boxplot in Fig. 4 illustrates the distribution of aggregated rankings across various car manufacturers, highlighting substantial variability in how different brands perform under the composite stakeholder model. Brands like Mitsubishi and Peugeot show consistently high rankings with minimal variation, suggesting broad agreement across stakeholders about their strong performance in green car criteria.

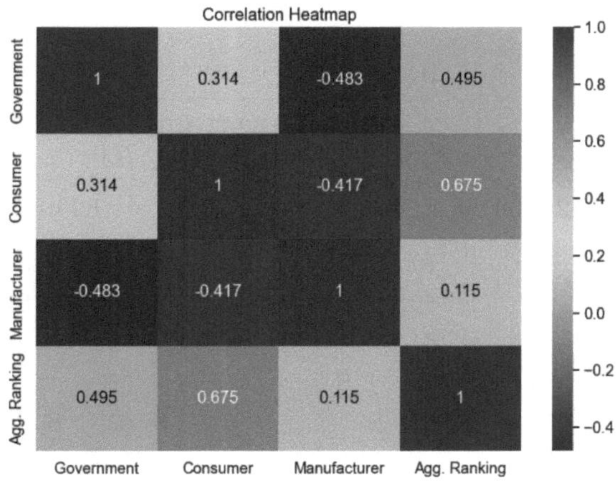

Fig. 3. Correlation heatmap between government, consumer, manufacturer, and aggregated rankings.

In contrast, brands such as Toyota, BMW, and Lexus display wide interquartile ranges, indicating greater disagreement or inconsistency in stakeholder evaluations. Notably, Porsche ranks low overall with very tight dispersion, implying uniform stakeholder consensus on its weak alignment with green priorities. Meanwhile, KuoZui Motors and Nissan cluster around mid-range values, suggesting moderate and more stable perceptions. This visualisation underscores not just which manufacturers perform better in the aggregate but also where stakeholder opinions diverge most sharply.

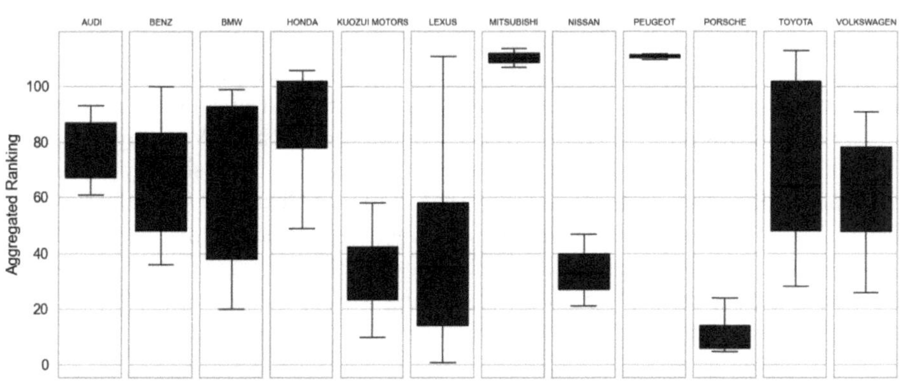

Fig. 4. Boxplots illustrating distribution of aggregated rankings across car brands.

Heatmap of Percentile-Based Performance Concentration. The heatmap in Fig. 5 visualises how each brand's models are distributed across percentile groups. Darker cells indicate a higher concentration of vehicles in that rank segment. This visual quickly reveals concentration patterns—such as LEXUS, which has a broad spread across all ranks, and PORSCHE, which is heavily skewed towards the lowest-performing 20%.

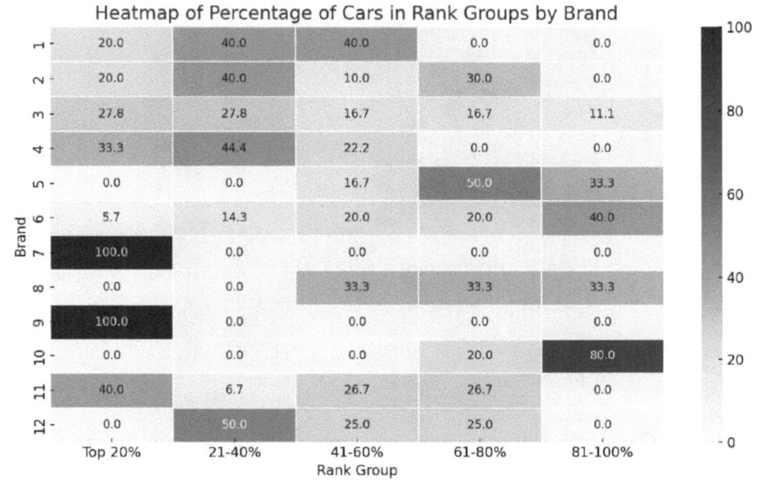

Fig. 5. Heatmap of vehicle distribution by rank percentile across brands.

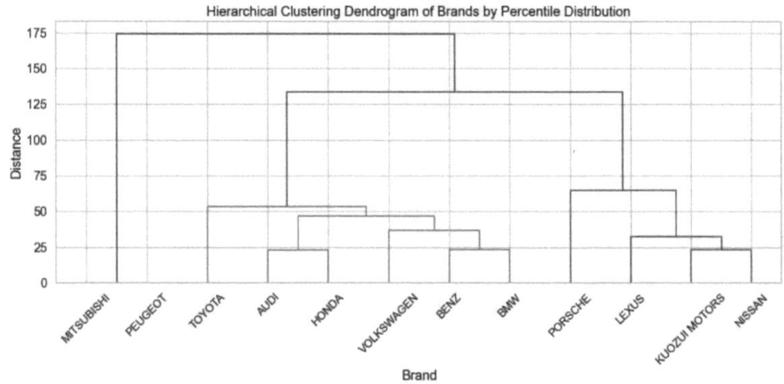

Fig. 6. Dendrogram of brand clusters based on vehicle rank distributions.

Cluster Analysis. Figure 6 presents a dendrogram based on hierarchical clustering of brands using their rank percentile distributions. It reveals natural groupings among brands. For instance, MITSUBISHI and PEUGEOT cluster tightly due to their 100% concentration in the top 20%. In contrast, KUOZUI MOTORS, LEXUS, and PORSCHE cluster together, sharing higher concentrations in the lower-performance percentiles (61–100%). These clusters suggest differences in market positioning or strategic focus across manufacturers.

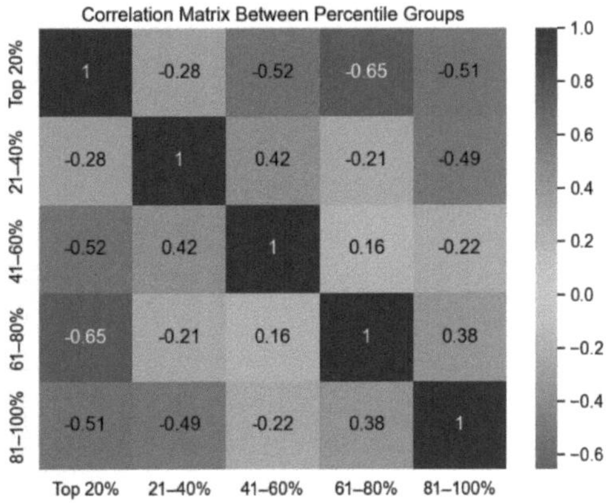

Fig. 7. Correlation matrix between vehicle rank percentile groups.

Correlation Matrix Between Percentile Groups. The correlation matrix in Fig. 7 explores relationships between percentile groups. Strong negative correlations between the Top 20% and Bottom 20% indicate that brands with many high-performing vehicles tend to have fewer low performers. Meanwhile, positive correlations among mid-tier groups (21–40%, 41–60%, 61–80%) suggest that brands often have a consistent spread within the mid-range. This insight may inform strategies aimed at shifting distributions upwards or consolidating mid-tier offerings.

5 Discussion

In this section, we briefly report on the strengths and limitations of the current study and outline avenues for future work. The use of DRSA, selected for its ability to model governmental preferences through qualitative, non-compensatory structures, aligns with public-sector decision-making. DRSA's transparent if-then rules support clear policy justifications and reduce cognitive load, making it suited to green car policy-making. The squared Euclidean distance in SRAM offers computational efficiency and emphasises outliers, valuable in aggregating stakeholder preferences.

However, the reliance on proxies such as Total Sales for consumers and Profit for manufacturers is a limitation, as they miss nuanced factors like brand loyalty, charging infrastructure, or market positioning. The government model relies on confidential technical parameters, limiting full reproducibility. These aspects could be refined in future research.

While the SRAM model currently uses equal weighting for stakeholders, future studies could explore differential weighting to reflect specific policy priorities. Incorporating factors like long-term maintenance costs and production constraints would also improve the model's realism.

This work provides a solid foundation, and future research will build upon it by addressing these limitations and enhancing the model's complexity in policy-making.

6 Conclusion

This paper presents a two-phase approach to model green car adoption, integrating the perspectives of government, consumers, and manufacturers. By developing individual preference models in Phase 1—based on consumer sales, manufacturer profit, and government policies—and aggregating these rankings in Phase 2 using SRAM, the model reflects collective preferences. The integration of environmental, economic, and social factors ensures the model adheres to the Triple Bottom Line (TBL) principles, providing a balanced foundation for policy-making.

The presented approach offers a robust framework for decision-makers to evaluate green car adoption. It facilitates informed policy formulation and aligns diverse stakeholders' goals in the green transition. Future work could expand this model by incorporating factors like technological advancements or consumer behaviour shifts to enhance its applicability and predictive power. This approach lays the groundwork for more effective, sustainable, and inclusive green car policies.

References

1. Ahn, H., Gümüş, M., Kaminsky, P.: Pricing and manufacturing decisions when demand is a function of prices in multiple periods. Oper. Res. **55**(6), 1039–1057 (2007). https://doi.org/10.1287/opre.1070.0411
2. Atteya, T., Chakhar, S., Labib, A., Cox, A., Ishizaka, A.: Estimating relative importance of criteria by post-processing dominance-based rough set approach's outputs. Eur. J. Oper. Res. **315**(3), 1096–1122 (2024). https://doi.org/10.1016/j.ejor.2023.12.027
3. Chakhar, S., Ishizaka, A., Labib, A., Saad, I.: Dominance-based rough set approach for group decisions. Eur. J. Oper. Res. **251**(1), 206–224 (2016). https://doi.org/10.1016/j.ejor.2015.10.060
4. Chakhar, S., Ishizaka, A., Thorpe, A., Cox, J., Nguyen, T., Meech, L.: Calculating the relative importance of condition attributes based on the characteristics of decision rules and attribute reducts: application to crowdfunding. Eur. J. Oper. Res. **286**(2), 689–712 (2020). https://doi.org/10.1016/j.ejor.2020.03.039
5. Chakhar, S., Saad, I.: Dominance-based rough set approach for groups in multicriteria classification problems. Decis. Support Syst. **54**(1), 372–380 (2012). https://doi.org/10.1016/j.dss.2012.05.050

6. Egaji, A., Chakhar, S., Brown, D.: An innovative decision rule approach to TYRE pressure monitoring. Expert Syst. Appl. **124**, 252–270 (2019). https://doi.org/10.1016/j.eswa.2019.01.051
7. Elkington, J.: Towards the sustainable corporation: win-win-win business strategies for sustainable development. Calif. Manage. Rev. **36**(2), 90–100 (1994). https://doi.org/10.2307/41165746
8. Greco, S., Matarazzo, B., Słowiński, R.: Rough sets theory for multicriteria decision analysis. Eur. J. Oper. Res. **129**(1), 1–47 (2001). https://doi.org/10.1016/S0377-2217(00)00167-3
9. Guajardo, J.A., Cohen, M.A., Netessine, S.: Service competition and product quality in the U.S. automobile industry. Manage. Sci. **62**(7), 1860–1877 (2016). https://doi.org/10.1287/mnsc.2015.2195
10. Hendel, I., Nevo, A.: Measuring the implications of sales and consumer inventory behavior. Econometrica **74**(6), 1637–1673 (2006). https://doi.org/10.1111/j.1468-0262.2006.00721.x
11. Hu, Q., Chakhar, S., Siraj, S., Labib, A.: Spare parts classification in industrial manufacturing using the dominance-based rough set approach. Eur. J. Oper. Res. **262**(3), 1136–1163 (2017). https://doi.org/10.1016/j.ejor.2017.04.040
12. Hwang, C.L., Yoon, K.: Multiple Attribute Decision Making: Methods and Applications. Springer-Verlag, New York (1981)
13. Legay, C., Cloutier, G., Chakhar, S., Joerin, F., Rodriguez, M.: Estimation of urban water supply issues at the local scale: a participatory approach. Clim. Change **130**(4), 491–503 (2015). https://doi.org/10.1007/s10584-015-1366-6
14. Lin, Y.l., Chakhar, S., Yang, R.: Sequential rank aggregation method. In: 2018 IEEE International Conference on Systems, Man, and Cybernetics (SMC), pp. 3194–3200 (2018). https://doi.org/10.1109/SMC.2018.00541
15. North, P., Nurse, A., Barker, T.: The neoliberalisation of climate? Progressing climate policy under austerity urbanism. Envir. Plan. Econ. Space **49**(8), 1797–1815 (2017). https://doi.org/10.1177/0308518X16686353
16. Saad, I., Chakhar, S.: A decision support for identifying crucial knowledge requiring capitalizing operation. Eur. J. Oper. Res. **195**(3), 889–904 (2009). https://doi.org/10.1016/j.ejor.2007.11.021
17. Saaty, T.L., et al.: The analytic hierarchy process (1980)
18. Slattery, C., Zidar, O.: Evaluating state and local business incentives. J. Econ. Perspectives **34**(2), 90–118 (2020). https://doi.org/10.1257/jep.34.2.90
19. Słowiński, R., Greco, S., Matarazzo, B.: Rough Sets in Decision Making, pp. 7753–7787. Springer New York, New York, NY (2009). https://doi.org/10.1007/978-0-387-30440-3_460
20. Souissi, M., Zavadskas, E., Chakhar, S.: Bivariate grid scale based multiple attribute evaluation technique (GAMETE) with incomplete information on weights. Technol. Econ. Dev. Econ. **31**(4), 1206–1236 (2025). https://doi.org/10.3846/tede.2025.24346
21. Szeląg, M., Greco, S., Słowiński, R.: Rule-based approach to multicriteria ranking. In: Multicriteria Decision Aid and Artificial Intelligence, chap. 6, pp. 127–160. John Wiley & Sons, Ltd (2013). https://doi.org/10.1002/9781118522516.ch6

Supporting Green Transition in Small Enterprises with a Decision Support System: A Multi-criteria Waste Management Model

Joyce Abreu Maia, Lucia Reis Peixoto Roselli(✉), and Adiel Teixeira de Almeida

Center for Decision Systems and Information Development (CDSID), Universidade Federal de Pernambuco, Recife, PE 50630-970, Brazil

joyce.maia@ufpe.br, {lrpr,almeida}@cdsid.org.br

Abstract. The rapid expansion of urban areas and the mechanization of agriculture, accelerated by globalization, have led to a substantial increase in waste generation world-wide. Effective management of this waste is complex due to its varied composition, which necessitates specific treatment methods. This complexity renders the management of solid waste a significant environmental challenge, particularly in rapidly growing sectors such as Brazil's automotive industry. A 9% increase in production in 2024 has directly amplified waste from associated services, including automotive detailing studios. In this context, multi-criteria decision support modeling offers a valuable tool for recommending optimal disposal methods. This study proposes a model for selecting a disposal method for a Brazilian automotive detailing microenterprise by integrating Value-Focused Thinking with the FITradeoff method. On structuring the problem, thirteen alternatives were identified and evaluated against seven criteria. The FITradeoff method was applied to preference modeling from two distinct perspectives. The analysis yielded a final recommendation to contract a specialized recycling company. This approach demonstrates potential for enhancing decision-making processes in waste management.

Keywords: Waste Management · Problem Structuring · FITradeoff Method · Value-Focused Thinking · Multi-Criteria Decision-Making/Aiding (MCDM/A)

1 Introduction

Rapid urban growth and the mechanization of agriculture, intensified by globalization, have resulted in generating high volumes of waste on a global scale [1]. Managing this flow is particularly complex because the composition of these wastes varies according to their original transformation process, which requires specific treatments for each type [2]. This complexity turns solid waste management (SWM) into a major environmental challenge, particularly in developing countries [3].

In 2024, Brazil's vehicle production grew by approximately 9% compared to the previous year, reaching 2.5 million units [4]. This growth in the national fleet directly

A. P. Cabral Seixas Costa et al. (Eds.): ICDSST 2026, LNBIP 582, pp. 149–162, 2026.
https://doi.org/10.1007/978-3-032-25146-6_10

results in companies generating an increase in waste in the production chain, with automotive detailing studios being a notable example. These establishments offer cleaning and disinfection services that use materials that generate a significant amount of solid waste.

To regulate waste management in Brazil, the National Solid Waste Policy (PNRS in Portuguese) was established by Law No. 12.305/2010 [5]. This legislation introduces shared responsibility for the life cycle of products, and thus requires manufacturers, importers, distributors, retailers, consumers, and municipal waste management services to act in an integrated manner to minimize waste generation and to reduce impacts on public health and the environment.

Furthermore, the PNRS defines environmentally sound final disposal as a set of practices that include reuse, recycling, composting, material recovery, energy recovery, and finally, the disposal of non-recyclable waste in sanitary landfills, provided that specific technical standards are observed at each stage [5].

For automotive detailing companies, the decision on which waste disposal method to apply is crucial to ensure compliance with environmental legislation while guaranteeing an environmentally sound final destination. Given this difficulty, the use of Problem Structuring Methods (PSMs) combined with multicriteria modeling can enable an appropriate choice, as they emphasize the need to acquire a comprehensive understanding of the context and complexity of the multiple decision problems that are already known to exist [6].

In this scenario, considering that the decision on which waste management method to adopt is crucial to ensure more sustainable disposal, and knowing that problem structuring methods can be used in conjunction with decision models to ensure a sound recommendation, this article aims to support the transition of disposal methods in a Brazilian company by constructing a multicriteria decision model.

Although this combination of methods has been successfully used in the context of waste management, its use in micro-enterprises is quite limited. Thus, it is considered that the main contribution of this article is to demonstrate how the integration of these two tools can provide a robust decision model for a Brazilian automotive detailing company, filling the gap in the application of this combination to waste-generating micro-enterprises.

2 FITradeoff Method

The FITradeoff method was created by De Almeida et al. [7] for selection problems, as an improvement to the traditional Tradeoff procedure [8]. It uses partial information, which enables a recommendation to be obtained without needing to find the exact point of indifference between two consequences.

This method was later expanded to ranking [9], classification [10], and portfolio problematics by using a cost-benefit approach [11]. The method allows two distinct approaches to be used to obtain a recommendation for the decision-maker (DM): elicitation by decomposition and holistic evaluation. These approaches can be applied in combination [7] or independently [12].

In the elicitation by decomposition approach, a linear programming problem is set up. This problem maximizes the global value of the alternatives and has as constraints the order of scale constants defined by the DM and his/her answers to the questions [7]. In holistic evaluation, the DM directly compares two alternatives, choosing the one he/she considers better or eliminating the one that is worse. In both cases, a new equation is added to the constraints of the linear programming problem: the alternative considered worse has a lower global value than the one that is considered better [12].

For ordering problems, unlike what occurs in choice problems, the method performs comparisons between alternatives seeking to identify dominance relations, and based on these, to determine the positions in the ranking that the alternatives should occupy. This is accomplished by solving the following linear programming problem [9]:

$$\max D(A_i, A_k) = \sum_{j=1}^{m} w_j v_j(A_i) - \sum_{j=1}^{m} w_j v_j(A_k) \tag{1}$$

s.t.

$$w_1 > w_2 > \cdots > w_m | \sum_{j=1}^{m} w_j = 1 \tag{2}$$

$$w_j v_j(x') > w_{j+1} \quad j = 1\ to m - 1 \tag{3}$$

$$w_j v_j(x'') < w_{j+1} \quad j = 1\ to m - 1 \tag{4}$$

$$w_j \geq 0,\ j = 1 \cdots m \tag{5}$$

For ranking problems, the objective function seeks to maximize the difference between the global values of two analyzed alternatives (1). The model's first constraint is the order of the scale constants defined by the DM (2), and further constraints are added to the model based on the DM's answers to the questions asked (3 and 4). For each criterion, a dominance matrix is generated, which is later used to determine the position of the alternatives. The process is concluded when a complete ranking is obtained or when the DM chooses to end the process [9].

The method is widely used in conjunction with problem structuring methods. Within the context of waste management, FITradeoff has been applied to support the management of waste electrical and electronic equipment (WEEE) collection systems by the public sector in Brazil [16]

3 Case Study

The model was applied to a Brazilian micro-enterprise in automotive detailing, the structure of which is highly centralized, as all contractual decisions are made by the majority partner, who holds the authority to veto decisions made and who will be held accountable for the consequences of those decisions.

Although the majority partner acts as the DM in this process, he does not operate in isolation, being influenced by the operators (the professionals who carry out the company's day-to-day activities). Because they have direct contact with the production process, these operators can provide important information regarding what waste disposal companies are the most suitable.

In addition to the DM, the minority partner and the operators act as stakeholders in this process, and the researcher acts as the analyst, who provides the necessary methodological support for obtaining and analyzing the model within the company.

3.1 Problem Structuring

The problem was structured using the VFT (Value-Focused Thinking) problem structuring method.

Definition of the Objectives

The values of a DM can be elicited by identifying their objectives, a process that necessitates a structured approach within the context of a strategic decision [14]. Following the Value-Focused Thinking (VFT) methodology, which involves converting a preliminary list into a standardized set of objectives [15], this study used the value-elicitation devices proposed by [13]. These devices (including creating a wish list, imagining hypothetical alternatives, analyzing problems and shortcomings, and assuming different perspectives) are designed to construct a comprehensive hierarchy of values. For the correct implementation of the VFT, questions were posed to the decision maker in each of the devices applied, forming 26 possible objectives.

Consequently, four devices were implemented: a Wishlist, Hypothetical Alternatives, Problems and Shortcomings, and Different Perspectives. The Wishlist established the foundational reasons for selecting a waste disposal method, later refined by using unrestricted reflection to capture implicit and aspirational goals. The Hypothetical Alternatives device required the DM to describe ideal and unacceptable options, with the analysis focusing on the rationale behind these judgments to reveal core values.

The Problems and Shortcomings device helped pinpoint operational concerns and potential improvements in the waste disposal process. The decision-maker was asked to think about which problems they could identify in the organization under study. The problems identified are presented in full in Fig. 1. Meanwhile, the Different Perspectives device sought novel insights by asking the DM to consider the employees' viewpoint. The decision-maker imagined themselves as an employee of the organization, and from this exercise, four statements were collected. After applying the devices, all the collected statements were gathered and presented to the decision maker. The full set of objectives derived from this process is summarized in Fig. 1

Through the use of the devices, it was possible to identify 26 statements from the decision-maker, which, after redundancy analysis, resulted in removing 13 objectives. The second step of VFT consists of distinguishing between fundamental objectives and means objectives. If an objective is considered essential in itself, representing a fundamental interest of the organization, it is classified as a fundamental objective. If its importance lies in serving as a means to achieve another objective, it is classified as a means objective. This separation leads to the structuring of objective networks: the fundamental objectives network, which serves as the basis for defining the criteria of the problem, and the means objectives network, which helps identify the alternatives [13].

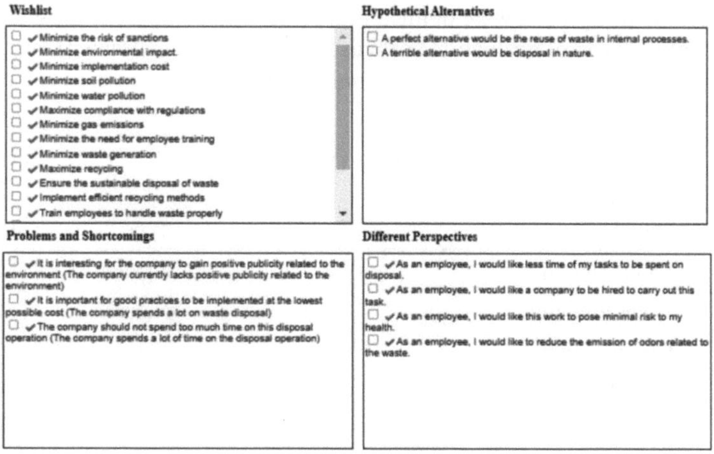

Fig. 1. Objectives found by using devices

These objectives were properly documented in the DSS. After eliminating the redundant objectives, those remaining were classified into means-objectives and fundamental objectives. Subsequently, the hierarchy of fundamental objectives was constructed, as presented in Fig. 2.

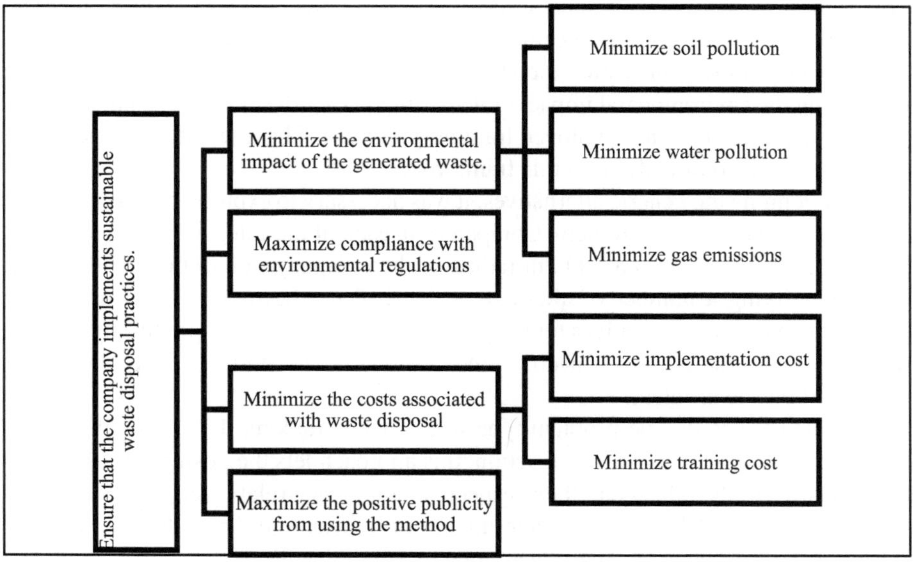

Fig. 2. Hierarchy of fundamental objectives

Setting the Criteria

Based on the hierarchy of higher- and lower-level fundamental objectives, the criteria to measure each final objective were defined (Table 1). Each criterion is presented with its code, description, type of evaluation scale used (natural or constructed), and the DM's preference direction (maximization or minimization).

Table 1. Criteria

Code	Criteria	Type	Direction
SOILP	Soil Pollution	Natural	Min
WATP	Water Pollution	Natural	Min
GASEM	Gas Emission	Constructed	Min
ENVREG	Compliance with environmental regulations	Constructed	Max
IMPCO	Implementation cost	Natural	Min
TRACO	Training Cost	Natural	Min
POPUBL	Positive publicity generated by using the method	Constructed	Max

Setting the Alternatives

The process first entails defining criteria from the DM's fundamental objective hierarchy [13], and then identifying the means to accomplish these ends. These means objectives provide the reference point for generating generic alternatives, which subsequently guide the creation of specific, actionable alternatives for the given problem.

By using iterative interviews, the DM's central means objectives were crystallized: minimizing the environmental impact of waste, increasing the efficiency of the disposal process, and strengthening employee training programs. The alternatives presented in Fig. 3 were derived directly from this framework.

After defining the generic alternatives, it was necessary to explore the entire space of possible solutions for this problem. Companies in each of the categories were surveyed, provided they were within a 300 km radius of the company in question. This process led to identifying a number of alternatives that had not been considered previously, reaching a total of 13 alternatives for the problem, including also the possibility of waste management by the company itself. Table 2 shows a description of the alternatives considered for the problem.

For this case study, the problem type selected was ranking. Even though the DM wishes to choose only one disposal action, this approach lets the position of each alternative relative to the others and their respective dominance relationships be visualized, which provides the DM with a greater amount of information. Once the problem type was defined, a consequence matrix was constructed (Table 3).

In the next section the problem was solved using the FITradeoff method [7].

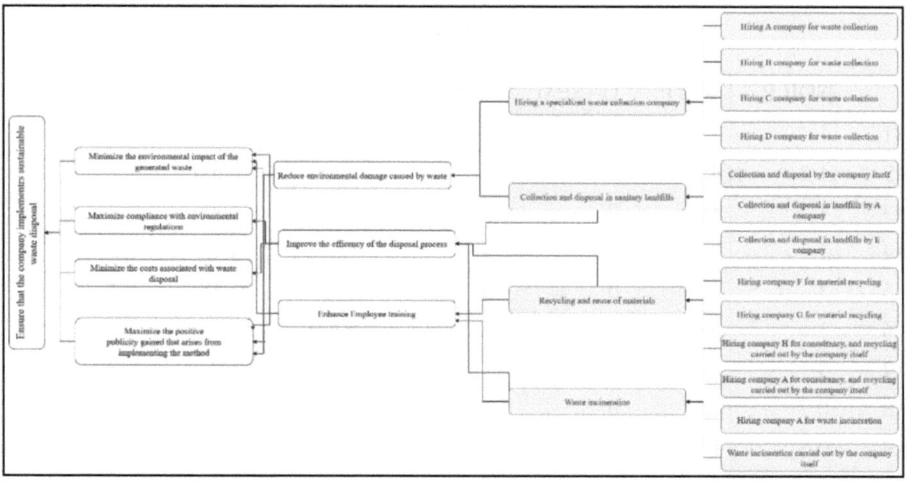

Fig. 3. Alternatives

Table 2. Description of the alternatives

Alternative	Description
A1	Hiring company A for waste collection
A2	Hiring company B for waste collection
A3	Hiring company C for waste collection
A4	Hiring company D for waste collection
A5	Collection and disposal in landfills by the company itself
A6	Collection and disposal in landfills by company A
A7	Collection and disposal in landfills by company E
A8	Hiring company F for material recycling
A9	Hiring company G for material recycling
A10	Hiring company H for consultancy, and recycling carried out by the company itself
A11	Hiring company A for consultancy, and recycling carried out by the company itself
A12	Hiring company A for waste incineration
A13	Waste incineration carried out by the company itself

Table 3. Consequence Matrix

	SOILP	WATP	GASEM	ENVREG	IMPCO	TRACO	POPUBL
A1	1	3	0	1	7000	1500	3
A2	1.5	4	0	1	6500	1200	3
A3	1.2	3.5	0	1	7200	1600	2
A4	1.3	3	0	1	6800	1400	2
A5	3	6	0	0	5000	1000	1
A6	3	5	0	1	5200	1100	2
A7	3	6	0	1	5300	1150	2
A8	0.7	1.2	0	1	8500	1900	4
A9	0.5	1	0	1	9000	2000	4
A10	1	2	0	1	10000	2500	5
A11	1.2	2.2	1	1	9500	2300	4
A12	2	5	1	1	6000	1500	3
A13	2.5	4	1	0	5500	1200	2

4 Problem Solving with FITradeoff

Considering that the DM demonstrated compensatory rationality after preference modeling, a method had to be selected that imposed minimal cognitive load on the DM, while also integrating a DSS to enhance the process.

Thus, the FITradeoff method was adopted, as it offers the axiomatic rigor of the traditional tradeoff procedure [8] combined with greater flexibility and interactivity. Additionally, tests were conducted to assess the presence of preferential independence among the alternatives.

To apply the FITradeoff method, the DSS at www.fitradeoff.org was accessed. After defining the values for the matrix, an intra-criteria evaluation was carried out using the DSS itself. Upon completion of the elicitation process, the value functions were displayed (Fig. 4).

After the intra-criteria evaluation, where the shapes of the marginal value functions of the criteria were obtained, this stage is concluded and the next one begins, in which the order of the scaling constants is established. This order is obtained in the DSS, through the display of comparisons for the decision-maker to choose. After the comparisons, it was possible to define the order, as shown in (6).

$$K_{ENVREG} > K_{POPUBL} > K_{IMPCO} > K_{TRACO} > K_{WATP} > K_{SOILP} > K_{GASEM} \quad (6)$$

The ordering of the scale constants alone was sufficient to determine the dominance relationships among some alternatives, as shown in Fig. 5.

Considering that the relationships established using only the order of the scaling constants are insufficient to determine the multiple positions in the ranking, decomposition-based elicitation rounds were conducted to establish them.

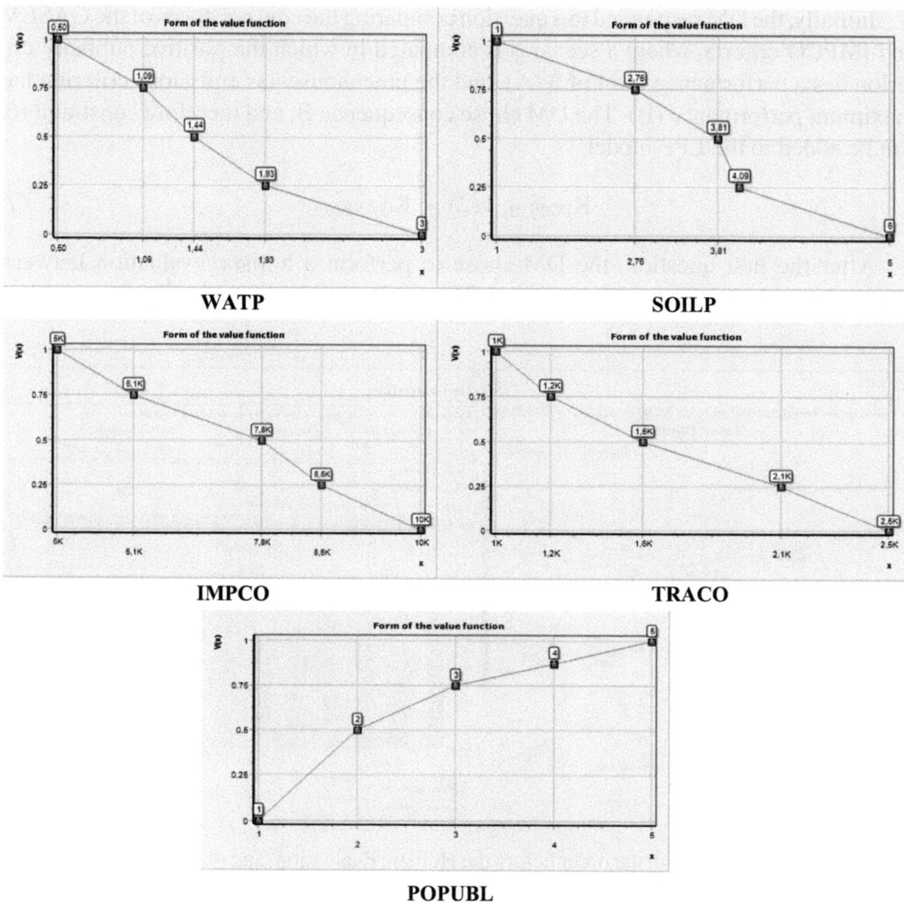

Fig. 4. Value Functions

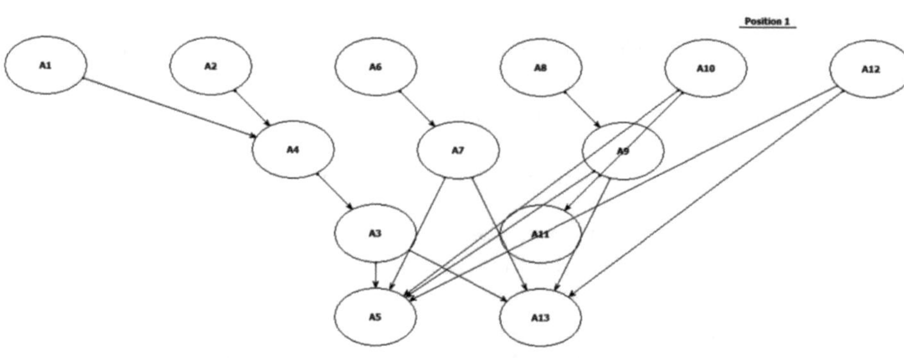

Fig. 5. Hasse Diagram with the order before ordering of scaling constants

Initially, the DM responded to a question comparing the consequences of the GASEM and IMPCO criteria, where a scenario is compared in which the positive publicity criterion has a performance level of 2 (A), and the greenhouse gas emissions criterion has maximum performance (B). The DM chose consequence B, and therefore constraint (6) can be added to the LPP model.

$$K_{POPUBL} v(2) < K_{GASEM} \quad (7)$$

After the first question, the DM chose to perform a holistic evaluation between alternatives A1 and A2, which were tied. This evaluation is shown in Fig. 6.

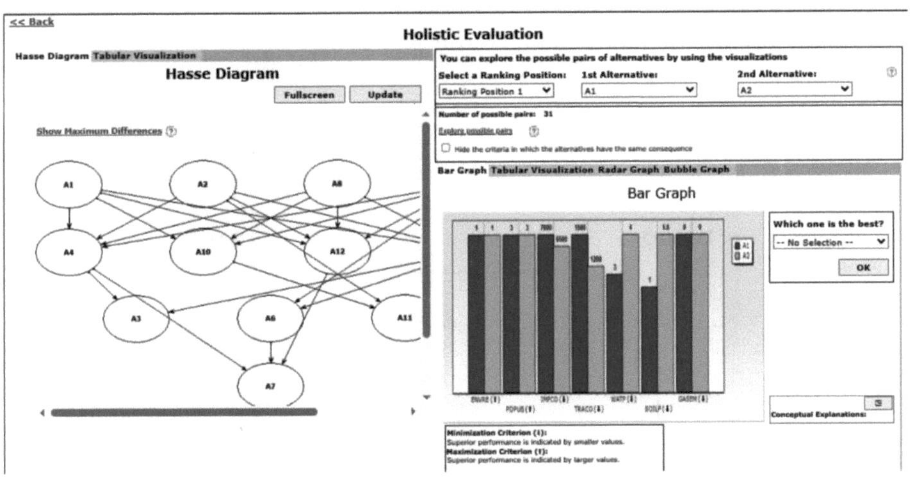

Fig. 6. Hasse Diagram with the order before the Holistic Evaluation, and the bar graph to compare A1 and A2 in the Holistic Evaluation

After the holistic evaluation, 7 positions remained in the ranking, and the DM chose to conduct another round of decomposition questions. The second question compared a scenario where the IMPCO criterion has a performance of 5.2 (A), and the TRACO criterion has maximum performance (B). The DM chose scenario A, which added the following constraint to the linear programming problem.

$$k_{IMPCO}.v(5, 2) > k_{TRACO} \quad (8)$$

This constraint was enough to establish 5 more positions in the ranking, completing the process. The final ranking is shown in Table 4.

The alternative chosen was A8 (Hiring company F for material recycling), followed by A9 (Hiring company G for material recycling) and A1 (Hiring company A for waste collection). This order reflects the DM's opinions and preferences.

Table 4. Final ranking

Ranking:	
1	[A8]
2	[A9]
3	[A1]
4	[A2]
5	[A10]
6	[A4]
7	[A3]
8	[A6]
9	[A7]
10	[A12]
11	[A11]
12	[A5, A13]

5 Discussion

The integration of decomposition and holistic evaluation paradigms has proven its relevance by optimizing the process, thereby significantly reducing the number of required cycles. Through the combined application, the decision-making framework was efficiently conducted with only two decomposition iterations and a single holistic evaluation. In contrast, the exclusive use of decomposition would generate a substantially larger volume of questions, as evidenced by Table 5.

To test the robustness of the solution obtained, a Sensitivity Analysis was performed directly in the DSS. The procedure consisted of creating different scenarios by varying the results of the alternatives under the selected criteria. In this study, the criteria "Implementation cost", "Training cost", "Soil pollution", and "Water pollution" were adjusted by $\pm 10\%$, while the criterion "Positive publicity resulting from the method" was varied at one level. The results of this analysis are presented in Fig. 7.

The results indicate a robustness of over 80% for all alternatives, which suggests that the preference order remains consistent across most scenarios tested and reinforces the reliability of the proposed solution. Given that the software recommended a priority order starting with hiring Company F for material recycling, it is suggested that the automotive detailing business in Brazil adopt this alternative as its primary action, as the analysis demonstrated that Company F performs well in relevant criteria such as positive publicity and compliance with environmental legislation.

Furthermore, it is recommended that the microenterprise prioritize formalizing contracts with specialized companies to ensure waste is managed according to best environmental practices and local regulations, while also establishing monitoring and evaluation mechanisms to measure the effectiveness of the adopted method and identify potential adjustments.

Table 5. Summary of questions answered in the decomposition process

Cycle	Consequence A	Consequence B	Answer	Number of levels
0			–	1
1	2 of POPUBL	Best of GASEM (0)	Consequence B	3
2	3 of POPUBL	Best of IMPCO (5 K)	Consequence B	4
3	4 of POPUBL	Best of IMPCO (5 K)	Consequence B	4
4	5.6 K of IMPCO	Best of TRACO (1 K)	Consequence A	5
5	6.1 K of IMPCO	Best of TRACO (1 K)	Consequence A	7
6	1.1 K of TRACO	Best of WATP (1)	Consequence B	9
7	1.1 K of TRACO	Best of WATP (1)	Consequence B	9
8	1.38 of WATP	Best of SOILP (0,50)	Consequence A	11
9	0.70 of SOILP	Best of GASEM (0)	Consequence B	11
10	6.2 K of IMPCO	Best of TRACO (1 K)	Consequence A	12

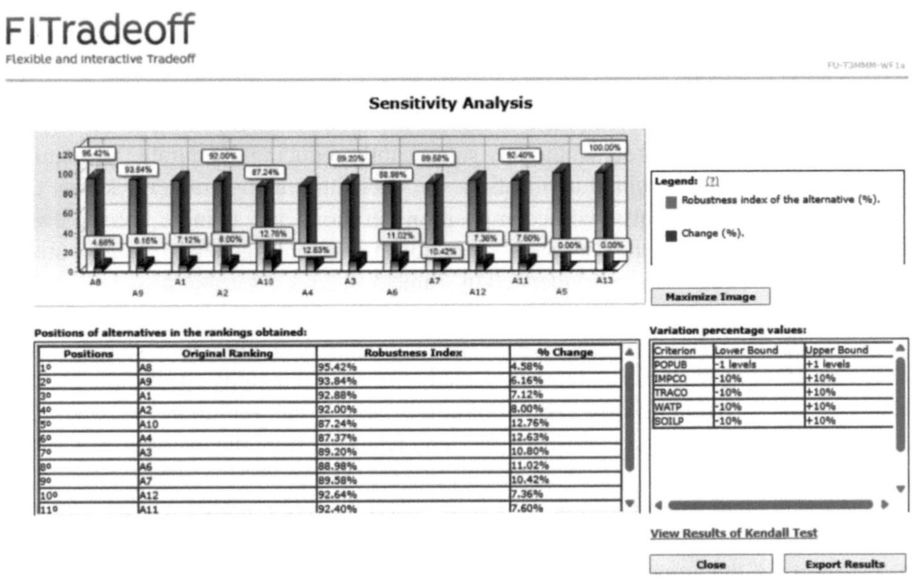

Fig. 7. Results of the Sensitivity Analysis

Although the initial decision points to hiring Company F, the business should remain open to revisiting available options in case there are significant changes in external or internal conditions, such as cost variations or new regulations. During implementation, the transition to the new waste disposal method should be structured, including training operators on specific procedures to ensure efficiency in waste separation and storage, as well as establishing a registration system to document all stages from collection to final destination, which will provide greater internal control and serve as evidence of compliance during inspections.

Finally, the company should adopt communication strategies to promote its sustainable actions and highlight efforts to minimize environmental impact, for example via marketing campaigns that emphasize environmental responsibility, thereby enhancing its appeal to environmentally-aware consumers.

6 Conclusions

Applying problem structuring methods like Value-Focused Thinking proved essential for systematically understanding and framing the challenges faced by the automotive detailing microenterprise. These approaches enabled gaps to be identified in its current waste disposal practices, established clear criteria for evaluating alternatives, and facilitated selecting the option most aligned with the DM's objectives.

The recommendation of alternative A8 as the priority solution reflects a robust decision-making process, one that effectively balanced costs, environmental impact, and corporate image benefits. Beyond enabling the selection of a viable alternative, the multi-criteria analysis offered valuable insights into the importance of operator training and of formalizing environmental management practices within the company.

Implementing these recommendations will require paying careful attention to training, monitoring, and communication. For long-term sustainability, the microenterprise must adopt an integrated, systematic approach to these changes. A genuine commitment to sound environmental practices will yield benefits that extend beyond regulatory compliance, and will strengthen the company's reputation and attract environmentally-aware clients.

However, the limitations of this study must be acknowledged. As this is research based on a single case, the results obtained reflect the specific reality of this one microenterprise and cannot be directly generalized. Thus, the need for future studies that carry out validation in different contexts is recognized, allowing the identification of decision patterns common to other small waste generators. Ultimately, this case serves as a practical illustration of how structured decision-making methods can be applied to promote sustainability by aligning objectives with legislation and best practices.

Acknowledgments. The authors would like to thank the Coordination for the Improvement of Higher Education Personnel (CAPES) and the Brazilian National Council for Scientific and Technological Development (CNPq) for the financial support.

Disclosure of Interests The authors have no competing interests to declare that are relevant to the content of this article.

References

1. Bilgilioglu, S.S., Gezgin, C., Orhan, O., Karakus, P.: A GIS-based multi-criteria decision-making method for the selection of potential municipal solid waste disposal sites in Mersin, Turkey. Environ. Sci. Pollut. Res. **29**(4), 5313–5329 (2021). https://doi.org/10.1007/s11356-021-15859-2

2. Bouyakhsass, R., et al.: Wastewater and solid waste environmental degradation cost in Mohammedia city. Water Air Soil Pollut. **234**(6), 234–344 (2023). https://doi.org/10.1007/s11270-023-06367-9
3. Saha, P., Roy, M.: Identification of municipal solid waste disposal sites for sustainable solid waste management in Cooch Behar Municipality, West Bengal: a RS-GIS based MCDA approach. Waste Manag. Bull. **2**(3), 127–144 (2024). https://doi.org/10.1016/j.wmb.2024.07.004
4. ANFAVEA: Associação Nacional dos Fabricantes de Veículos Automotores, 2025. Indústria automobilística brasileira Brazilian automotive industry. Available at: https://anfavea.com.br/site/wp-content/uploads/2025/06/DIGITAL-ANUARIO-2025ALT.CAP_.4_compressed.pdf. Accessed 4 Dec 2025
5. Brazil: Lei no. 12.305, de 02 de agosto de 2010 (Law no. 12.305, August 2nd, 2010). In: Política Nacional dos Resíduos Sólidos (National Policy of Solid Waste) (2010). Available at: https://www.planalto.gov.br/ccivil_03/_ato2007-2010/2010/lei/l12305.htm. Accessed 6 Dec 25
6. Vaz, C.S.D., Ferreira, F.A.F., Dabić, M., et al.: Measuring SME propensity for resilience using problem structuring methods and TOPSIS. Rev. Manag. Sci. (2025). https://doi.org/10.1007/s11846-025-00958-y
7. de Almeida, A.T., et al.: A new method for elicitation of criteria weights in additive models: flexible and interactive tradeoff. Eur. J. Oper. Res. **250**(1), 179–191 (2016). https://doi.org/10.1016/j.ejor.2015.08.058
8. Keeney, R.L., Raiffa, H.: Decisions with Multiple Objectives: Preferences and Value Tradeoffs. Cambridge University Press (1976)
9. Frej, E.A., de Almeida, A.T., Costa, A.P.C.S.: Using data visualization for ranking alternatives with partial information and interactive tradeoff elicitation. Oper. Res. **19**(4), 909–931 (2019). https://doi.org/10.1007/s12351-018-00444-2
10. Kang, T.H.A., Frej, E.A., de Almeida, A.T.: Flexible and interactive tradeoff elicitation for multicriteria sorting problems. Asia-Pacit. J. Oper. Res. **37**(05), 1–22 (2020). https://doi.org/10.1142/s0217595920500207
11. Frej, E.A., Ekel, P., de Almeida, A.T.: A benefit-to-cost ratio based approach for portfolio selection under multiple criteria with incomplete preference information. Inf. Sci. **545**, 487–498 (2021). https://doi.org/10.1016/j.ins.2020.08.119
12. de Almeida, A.T., Frej, E.A., Roselli, L.R.P.: Combining holistic and decomposition paradigms in preference modeling with the flexibility of FITradeoff. Cent. Eur. J. Oper. Res. **29**(1), 7–47 (2021). https://doi.org/10.1007/s10100-020-00728-z
13. Keeney, R.L.: Value-Focused Thinking: A Path to Creative Decisionmaking. Harvard University Press, Cambridge (1992)
14. Keeney, R.L.: Value-focused thinking: identifying decision opportunities and creating alternatives. Eur. J. Oper. Res. **92**(3), 537–549 (1996). https://doi.org/10.1016/0377-2217(96)00004-5
15. Sheng, H., Nah, F.F.H., Siau, K.: Strategic implications of mobile technology: a case study using value-focused thinking. J. Strateg. Inf. Syst. **14**(3), 269–290 (2005). https://doi.org/10.1016/j.jsis.2005.07.004
16. de Fernandes, C.H.A., Silva, L.C., Guarnieri, P., de Vieira, B.O.: Multicriteria model proposition to support the management of systems of e-waste collection. Logistics. **5**(3), 1–20 (2021). https://doi.org/10.3390/logistics5030060

A K-Medoids-Based Segmentation Model for Incentive Allocation in the Renewable Energy Sector

Bilal En-Nouaary[1(✉)], Aziz Hmioui[1], and Nadine Kafa[2]

[1] Ecole Nationale de Commerce et de Gestion de Fès, Sidi Mohamed Ben Abdellah University, Fez, Morocco
bilal.ennouaary@usmba.ac.ma
[2] CESIT, KEDGE Business School, Talence, France

Abstract. The effective allocation of incentives remains a major challenge for policymakers seeking to accelerate renewable energy deployment while balancing economic, social, and environmental objectives. This study proposes a a policy decision-support tool guiding incentive selection under energy transition. Using k-medoids clustering, RE projects are segmented based on generation performance indicators, namely installed capacity and reliability indices, while a triple bottom linebased scoring scheme evaluates economic, social, and environmental sustainability.

Applied to UK RE projects dataset, the model reveals distinct and policy-relevant segments characterized by scale, reliability, and sustainability heterogeneity. The results reveal ten distinct project segments formed based on scale and reliability, where each segment include projects with different sustainability profiles. Based on these preliminary findings, the study outlines exploratory, segment-specific incentive orientations aimed at improving policy incentive strategies. The proposed model represents one of the first machine-learningdriven models for renewable energy incentive allocation.

Keywords: Decision Support Systems · Data-Driven Policy Design · Incentive allocation · Machine Learning · K-medoids · Renewable Energy

1 Introduction

In the past two decades, the large-scale deployment of renewable energy (RE) projects has been unprecedented. They have been the key to meeting rising global energy needs while gradually displacing fossil fuelbased sources. Nevertheless, their diffusion continues to be constrained by several barriers. Technologically, the intermittent and variable nature of RE generation, particularly wind and solar energies, introduces unpredictability, complicating real-time energy equilibrium in the power grid, which is not always adequately adapted to accommodate prosumers injecting surplus electricity [1]. Economically, unclear and uncertain

A. P. Cabral Seixas Costa et al. (Eds.): ICDSST 2026, LNBIP 582, pp. 163–176, 2026.
https://doi.org/10.1007/978-3-032-25146-6_11

fiscal frameworks, long payback periods, and limited access to financing discourage private investors from engaging in RE projects [2]. Environmentally and socially, renewable projects tend to require large land and water resources [3] while creating fewer jobs [4], which further weakens their competitiveness compared to fossil-based alternatives [5]. These asymmetries underline the need for a segmentation decision support tool to inform incentive mechanisms in the RE sector.

Project segmentation for incentives allocation, in particular, remains a complex challenge in the RE field, as it requires integrating production data with sustainability dimensions. Recent literature has proposed various methodologies to address this problem, including economic models [6], epidemic models [2], game theory-based models [7], forecasting accuracy-based models [8], and contribution distribution models [9]. Yet, the challenge goes beyond interpretability, strategic interactions, and forecasting: effective segmentation must process large volumes of production data to reveal patterns, adapt to evolving data sets to continuously and accurately predict project segments, and also reconcile sustainability dimensions. This work therefore addresses the question of how RE energy project segmentation can support policymakers in selecting tailored incentive schemes across a broad portfolio of instruments, including feed-in tariffs, feed-in premiums, grants, and tax reductions. Positioned within the broader agenda of evidence-based energy transition governance, the study responds to the increasing complexity and data richness of modern energy systems. As such systems evolve, policymaking must move beyond generic incentive mechanisms toward empirically grounded, project-specific allocation strategies.

This study builds on previous work to improve incentive allocation for RE by addressing key limitations identified in the literature. While prior models have employed diverse approaches, these methods have largely treated utility-scale projects and RE communities as distinct domains [6–8] and have prioritized economic criteria at the expense of broader sustainability considerations [10]. Moreover, despite growing interest in data-driven policymaking, no existing study has implemented an unsupervised machine learning (ML) approach to segment RE projects for incentive allocation. By introducing a novel ML-based segmentation model that incorporates both generation performance and sustainability indicators, this study extends prior work and offers policymakers a more holistic policy decision-support tool to guide incentive selection under energy transition.

2 Literature Review

The design of effective incentive models plays a critical role in accelerating the deployment of RE by addressing unpredictability, high investment cost, limited financing, and ultimately the fragile cost competitiveness of RE projects against fossil energy projects. This review of the literature examines a range of methodologies to optimize incentive allocation for RE projects, ranging from economic models [6], epidemic models [2], game theory [7], forecasting accuracy-based models [8], and contribution distribution models [9]. These contributions can be broadly divided into two main categories: those focusing on utility-scale RE projects and those targeting RE communities.

In the first category, Mostafayi Darmian et al. [10] optimized government incentives and private investment for utility-scale RE projects using a hybrid multicriteria and bi-level optimization model, showing that high-risk technologies such as hydropower require stronger incentives, though the analysis was limited to market-traded firms. Similarly, Kong et al. [8] linked incentives to forecasting accuracy, encouraging improved prediction and reduced reliance on energy storage systems in the South Korean market. From a spatial perspective, Piel et al. [11] demonstrated that location-based incentives can promote geographically balanced wind deployment and reduce generation variability, albeit with a focus restricted to wind energy.

The second category focuses on RE communities, where incentives aim to foster participation and cooperation. Bilardo [6] proposed a dynamic incentive model rewarding prosumers based on loadgeneration alignment, while Lilliu et al. [7] introduced a game-theoretic peer-to-peer trading mechanism that reduced costs and enhanced self-consumption. Extending this line of work, Wang et al. [1] combined motivational psychology with coalition game theory, showing that incentive-based coordination can improve participation and overall grid performance.

Although the reviewed studies underscore diversified approaches in incentive allocation, none of the developed models integrate all sustainability criteria. Moreover, they focus predominantly on economic and financial mechanisms. To improve the allocation of incentives, it is essential to recognize that RE projects differ significantly in scale, technology, and production factors, as well as in economic, social, and environmental profiles. The segmentation of these projects into homogeneous clusters with similar characteristics is required to design tailored incentive schemes that more precisely address each segment's needs.

Despite the growing body of literature on RE incentive allocation, several critical gaps remain. First, existing studies predominantly address either utility-scale projects or community-based initiatives in isolation, without proposing a unified framework capable of accommodating both scales simultaneously. Second, most incentive models are narrowly focused on economic or financial mechanisms, overlooking other policy-relevant instruments such as fiscal, R&D, or institutional incentives. Third, sustainability considerations are only partially integrated. Finally, no prior study has adopted a dedicated ML-based clustering approach for incentive allocation. To address these gaps, this study proposes an integrated methodology that leverages unsupervised ML to segment RE projects into homogeneous groups based on generation performance and all triple bottom line (TBL) sustainability dimensions.

Several unsupervised clustering paradigms have been proposed in the literature, including Bayesian clustering and ML-based approaches. Bayesian methods offer the advantage of explicitly accounting for uncertainty; however, they require careful specification of prior assumptions and typically yield probabilistic outputs rather than explicit segments [12]. As a result, their applicability to incentive allocation remains limited where clear and operational project categories are required. Recent advances in ML have also popularized deep learning (DL)based clustering techniques. While DL methods are capable of learning complex nonlinear representations, they typically demand large datasets, involve high computational costs, and produce latent embeddings that are difficult to translate into transparent incentive segments [13]. Given the moderate size of RE project datasets and the need for interpretability, such approaches would constitute methodological over-engineering in the present context. Consequently, more conventional ML clustering techniques, such as shallow neural networks (SNNs), DBSCAN, hierarchical clustering, and partitioning methods could represent a more suitable alternative, boasting strengths such as topology preservation, flexibility in cluster shapes, or dendrogramic representations. However, these methods exceed the requirements of the present study, which focuses on univariate partitioning clustering. Their adoption would therefore add unnecessary methodological complexity.

Partitioning algorithms remain widely used due to their simplicity, reproducibility, and low interpretational burden. Among them, k-means has become the de facto baseline in segmentation studies [14]. In the energy domain, Scaramuzzino et al. [14] applied k-means to classify EU28 and Swiss regions based on renewable potential and socioeconomic factors, yielding policy-relevant insights. However, k-means is sensitive to outliers and noise, and its reliance on abstract centroids limits the association of clusters with representative real-world cases [15]. To overcome these limitations, this study adopts k-medoids clustering as the core ML method. Unlike k-means, k-medoids selects actual data points as cluster centers, enhancing robustness to outliers and improving interpretability [16]. By minimizing pairwise dissimilarities rather than squared distances, k-medoids yields more stable and representative clusters, enabling a direct linkage between RE project segments and tangible reference cases.

3 Methodology

Figure 1 presents the proposed policy decision-support tool guiding incentive selection under energy transition, which adopts a hybrid architecture that combines k-medoidsbased clustering with sustainability scoring to inform incentive allocation for RE projects. The model is designed to generate interpretable categories that can be directly used by policymakers for regulatory and incentive design purposes. The two components operate in parallel. The ML branch derives a data-driven segmentation of projects by clustering generation performance indicators, while the sustainability branch evaluates each project's profile across the TBL dimensions. The outputs of these two branches are subsequently integrated to inform incentive allocation.

3.1 Data Collection

The dataset used in this study consists of detailed project-level information, including installed capacity, technology type, location, and development status.

Additional indicators related to generation performance, such as capacity factor, intermittency, and reserve contribution, were computed or assigned based on project characteristics. Economic, social, and environmental variables were also incorporated to reflect broader sustainability dimensions, including levelized cost of electricity (LCOE), social acceptability, land use, and water use. Once consolidated, the dataset supports two parallel processes: generation performance-based clustering and TBL-based sustainability scoring.

3.2 Generation PerformanceBased Clustering

The generation performance branch of the model operates on a set of project-level indicators that include installed capacity, capacity factor, intermittency index, and reserve contribution. Installed capacity serves as a proxy for project scale, reflecting the overall magnitude of generation infrastructure. Capacity factor captures the effective utilization of this capacity over time, providing insights into operational efficiency and technological reliability. The intermittency index measures the variability in power output, indicating the predictability and stability of the energy supply. The reserve contribution index quantifies the extent to which a project's output coincides with periods of peak system demand, serving as a proxy for the project's contribution to grid adequacy and its potential to reduce reliance on flexible backup sources. In parallel, the intermittency index was derived from the variability of hourly generation, measuring the consistency and predictability of power output. Projects with lower intermittency scores are generally more stable and reliable from a grid integration perspective.

A preprocessing phase was carried out to ensure consistency and completeness across all indicators. The final input for clustering was composed of two features: installed capacity and a composite reliability index, constructed as the average of the normalized capacity factor, intermittency, and reserve contribution scores. This bivariate configuration enables the model to capture both scale-related and reliability aspects of projects, ensuring that segmentation outcomes account for both infrastructure magnitude and performance quality without resorting to high-dimensional or opaque feature spaces. To determine the optimal number of clusters, both the elbow and silhouette methods are employed. The elbow method relies on identifying the point at which additional clusters no longer yield substantial improvements in within-cluster compactness, offering a simple yet widely accepted visual heuristic [17]. In parallel, the silhouette method evaluates how well each observation lies within its assigned cluster compared to neighboring clusters, thus quantifying the coherence and separation of the clustering structure. The use of both techniques balances intuitive interpretability with statistical robustness.

Then, K-medoids clustering is adopted to segment RE projects based on their generation performance profiles. To evaluate the quality of the generated clustering, both the silhouette coefficient and the Between-Within Proportion (BWP) index are used. While higher silhouette coefficient values indicate better-defined partitions, The BWP index assesses the ratio of between-cluster variance to within-cluster variance, thereby quantifying the extent to which the clustering

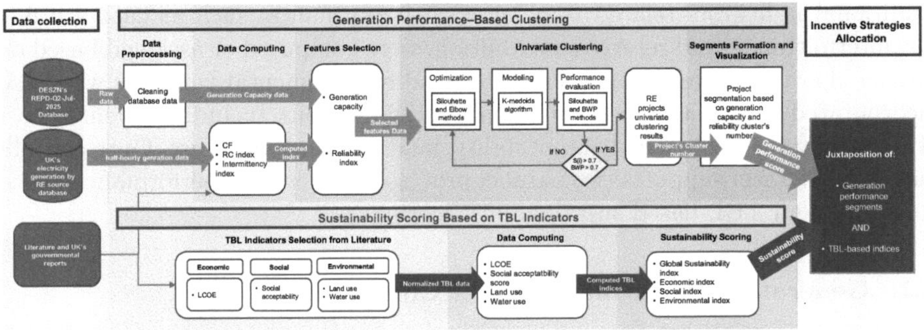

Fig. 1. The Methodology of the Developed Policy Decision-Support Tool Guiding Incentive Selection under Energy Transition

maximizes inter-group distinction while minimizing intra-group dispersion. A silhouette and BWP threshold of 0.7 is enforced, below which the clustering process is considered suboptimal and must be re-initiated from the cluster optimization phase to refine segmentation accuracy.

Following the clustering phase, each project is assigned to one of the k clusters for each generation performance feature. Clusters are ranked numerically from 1 to k, where 1 represents the lowest performance group for a given indicator. For example, a project assigned to cluster 1 in the installed capacity dimension is interpreted as a small-scale project, while higher-numbered clusters represent progressively larger scales. This ordinal ranking preserves the relative positioning of projects along each feature and supports a meaningful interpretation of segmentation results.

To support the formation of generation-based project segments, the cluster assignments from the installed capacity and reliability features are juxtaposed in a two-dimensional space. This visual mapping enables the identification of meaningful groupings that reflect both the scale and operational reliability of projects. These visually formed segments then serve as the foundation for the subsequent incentive allocation phase, where sustainability indices based on the TBL approach are introduced.

3.3 Sustainability Scoring Based on TBL Indicators

In parallel to the generation performance clustering, the model integrates a scoring branch that evaluates RE projects according to the TBL approach, encompassing economic, social, and environmental dimensions. This approach ensures that segmentation is not based solely on technical or capacity-related indicators but also reflects broader sustainability considerations aligned with public policy objectives. The economic criterion is captured by the LCOE. Lower LCOE values signal more economically viable projects. The social criterion is proxied by a social acceptability score. The environmental criterion includes two resource-based indicators: land use, and water use. These indicators favor technologies with lower spatial and hydrological footprints.

All four indicators are normalized using minmax scaling to ensure comparability across different units and ranges. The economic, social, and environmental indices are then derived directly from the respective normalized indicators. The overall sustainability score is computed as the unweighted arithmetic average of the three dimensions, providing a global sustainability index that reflects the composite performance of each project across the TBL criteria. These normalized indices serve as sustainability scores that are incorporated after the visual segmentation of projects based on generation performance.

3.4 Incentive Strategies Allocation

For each segment defined through the generation performance-based clustering, projects are subsequently assessed through the lens of sustainability. Within a given segment, projects are analyzed across economic, environmental, and social dimensions. It is at this stage that incentive strategies are formulated, aligning support mechanisms with the specific sustainability strengths and weaknesses observed within each segment. This phase is particularly critical, as the generation performance segmentation allows for the identification of high-performing and low-performing project groups. In this way, the segmentation outcomes serve as a diagnostic phase, helping policymakers to tailor incentive strategies according to the actual needs and sustainability value of each project.

4 Application of the Proposed Model to UK RE Sector

In this section, results generated from the application of the proposed k-medoids segmentation model are highlighted and discussed to inform potential incentive strategies for RE policymakers. The model allows policymakers to match incentive instruments with the distinct strengths and weaknesses of each project in each segment, ensuring targeted and efficient allocation of public support. The model was tested using open-source datasets from the UK renewable energy sector.

4.1 Dataset Description

The cleaned and consolidated dataset comprises 7,891 renewable energy projects across all 12 UK administrative areas: nine English regions and three devolved nations, and an additional offshore category. These projects span 20 distinct technology types, grouped into eight broader categories to facilitate analysis and interpretation. Installed capacities range from 0.1 MW to 4,100 MW, reflecting the diversity of the UK's renewable energy portfolio.

Project-level installed capacity and land use data were obtained from the UK Renewable Energy Planning Database (REPD-Q2-Jul-2025) [18]. Capacity factor values were retrieved from official datasets published by the Department for Energy Security and Net Zero [18], while hourly electricity generation and demand data, used to compute reserve contribution and intermittency indices, were sourced from the National Energy System Operator (NESO) [19]. Economic viability was assessed using levelized cost of electricity (LCOE) figures published in the UK government's 2023 electricity generation cost report. Social acceptability scores were constructed from the DESNZ Public Attitudes Tracker - Spring 2025) [18], and environmental resource intensity was measured through land use estimates validated by Troldborg et al. [20] and water use data drawn from Clark et al. [21].

4.2 Univariate Clustering of Generation Performance Features

To identify distinct project groups based on generation performance, univariate k-medoids clustering was applied separately to two key indicators: installed capacity and reliability index. Installed capacity serves as a structural proxy for project scale, given the well-established distinction between utility-scale and community-scale RE projects, as these categories differ not only in investment requirements and grid impact but also in the form and volume of incentives they typically receive. The reliability index captures the operational stability and consistency of electricity generation over time, since RE technologies differ significantly in intermittency and generation variability. This indicator reflects a project's capacity to contribute to system adequacy and reduce balancing costs, which are key concerns for energy planners. The dimensionality of the segmentation was deliberately limited to these two indicators to preserve transparency and interpretability for policymakers. Accordingly, four clusters were retained for each feature, offering an interpretable clustering into small-, mid-, large-, and mega-scale projects, and unstable, limited reliability, reliable, and highly reliable projects.

The quality of the clustering results was assessed using the silhouette and the BWP coefficients. For the installed capacity clustering, the model achieved an silhouette coefficient of 0.80 and a BWP coefficient of 0.87, indicating well-defined and internally coherent clusters with a clear separation between scale categories. The reliability-based clustering exhibited even stronger performance, with an silhouette coefficient of 0.93 and a BWP coefficient of 0.99.

After applying univariate clustering to both the installed capacity and reliability index, the juxtaposition of the resulting four-by-four cluster structure revealed ten distinct project segments. These segments were visualized on a two-dimensional axis, with capacity clusters on the x-axis and reliability clusters on the y-axis. Each segment was then interpreted and labeled according to the combined performance characteristics it represents. For instance, segments such as *highly reliable large-scale*, *mid-scale reliable*, or *small-scale unstable* reflect clear distinctions in both scale and reliability. The resulting visual map is presented in Fig. 2, where cluster boundaries (represented by the horizontal and vertical red lines) correspond to the optimal thresholds derived from the clustering phase.

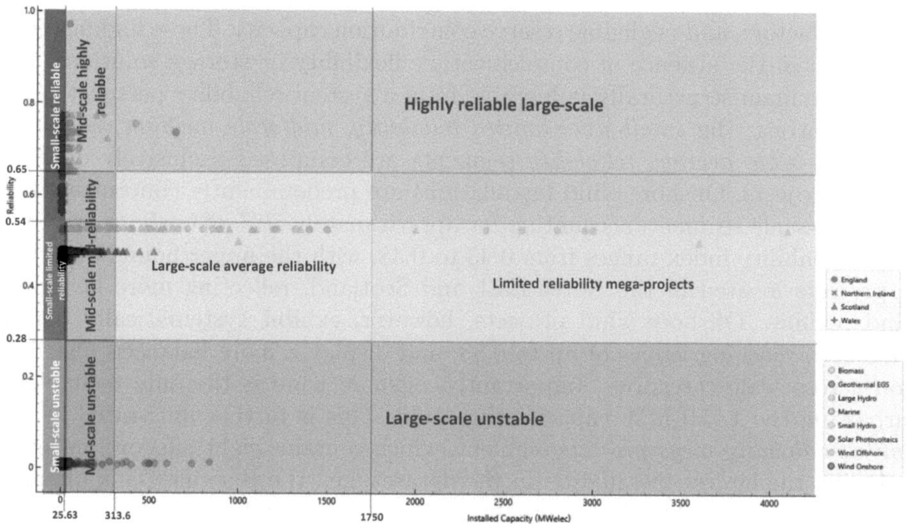

Fig. 2. Segmentation Results Based on the Generation Performance Clustering

4.3 Segments Formation and Visualization

While the initial clustering procedure yielded a theoretical matrix of 16 segments based on the 4×4 combinations of capacity and reliability clusters, only 10 segments were ultimately retained. This reduction was motivated by analytical and policy-oriented considerations. First, some segments were found to be sparsely populated or even empty, such as the *highly reliable large-scale, large-scale average reliability, and large-scale unstable* categories, making them statistically unstable and of limited practical relevance. Removing or merging such segments improves the interpretability of the model and avoids overfitting. Second, certain adjacent clusters exhibited highly similar profiles; for example, projects falling into capacity cluster 1 and reliability cluster 3 were nearly indistinguishable from those with similar small capacity and reliability cluster 4. That's why they are unified in the *small-scale reliable* projects segment. Finally, although the visual boundary between capacity clusters 1 and 2 appeared somewhat blurred, it was considered important to isolate small-scale projects (below 25.63 MW) as a distinct group. These installations are often associated with industrial or commercial developments, implying distinct operational contexts and support needs that justify tailored policy treatment.

The segmentation results reveal a clear and policy-relevant differentiation of RE technologies across reliability and scale classes. First, *the small-scale unstable, mid-scale unstable, and large-scale unstable* segments are populated exclusively by PV projects. Nearly 90% of these PV installations are concentrated in the *small-scale unstable* segment. Across these three segments, the PV reliability index is extremely low, ranging from 0 to 0.01. This weak performance is primarily attributable to the combined effects of low operational stability, low capacity factors, and negligible reserve contribution capacity. These findings confirm that, in the absence of complementary flexibility or storage solutions, PV projects remain structurally vulnerable from a system reliability perspective.

In contrast, the *small-scale limited reliability, mid-scale medium reliability, and large-scale average reliability* segments are composed exclusively of wind energy projects. Onshore wind installations are predominantly concentrated in the small-scale segment, accounting for approximately 70% of onshore projects. Their reliability index ranges from 0.43 to 0.48, with the upper bound observed in projects located in the North-East and Scotland, reflecting more favorable wind regimes. Offshore wind projects, however, exhibit systematically higher reliability, reaching values of up to 0.53, and display a more balanced distribution across scale categories. Importantly, offshore wind is the only technology surpassing the 1,750 MW capacity threshold. This is further illustrated in the *limited reliability mega-projects* segment, which contains eight offshore projects, including the largest installation in the dataset, i.e., the Berwick Bank offshore wind farm with 4,100 MW, highlighting the strategic role of offshore wind in high-capacity deployment.

The transition toward higher reliability segments is characterized by increased technological diversity. In the s*mall-scale reliable* segment, a heterogeneous mix of biomass, small and large hydropower, and marine energy projects is observed, although biomass projects dominate this class. Biomass installations exhibit reliability indices ranging from 0.65 to 0.75, with the highest values associated with dedicated biomass facilities and energy-from-waste incineration plants, reflecting their dispatchability and strong contribution to system operations. A similar but more pronounced pattern emerges in the *mid-scale highly reliable* segment, where biomass, hydropower, and marine projects coexist alongside the sole geothermal installation in the dataset. The Tregath business park station road geothermal development, with an installed capacity of 49 MW, achieves the highest reliability index overall (0.97), emphasizing the exceptional stability and baseload characteristics of geothermal energy.

Finally, the *highly reliable large-scale* segment is exclusively composed of biomass projects, limited to five installations, with the largest reaching 645 MW. Their presence in this segment confirms the scalability of biomass technologies, while also highlighting their role as stable renewable resources within the energy mix.

4.4 Sustainability Scoring

Beyond reliability, the sustainability scoring results provide an additional and complementary decision layer by shifting the focus from generation performance to long-term economic, social, and environmental value. Rather than applying uniform sustainability requirements across all projects, the analysis highlights the necessity of assessing project-specific sustainability needs.

In this respect, large-scale segments and PV-dominated segments emerge as priority targets for environmental and social sustainability incentives. Large-scale projects, by virtue of their spatial footprint and systemic influence, generate more pronounced environmental externalities and social interactions, particularly in terms of land use, local acceptance, and regional development effects. Similarly, PV projects, while representing the majority of installations in the dataset, are predominantly concentrated in low-reliability segments, making their cumulative environmental and social impacts more significant than their individual contribution would suggest.

Conversely, high-reliability segments face different constraints, primarily related to cost structures, capital intensity, and long-term economic viability. For these segments, sustainability scoring indicates that economic sustainability incentives are more appropriate than additional environmental or social requirements. Supporting these projects through economically focused instruments helps preserve their role as stabilizing assets within the energy mix.

4.5 Potential Incentive Allocation Strategies

To translate the segmentation and sustainability scoring into actionable guidance, several preliminary policy recommendations are proposed. R&D incentives and soft loans may benefit small-scale, high-reliability projects by supporting technological upgrades and gradual capacity expansion. For high-sustainability, high-reliability segments, ensuring fiscal and regulatory stability appears more impactful than additional financial incentives. Feed-in Tariffs could remain suitable for mid and high-reliability projects facing market-related risks that hinder economic viability, as guaranteed pricing mechanisms may reduce uncertainty and facilitate threshold crossing. In contrast, Feed-in Premiums may be better suited for high-reliability projects with strong sustainability performance, as they allow market participation while potentially rewarding higher generation performance and sustainability outcomes. Finally, fiscal incentives and grants could be considered for large-scale and PV-dominated segments, particularly where projects are associated with significant environmental or social externalities. When conditionally linked to incremental improvements in environmental mitigation, social value creation, and the adoption of green supply chain management practices [22], such instruments may contribute to improving sustainability performance while preserving system reliability. Particularly, this approach could help reduce PV cumulative environmental impacts without directly incentivizing additional capacity deployment.

5 Conclusions

This study proposes an integrated policy decision-support tool guiding incentive selection under energy transition by jointly addressing generation performance and sustainability dimensions. By combining unsupervised ML, through k-medoids clustering, with a TBLbased sustainability scoring approach, the model enables the segmentation of heterogeneous RE projects into homogeneous and interpretable groups. The results demonstrate that RE projects differ substantially in scale, reliability, and sustainability profiles, and that these differences have direct implications for incentive design. The empirical application to the UK RE portfolio highlights clear structural patterns across technologies and segments, notably the concentration of PV projects in low-reliability classes, the scale-dependent performance of wind technologies, and the stabilizing role of biomass, hydro, marine, and geothermal projects in high-reliability segments. Building on these findings, the study advances exploratory, segment-specific incentive orientations that reflect the structural heterogeneity of RE projects and their differentiated contributions to system performance and sustainability. Rather than prescribing fixed policy instruments, these orientations emphasize alignment between incentive types and dominant segment needs: environmental and social instruments appear more pertinent where scale and diffusion amplify externalities, while economic and stability-oriented mechanisms are more suitable for segments providing firm system stability value. This calibrated approach acknowledges that incentive effectiveness is inherently conditional on project scale, reliability, and sustainability profile and that uniform instruments risk diluting policy impact.

Overall, this study contributes to the advancement of evidence-based energy transition governance by operationalizing a data-driven and sustainability-informed approach to incentive design. In contrast to uniform or historically embedded policy instruments, the proposed policy decision-support tool demonstrates how empirical generation patterns and sustainability profiles can be systematically translated into differentiated and targeted policy incentive schemes across the broad panoply of instruments, such as FiTs, R&D support, and soft loans.

These considerations highlight several limitations of the present study. First, the proposed incentive orientations should be interpreted as analytical signposts intended to support adaptive and evidence-informed policy design, rather than as definitive or universally prescriptive solutions. The policy decision support tool is conceived as an exploratory, non-predictive decision-support tool whose applicability remains subject to further empirical validation. Second, the selection of the clustering features depends on methodological choices; the model deliberately adopts a simplified structure to preserve interpretability for policymakers, at the expense of capturing all system complexities. In addition, the results are contingent on the quality, granularity, and completeness of project-level data: in data-scarce contexts, key indices may require approximation, potentially affecting segmentation precision.

Future research should therefore address these limitations by expanding the dataset and integrating additional performance and grid-related indicators. In the UK context specifically, further work could tailor the model to existing incentive schemes and regulatory constraints. Moreover, involving UK policymakers in the calibration of the sustainability scoring phase, through the structured integration of their qualitative judgments, would enhance the contextual relevance, legitimacy, and operational applicability of the proposed incentive recommendations.

Acknowledgement. Bilal En-Nouaary would like to acknowledge the financial support provided by the CNRST under the "PhD-Associate Scholarship PASS" program, which has enabled the accomplishment of this work.

References

1. Wang, J., Chen, H., Cao, Y., Wang, C., Li, J.: An integrated optimization framework for regional energy planning with a sustainability assessment model. Sustain. Prod. Consumption **36**, 526–539 (2023). ISSN 2352–5509. https://doi.org/10.1016/j.spc.2022.08.032
2. Reinoso-Avecillas, F.Z., Moreno-Gamboa, F., Nieto-Londoño, C.:. Renewable energy technology diffusion model: evaluation of financial incentives in the electricity market of ecuador. Int. J. Thermofluids **24**, 100884, (2024). ISSN 2666–2027. https://doi.org/10.1016/j.ijft.2024.100884
3. Asdrubali, F., Baldinelli, G., D'Alessandro, F., Scrucca, F.: Life cycle assessment of electricity production from renewable energies: review and results harmonization. Renew. Sustain. Energy Rev. **42**, 1113–1122 (2015). ISSN 1364–0321. https://doi.org/10.1016/j.rser.2014.10.082
4. Kim, J., Mohommad, M.A.: Jobs Impact of Green Energy. Working Paper WP/22/101, May 2022
5. En-Nouaary, B., Hmioui, A.: Towards sustainability in incentive allocation for renewable energy projects: a hybrid AHP-ML methodology. In: 2025 International Conference on Artificial Intelligence, Computer, Data Sciences and Applications (ACDSA), pp. 1–6, August 2025
6. Bilardo, M.: A fair dynamic incentive allocation method for virtual energy sharing in renewable energy communities that rewards members' virtuosity and engagement. Renew. Energy **255**, 123756 (2025). ISSN 0960–1481. https://doi.org/10.1016/j.renene.2025.123756
7. Lilliu, F., Recupero, D.R., Vinyals, M., Denysiuk, R.: Incentive mechanisms for the secure integration of renewable energy in local communities: a game-theoretic approach. Sustain. Energy Grids Networks **36**, 101166 (2023). ISSN 2352–4677. https://doi.org/10.1016/j.segan.2023.101166
8. Kong, J., Oh, S., Kang, B.O., Jung, J.: Development of an incentive model for renewable energy resources using forecasting accuracy in South Korea. Energy Sci. Eng. 10, November 2021. https://doi.org/10.1002/ese3.1020
9. Casalicchio, V., Manzolini, G., Prina, M.G., Moser, D.: From investment optimization to fair benefit distribution in renewable energy community modelling. Appl. Energy **310**, 118447 (2022). ISSN 0306–2619. https://doi.org/10.1016/j.apenergy.2021.118447
10. Darmian, S.M., Tavana, M., Ribeiro-Navarrete, S.: An investment evaluation and incentive allocation model for public-private partnerships in renewable energy development projects. Socio-Econ. Plan. Sci. **95**, 101953 (2024). ISSN 0038–0121. https://doi.org/10.1016/j.seps.2024.101953
11. Piel, J.-H., Hamann, J.F.H., Koukal, A., Breitner, M.H.: Promoting the system integration of renewable energies: toward a decision support system for incentivizing spatially diversified deployment. J. Manage. Inf. Syst. **34**(4), 994–1022 (2017). ISSN 0742-1222. https://doi.org/10.1080/07421222.2017.1394044. Publisher: Routledge _eprint: https://doi.org/10.1080/07421222.2017.1394044

12. Gronau, Q.F., Wagenmakers, E.-J.: Limitations of bayesian leave-one-out cross-validation for model selection. Comput. Brain Behav. **2**(1), 1–11 (2019). ISSN 2522-087X. https://doi.org/10.1007/s42113-018-0011-7
13. Alzubaidi, L., et al.: Review of deep learning: concepts, CNN architectures, challenges, applications, future directions. J. Big Data **8**(1), 53 (2021). ISSN 2196–1115. https://doi.org/10.1186/s40537-021-00444-8
14. Scaramuzzino, C., Garegnani, G., Zambelli, P.: Integrated approach for the identification of spatial patterns related to renewable energy potential in European territories. Renew. Sustain. Energy Rev. **101**, 1–13 (2019). ISSN 13640321. https://doi.org/10.1016/j.rser.2018.10.024. arXiv:1911.01278 [cs]
15. Khan, I.K., et al.: Addressing limitations of the K-means clustering algorithm: outliers, non-spherical data, and optimal cluster selection. AIMS Math. **9**(9), 25070–25097 (2024). ISSN 2473–6988. https://doi.org/10.3934/math.20241222. Publisher: American Institute of Mathematical Sciences
16. He, Y., Xu, Z., Liu, N.: Research on k-medoids algorithm with probabilistic-based expressions and its applications. Appl. Intell. **52**(10), 12016–12033 (2022). ISSN 0924-669X. https://doi.org/10.1007/s10489-021-02937-8. Publisher: Springer
17. de Rubens, G.Z.: Who will buy electric vehicles after early adopters? Using machine learning to identify the electric vehicle mainstream market. Energy **172**, 243–254 (2019). ISSN 0360–5442. https://doi.org/10.1016/j.energy.2019.01.114
18. GOV.UK. Renewable Energy Planning Database: quarterly extract, November 2025. https://www.gov.uk/government
19. NESO. Neso (2026). https://www.neso.energy/search
20. Troldborg, M., Heslop, S., Hough, R.L.: Assessing the sustainability of renewable energy technologies using multi-criteria analysis: suitability of approach for national-scale assessments and associated uncertainties. Renew. Sustain. Energy Rev. **39**, 1173–1184 (2014). ISSN 1364–0321. https://doi.org/10.1016/j.rser.2014.07.160
21. Clark, C.E., Harto, C.B., Sullivan, J.L., Wang, M.Q.: Water use in the development and operation of geothermal power plants. Technical report, Argonne National Lab. (ANL), Argonne, IL (United States) (2010)
22. Beniaich, R., En-Nouaary, B., Hmioui, A.: A comprehensive overview of green supply chain management practices in renewable energy systems (2024)

An Expert-Based Approach for Validating Renewable Energy Green Practices Key Performance Indicators in Morocco

Rania Beniaich[1(✉)], Aziz Hmioui[1], Nadine Kafa[2], and Bouchra Laarabi[3]

[1] Ecole Nationale de Commerce et de Gestion de Fès, Sidi Mohamed Ben Abdellah University, Fes, Morocco
rania.beniaich@usmba.ac.ma
[2] CESIT, KEDGE Business School, Talence, France
[3] Ecole Normale Supérieure, Mohammed V University, Rabat, Morocco

Abstract. Renewable energy (RE) systems are essential for mitigating climate change and reducing dependence on fossil fuels, yet they also generate environmental challenges, including land use, water consumption, waste generation, and impacts on biodiversity. In Morocco, the energy transition is increasingly integrating green design practices such as floating photovoltaics, solar PV carports, and agrivoltaics. However, their performance remains largely unevaluated. This study proposes a context-specific set of Key Performance Indicators (KPIs) to assess green practices in the RE sector across four sustainability dimensions: technological, social, environmental, and economic. The KPIs were first identified from the literature and subsequently validated using the Majority Judgment method with Moroccan experts. Results highlight the central importance of economic viability, with Levelized Cost of Energy (LCOE) identified as very relevant, alongside water consumption, reflecting Morocco's persistent water stress. The validated KPI framework provides an operational basis for future decision support systems supporting sustainable energy transitions.

Keywords: Decision Support Systems · Sustainability Assessment · Green Practices · Key Performance Indicators · Majority Judgment Method · Renewable Energy

1 Introduction

Global energy consumption continues to increase, despite ongoing international efforts to hinder it [1]. This expansion in energy use creates serious environmental problems. Among them, climate change and water scarcity are considered the most urgent to handle. In this context, renewable energy (RE) sources, such as solar, wind, hydro, biomass, geothermal, tidal, and wave energy, are widely promoted as a cleaner and more sustainable alternatives to fossil fuels, which are responsible for 90 % of greenhouse gas (GHG) emissions in the energy sector [1].

A. P. Cabral Seixas Costa et al. (Eds.): ICDSST 2026, LNBIP 582, pp. 177–190, 2026.
https://doi.org/10.1007/978-3-032-25146-6_12

However, despite the environmental advantages of the RE technologies, they also present important challenges that could potentially counterbalance their primary reason of existence i.e. environment protection.

Like any industrial process, the manufacturing stage often results in intensive energy consumption leading to significant GHG emissions. Beyond this, large land requirements, visual and noise pollution, high water consumption, effects on biodiversity, and significant waste generation after the end of life (EOL) of their life cycle are the main environmental downsides which must be handled to ensure their sustainability [2]. Green practices constitute a key response to these challenges, as they aim to optimize resource utilisation, reduce pollution and waste generation, and preserve all the forms of life in our ecosystem, with human beings as a priority. For instance, wind turbine blades, considered difficult to recycle at the end of their lifecycle, can be reused as public seating, playgrounds, or bike sheds [3]. Furthermore, vertical-axis wind turbines (VAWTs) are being designed to address both biodiversity conservation and noise reduction concerns [4]. Moreover, agrivoltaic systems, which combine solar panel installations with agricultural activities, allow for dual land use, reducing land occupation [5]. While These green practices are essential for enabling a cleaner energy future, their implementation alone is insufficient without properly evaluating and monitoring their performance to ensure continuous improvement and real environmental benefits.

In Morocco, the energy transition has become a strategic national priority over the past two decades. The country has implemented an ambitious National Energy Strategy (NES) aimed at diversifying the energy mix through the expansion of renewable energy capacity and reducing reliance on imported fossil fuels. With targets of reaching 52% renewable capacity by 2030 and 70% by 2050, substantial investments have been directed toward the exploitation of Morocco's renewable energy potential [6,7]. Today, Morocco's renewable energy landscape is gradually shifting from a focus on increasing installed capacity to integrating green practices in the sector. For example, small-scale Floating Photovoltaic (FPV) systems have been installed on agricultural reservoirs in the Sidi Slimane and Ben Guerir regions, representing the first applications of their kind in North Africa [8]. More recently, Morocco has taken a further step with the inauguration of a large-scale floating solar power plant installed on the Oued Rmel dam reservoir near Tangier, with an installed capacity of 13 MW, highlighting the growing adoption of green design practices in the country [9]. However, while these initiatives are beginning to emerge in practice, the evaluation of such green practices remains largely unexplored.

In the existing literature, despite the growing interest in green practices within the renewable energy sector such as recycling and reuse strategies, as well as green design solutions including FPV, solar PV carports, agrivoltaics, and vertical axis wind turbines, their evaluation remains fragmented. Most existing studies tend to focus on partial assessments, addressing only one dimension at a time. For example, several works like Alzahrani et al. [10] emphasize techno-economic performance, while others concentrate exclusively on environmental impacts, mainly through life cycle assessment (LCA) approaches like Cromratie Clemons et al. [11], or on social impacts with a focus on job creation and social acceptance [12,13]. However, an integrated evaluation framework that simultaneously covers the social, technical, economic, and environmental dimensions of green practices in the renewable energy sector is still lacking. This gap is even more pronounced in the Moroccan context, where sustainability assessments have largely focused on conventional renewable energy projects [6,14], while the evaluation of emerging green practices remains largely unexplored.

To the best of our knowledge, no study has yet proposed a comprehensive set of KPIs specifically dedicated to assess the performance of green practices in the renewable energy sector, covering the four dimensions of sustainability and being adapted to the Moroccan context. This paper addresses this gap by proposing a set of relevant, context-specific KPIs to assess the performance of green practices in the renewable energy sector across four dimensions: social, technical, economic, and environmental. By providing validated and reproducible indicators, this work supports evidence-based energy transition by enabling policymakers, project developers, and industry stakeholders to objectively evaluate green practices and guide informed decision-making in the renewable energy sector.

2 Literature Review

The deployment of renewable energy systems is widely promoted as a key pathway for reducing dependence on fossil fuels and mitigating climate change. However, renewable energy technologies are not impact-free across their life cycle and may generate significant environmental challenges, including intensive resource use, land occupation, water consumption, waste generation, biodiversity disturbance, and noise pollution [15].

In response, the literature increasingly emphasizes the adoption of green practices aimed at improving sustainability performance. These include green design solutions such as FPV systems that reduce land footprint and water evaporation [16], agrivoltaics that enable the coexistence of energy production and agriculture [1], and vertical axis wind turbines designed to mitigate noise pollution and biodiversity impacts [4]. Closed-loop practices such as recycling and reuse strategies have also gained attention, including silicon recovery from end-of-life PV panels [3] and the reuse of wind turbine components for secondary applications such as public infrastructure [17], contributing to waste reduction and circular economy objectives.

While these practices are increasingly promoted, their implementation alone is not sufficient. Evaluating their performance through appropriate indicators is essential to assess their technological, social, environmental, and economic contributions. Nevertheless, existing evaluation approaches remains fragmented. Many studies focus on techno-economic performance, prioritizing indicators such as "levelized cost of energy (LCOE)", "investment cost", "O&M cost", "energy yield", and "energy efficiency" [18,19]. Others adopt an economic-environmental perspective when assessing closed loop practices, highlighting indicators such

Table 1. Key Performance Indicators (KPIs) for assessing green practices in the renewable energy sector

Dimension	Key Performance Indicator	Description
Technological	Energy Efficiency (%)	How effectively the technology converts natural resources into usable energy [18].
	Energy Yield (kWh/kWp/year)	The total amount of energy produced by the project over a specific period [19].
Social	Jobs created (jobs-year)	The number of new employment opportunities generated directly or indirectly through green practices [6].
	Temperature reduction of the area (°C)	Decrease in ambient temperature due to green technologies such as PV trees that provide shading and reduce heat accumulation [20].
	Visual or aesthetic impact	Public perception of the visual integration and attractiveness of green technologies within their environment [21].
Environmental	GHG emissions (gCO_2eq/kWh)	Greenhouse gas emissions generated during manufacturing, including energy use and material processing [11].
	Energy consumption (kWh/kW)	Total energy consumed during manufacturing, including material processing [22].
	Water consumption (L/kWh)	Volume of water used during manufacturing and material treatment [22].
	Land use (m^2/MW)	Land area required for installation and operation of RE projects [6].
	Water savings (m^3/year)	Reduction in evaporation losses achieved through floating photovoltaic systems [16].
	Impact on biodiversity (m^2 or ha)	Effects on local ecosystems and natural habitats [6].
	Noise level (dB)	Operational noise disturbances generated by RE installations [11].
	Recyclability of components (%)	Extent to which materials can be recovered and reused at end of life [23].
Economic	Investment cost (USD/kW)	Initial expenditures including eco-materials, equipment, installation, engineering services, grid connection, and auxiliary construction [18].
	O&M cost (USD/kW-year)	Operating and maintenance costs required to sustain reliable plant performance [18].
	LCOE (USD/kWh)	Unit cost of electricity over the system lifetime, accounting for eco-friendly investments [6].
	Investment recovery (USD/kW)	Financial gains from surplus resources, waste disposal, and recycled materials [24].
	Waste disposal cost (USD/ton)	Total expenses related to waste management, including recycling [25].

as waste management cost, recyclability fee, and recycling rate [23,25]. Environmental assessments are commonly conducted through Life Cycle Assessment, using indicators such as "greenhouse gas emissions", "water consumption", "land use", "noise level", and "biodiversity impacts" [11,22]. Social aspects are addressed less frequently, mainly through indicators related to public acceptance like "visual impact", "job creation", and "local temperature reduction"

[20,21]. Overall, these studies provide valuable but partial insights, reflecting the absence of an integrated framework capable of capturing the multidimensional performance of green practices.

In Morocco, the energy transition has recently begun to integrate emerging green practices alongside conventional renewable energy projects. However, their sustainability performance has not yet been evaluated. Existing Moroccan studies primarily focus on conventional projects, as illustrated by En-Nouaary et al. [6], who assess sustainability using economic (LCOE, investment and O&M costs), technological (production potential, maturity, variability), socio-political (job creation, political support), and environmental indicators (land requirement, GHG emissions, ecosystem impact). While this provides important insights into conventional renewable systems, the evaluation of emerging green practices remains largely unexplored. To date, no study has proposed a comprehensive KPI framework dedicated to assessing green practices and adapted to the Moroccan context. To address this limitation, KPIs were identified from the international literature and structured into four sustainability dimensions: technological, social, environmental, and economic, as presented in Table 1. These KPIs were subsequently subjected to an expert-based validation process to ensure their relevance and contextual adequacy.

3 Methodology

The validation of the Key Performance Indicators (KPIs) identified from the literature constitutes an essential step in ensuring their relevance, robustness, and applicability within the Moroccan renewable energy context. This process was designed as a two-stage expert evaluation, as illustrated in Fig. 1, which presents the overall methodology applied for KPI validation. In this study, the KPI validation focuses on the three dominant renewable energy technologies: solar photovoltaic, wind, and hydropower, which together account for more than 90% of global installed renewable energy capacity and represent the most widely deployed technologies in Morocco [15,26].

In the first stage, all KPIs identified from the literature were assessed using a structured questionnaire based on the Majority Judgment method. Based on the results of this first evaluation, KPIs assessed as relevant or very relevant were directly retained without modification in the final KPI set. Conversely, KPIs judged as having low relevance were excluded from the final list, as they were considered insufficiently representative or not adapted to the Moroccan context.

In the second stage, KPIs with medium relevance were re-evaluated through semi-structured individual interviews with the same experts (approximately 20–30 min each). This step aimed to clarify divergent judgments, address ambiguities, and guide final decisions on KPI retention, modification, or consolidation.

The following subsections describe the main methodological stages of the KPI validation process, including the questionnaire design, the expert selection, data collection and the Majority Judgment procedure.

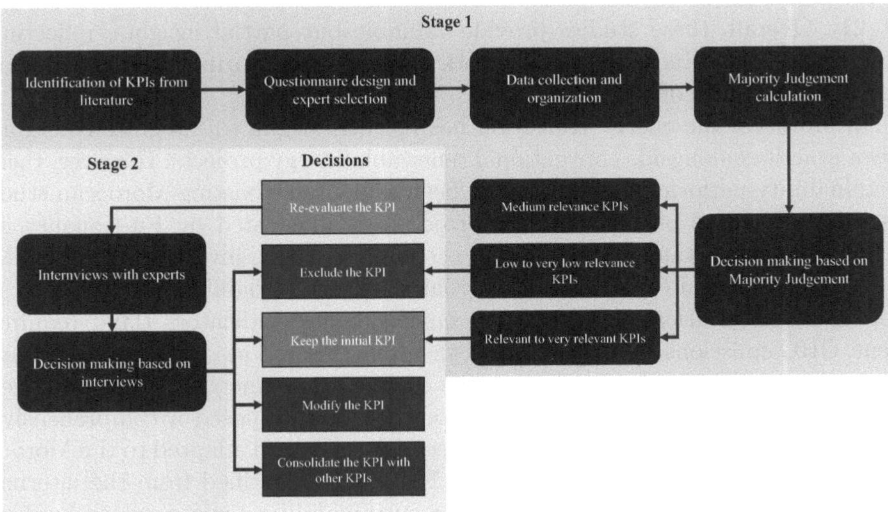

Fig. 1. The Methodology for KPIs Validation

3.1 Questionnaire Design and Expert Selection

The validation of the KPIs was conducted through an expert-based evaluation using a structured questionnaire. This approach was adopted to assess the relevance of each indicator within the Moroccan renewable energy context, where expert knowledge plays a key role in complementing and contextualizing findings from the literature. A panel of nine Moroccan experts was carefully selected for this purpose. All experts have more than five years of professional experience in the renewable energy sector, ensuring a sufficient level of technical expertise and practical knowledge. The panel includes specialists with experience ranging from project management, research and development to policy-making and industrial operations. As shown in Table 2, the experts come from a variety of organizational backgrounds, including public administrations, private consulting firms, and industrial companies active in renewable energy technologies such as solar, wind, and hydrogen. This diversity was intentionally sought in order to capture multiple perspectives and reduce potential institutional or sectoral bias in the evaluation process.

The questionnaire was designed and distributed individually to each expert. Each KPI was accompanied by a clear and concise definition to ensure a common understanding among experts and to minimize interpretation ambiguity. Experts were asked to assess the relevance of each KPI using an ordinal five-level scale: to reject, low relevance, medium relevance, relevant, and very relevant.

3.2 Data Collection and Majority Judgment Procedure

Following the distribution of the questionnaire, responses were collected from all nine experts, resulting in a complete and exploitable dataset. Each expert

Table 2. Profile of Moroccan experts involved in the KPI validation process

Expert	Position/Area of Expertise	Organization	Years of Experience
Expert 1	R&D Manager	IRESEN	12
Expert 2	RE Consultant	GOPA Tech GmbH	10
Expert 3	Director in RE and Energy Efficiency	Ministry of Energy	15
Expert 4	Senior Solar PV Project Manager	INNOVX	10
Expert 5	R&D Manager	Green Energy Park	5
Expert 6	Solar Energy and Energy Efficiency Specialist	Misen	12
Expert 7	Chief Executive Officer (CEO)	IOM Solutions	13
Expert 8	Solar Energy Specialist	MASEN	9
Expert 9	Solar, Wind, and Hydrogen Specialist	INNOVX	8

provided an independent evaluation of the relevance of the KPIs identified from the literature, using the predefined ordinal scale. The individual responses were compiled and organized in a structured data matrix, where rows represent KPIs and columns correspond to expert evaluations.

For each KPI, a frequency count was conducted for each relevance level to capture the distribution of expert judgments across the ordinal scale. This step enabled a transparent and systematic aggregation process, ensuring traceability from individual expert opinions to collective outcomes. The aggregation of expert judgments was conducted using the Majority Judgment (MJ) method, which consists of ordering the relevance ratings from the lowest to the highest level and determining the median value, which represents the majority judgment for the considered KPI.

The MJ method was originally proposed by Balinski and Laraki [27] as an alternative voting and evaluation procedure based on ordinal judgments rather than rankings. Since its introduction, the method has been applied in various decision-making and evaluation contexts, demonstrating strong robustness and resistance to extreme judgments Balinski and Laraki [27] The MJ procedure is particularly suitable for the present study, as it provides a reliable aggregation framework for qualitative evaluations involving heterogeneous expert profiles. Unlike mean-based aggregation approaches, MJ limits the influence of extreme values and minority opinions, thereby producing a collective judgment that more accurately reflects the central tendency of expert consensus. This characteristic is particularly relevant in sustainability assessments, where expert evaluations are often subjective and multidimensional.

4 Results and Discussion

The results of the KPI validation process are presented and discussed based on the majority judgment method. In line with the adopted two-stage framework, this section is organized into two subsections. The first reports the outcomes of the first stage of majority judgment evaluation, while the second presents the second stage required to re-evaluate exclusively KPIs with medium relevance.

4.1 The First Stage of the Validation Process

The first stage of the validation process relied on the Majority Judgment method to assess the relevance of the KPIs identified from the literature. While most KPIs were evaluated as relevant or very relevant by the expert panel, some KPIs were evaluated as having medium relevance, requiring further clarification through a second validation stage. However, *Temperature reduction of the area* was the only KPI judged as having low relevance. Figure 2 presents both the Majority Judgment outcomes and the count of expert ratings for each KPI, providing insight into the level of consensus and dispersion across expert evaluations.

Initial KPIs	No relevance	Low relevance	Medium relevance	Relevant	Very relevant	MJ Decision
Energy Efficiency	1	3	2	2	1	Medium relevance
Energy Yield	0	1	0	6	2	Relevant
Jobs Created	1	1	4	1	2	Medium relevance
Temperature reduction of the area	0	5	1	1	2	Low relevance
Visual or Aesthetic Impact	0	3	1	3	2	Relevant
GHG Emissions	0	0	3	3	3	Relevant
Water Consumption	0	1	0	2	6	Very relevant
Energy Consumption	0	1	2	3	3	Relevant
Land Use	0	1	3	5	0	Relevant
Water Savings	0	1	1	4	3	Relevant
Impact on Biodiversity	1	1	3	1	3	Medium relevance
Noise Level	0	2	3	3	1	Medium relevance
Recyclabiliy of Components	0	3	1	4	1	Relevant
Investment Cost	0	2	1	2	4	Relevant
O&M Cost	0	1	1	5	2	Relevant
Levelized Cost of Energy (LCOE)	0	1	0	3	5	Very relevant
Investment Recovery	0	1	1	3	4	Relevant
Waste Disposal Cost	1	3	3	0	2	Medium relevance

Fig. 2. Majority Judgment Results

In the technological dimension, *energy yield* received a majority judgment of relevant, while *energy efficiency* was assessed as having medium relevance. A strong consensus was observed for *energy yield*, highlighting its recognition as a robust and operational indicator of technological performance. While, *energy efficiency* did not achieve consensus, with expert ratings dispersed across medium, relevant, and low relevance, suggesting uncertainty regarding its added value and potential overlap with *energy yield*, as improvements in efficiency typically translate into increased energy production.

Within the social dimension, *jobs creation* received a majority judgment of medium relevance, while *visual impact* was assessed as relevant and *reduction of local ambient temperature* was judged as having low relevance. No consensus was observed for *visual impact*, with high dispersion in expert ratings, reflecting the subjective nature of social acceptance. In contrast, a moderate consensus emerged for *reduction of local ambient temperature*, with five out of nine experts rating it as having low relevance. This result is explained by the fact that this indicator is mainly associated with a specific green PV design solution, namely PV trees, which aim to provide shade and reduce ambient temperature [21]. However, in the Moroccan context, this technology remains limited and spatially dispersed, preventing measurable temperature effects at the zone level and reducing the practical relevance of this KPI.

Within the environmental dimension, *greenhouse gas emissions*, *energy consumption*, *land footprint*, *water savings*, and *recyclability of components* received majority judgments of relevant, while water consumption was rated as very relevant. Whereas *impact on biodiversity* and *noise level* were assessed as having medium relevance. A strong consensus was observed for water-related KPIs, namely *water consumption* and *water savings*, reflecting their perceived importance in evaluating green practices such as eco-friendly material use and FPV systems. This consensus can be explained by Morocco's persistent water stress and recurrent drought conditions, which have intensified pressure on water resources and made water efficiency a central concern in national planning [6]. In this context, experts appear particularly sensitive to KPIs capturing water efficiency and savings.

Finally, within the economic dimension, *investment cost*, *operation and maintenance cost*, and *investment recovery* received majority judgments of relevant, while *Levelized Cost of Energy (LCOE)* was rated as very relevant. In contrast, *waste disposal cost* was assessed as having medium relevance. A moderate consensus was observed for *investment cost*, *O&M cost*, and *investment recovery*, with six to seven out of the nine experts distributing their votes between relevant and very relevant, confirming their importance in economic performance assessment. A strong consensus emerged for *LCOE*, highlighting its central role in evaluating economic viability. Conversely, *waste disposal cost* did not achieve consensus and exhibited high dispersion in expert ratings, reflecting divergent views regarding its definition and scope and justifying further clarification in the second validation stage.

Based on these outcomes, the KPIs judged as relevant or very relevant were retained without modification, according to the adopted validation methodology. These include *energy yield*, *Visual or Aesthetic Impact*, *GHG Emissions*, *Water Consumption*, *Energy Consumption*, *Land Footprint*, *Water Savings*, *Recyclability of Components*, *Investment Cost*, *O&M Cost*, *LCOE*, and *Investment Recovery* (see Fig. 2). However, KPIs assessed as having low relevance were directly excluded from the initial list; accordingly, *reduction of local ambient temperature* was removed. Finally, KPIs that received a medium relevance judgment were subjected to a second validation stage to clarify divergent expert views and determine their retention, modification, or consolidation.

4.2 The Second Stage of the Validation Process

The second stage of the validation process was conducted exclusively for the KPIs that were assessed as having medium relevance. This stage consisted of targeted discussions with the same panel of experts, with the objective of clarifying divergent judgments and reaching final decisions regarding KPI refinement.

Within the technological dimension, experts agreed to remove *energy efficiency*, as it was considered largely overlapping with the KPI *energy yield*, which was already evaluated as relevant. In practice, higher *energy efficiency* typically leads to higher *energy yield*, meaning that both KPIs convey similar information. Experts therefore considered *energy yield* to be sufficiently representative on its own, providing a more direct and operational measure of energy performance, while *energy efficiency* did not add significant decision-making value.

In the social dimension, *jobs creation* was retained without modification, as it was considered the most representative indicator of the social dimension. Experts noted that this indicator is particularly relevant for large-scale RE projects, while its relevance is more limited for small-scale installations.

Regarding the environmental dimension, experts agreed to retain both *impact on biodiversity* and *noise level*, recognizing their essential role in assessing environmental impacts and evaluating green practices aimed at mitigating biodiversity loss and noise pollution. To facilitate practical assessment and improve consistency, it was suggested that these KPIs could be expressed in terms of affected surface area, allowing a spatial representation of environmental disturbances. In addition, while *recyclability of components* was confirmed as a relevant KPI, experts recommended complementing it with an environmental indicator such as *waste volume* in order to quantify the amount of scrap material and enable a comparison of waste levels before and after recycling.

Finally, within the economic dimension, discussions concerning the KPI *waste disposal cost* emphasized the need for clearer and more precise terminology when evaluating recycling as a green practice. Experts agreed that referring explicitly to *recycling cost* is more appropriate than using the broader notion of *waste disposal cost*.

Based on the two-stage validation process, a final set of KPIs was established following the retention, exclusion, and refinement decisions agreed upon with the expert panel. Figure 3 summarizes the KPI validation framework, showing the process from the initial set of KPIs identified from the literature to the final validated KPIs.

These findings are consistent with the results reported by En-Nouaary et al. [6], who demonstrated that Moroccan experts prioritize techno-economic and environmental performance indicators over social metrics in the evaluation of renewable energy projects. In particular, their study identified the Levelized Cost of Energy (LCOE) and ecosystem-related impacts (explicitly including water consumption) as the most influential criteria in project prioritization. This aligns closely with the present validation outcomes, where LCOE and water consumption were the only KPIs rated as very relevant, confirming their central role in sustainability-oriented decision-making. Conversely, En-Nouaary et al.

Dimension	Initial KPIs	First stage results	Second stage results	Final KPIs
Technological	Energy Efficiency	To re-evaluate	To remove due to overlap with energy yield	
	Energy Yield	To keep		Energy Yield (kWh/kWp/year)
Social	Jobs Created	To re-evaluate	To keep without modification as the most representative social KPI	Social Jobs created
	Temperature reduction of the area	To exclude		
	Visual or Aesthetic Impact	To keep		Visual or Aesthetic Impact
Environmental	GHG Emissions	To keep		Environmental GHG emissions (gCO2eq/kWh)
	Water Consumption	To keep		Water consumption (L/kWh)
	Energy Consumption	To keep		Energy Consumption (kWh/kW)
	Land Use	To keep		Land Use (m2/MW)
	Water Savings	To keep		Water savings (m3/year)
	Impact on Biodiversity	To re-evaluate	To keep, recommended to be operationalized through spatial metrics	Impact on Biodiversity (ha)
	Noise Level	To re-evaluate		Noise Level (dB)
	Recyclabiliy of Components	To keep	To keep, complemented with *waste volume*	Recyclability of components (%)
				Waste Volume (m3)
Economic	Investment Cost	To keep		Economic Investment Cost (USD/kW)
	O&M Cost	To keep		O&M Cost (USD/kW-year)
	Levelized Cost of Energy (LCOE)	To keep		LCOE (Levelized Cost of Energy) (USD/kWh)
	Investment Recovery	To keep		Investment Recovery (USD/kW)
	Waste Disposal Cost	To re-evaluate	To modify, reformulated as *recycling cost*	Recycling Cost (USD/ton)

Fig. 3. Validation Process from initial to final KPIs

[6] reported that job creation ranked among the 50% least prioritized indicators, which is coherent with the medium relevance and high dispersion observed for this KPI in the current study.

The strong relevance attributed to water-related KPIs can be explained by Morocco's persistent water stress and recurrent drought conditions. The country has experienced more than a decade of increasing rainfall variability and prolonged drought periods driven by climate change. These constraints have intensified pressure on water resources and have made water efficiency and water savings central concerns in national planning and project evaluation, particularly for energy systems that interact with water resources En-Nouaary et al. [6]. In this context, experts appear particularly sensitive to KPIs capturing water efficiency and water savings, especially when assessing green PV design practices such as floating photovoltaic systems and the use of eco-friendly materials.

Conversely, the low relevance attributed to the KPI "reduction of local ambient temperature" indicates that the usefulness of sustainability indicators strongly depends on the maturity of the technology and its actual level of deployment in the local context. In Morocco, PV trees aiming to reduce ambient temperature remain limited, which reduces the practical relevance of this indicator.

5 Conclusion

Renewable energy systems are essential to mitigate climate change mitigation and to reduce dependence on fossil fuels. However, they also present important challenges that could potentially counterbalance their primary reason of existence i.e. environment protection. Green practices, including eco-design and closed-loop practices, constitute a key response to these challenges.

In Morocco, despite the growing integration of green initiatives such as floating photovoltaics, agrivoltaics, and recycling strategies, their performance evaluation remains largely unexplored. This study addressed this gap by proposing and validating a context-specific set of Key Performance Indicators to assess the performance of green practices in the Moroccan renewable energy sector across four sustainability dimensions : technological, social, environmental, and economic.

A two-stage validation framework based on the Majority Judgment method was applied with Moroccan experts, leading to a final KPI framework through indicator retention, refinement, and exclusion. Results highlight the central importance of economic viability, with *Levelized Cost of Energy (LCOE)*identified as very relevant, alongside *water consumption*, reflecting Morocco's persistent water stress and the growing importance of resource efficiency in national energy planning. Environmental indicators including *greenhouse gas emissions, energy consumption, land footprint, water savings, recyclability of components, and waste volume* were assessed as relevant, confirming their importance for evaluating environmental performance. In contrast, social indicators such as *jobs creation* and *visual impact* exhibited higher dispersion, reflecting their dependence on project scale and the subjective nature of social acceptance. Several KPIs were refined for clarity such as *waste disposal cost*, while others such as *Temperature reduction of the area* were excluded due to limited applicability in the Moroccan context.

Beyond methodological contributions, this work provides an operational foundation for evaluating green practices throughout the renewable energy life cycle. By translating green practices into measurable and context-adapted sustainability indicators, the proposed KPI framework supports evidence-based energy transition by enabling policymakers, project developers, and industry stakeholders to assess performance, prioritize interventions, and guide planning and investment decisions. Future research will focus on integrating the validated KPIs into decision-support tools and applying them to real renewable energy project case studies in Morocco, thereby strengthening practical implementation and supporting data-driven strategies for sustainable energy transitions.

Acknowledgments. The authors sincerely thank all experts who contributed to this study through their valuable time and insightful evaluations. The authors also express their gratitude to Mr. El Ydrissi Massaab from INNOVX company for his support and assistance during the research process. The first author gratefully acknowledges the financial support provided by the CNRST under the PhD-Associate Scholarship PASS program, which has enabled us to pursue our research.

References

1. Ritchie, H., Roser, M.: CO_2 emissions. Our World Data (2024). https://ourworldindata.org/co2-emissions
2. Olabi, A.G., et al.: Renewable energy systems: comparisons, challenges and barriers, sustainability indicators, and the contribution to UN sustainable development goals. Int. J. Thermofluids **20**, 100498 (2023). ISSN 2666-2027. https://doi.org/10.1016/j.ijft.2023.100498
3. Wang, Y., Li, A., Zhang, S., Guo, B., Niu, D.: A review on new methods of recycling waste carbon fiber and its application in construction and industry. Constr. Build. Mater. **367**, 130301 (2023). ISSN 0950–0618. https://doi.org/10.1016/j.conbuildmat.2023.130301
4. Gemayel, D., Abdelwahab, M., Ghazal, T., Aboshosha, H.: Modelling of vertical axis wind turbine using large eddy simulations. Results Eng. **18**, 101226 (2023). ISSN 2590–1230. https://doi.org/10.1016/j.rineng.2023.101226
5. Tölgyesi, C., et al.: Ecovoltaics: framework and future research directions to reconcile land-based solar power development with ecosystem conservation. Biol. Conserv. **285**, 110242 (2023). ISSN 0006–3207. https://doi.org/10.1016/j.biocon.2023.110242
6. En-Nouaary, B., Beniaich, R., Hmioui, A.: Prioritizing sustainability of renewable energy projects in Morocco. Int. J. Energy Sector Manage. **19**(3), 612–631 (2024). ISSN 1750–6220. https://doi.org/10.1108/IJESM-06-2024-0014
7. Masen, September 2025. https://www.masen.ma/fr
8. Energy handle (2026). https://energyhandle.com/en/project/first-floating-solar-pv-plant-in-north-africa-2/
9. Allali, R.: Tanger med launches a floating solar park, July 2024
10. Alzahrani, K.M., Hee, J.L., Elsakka, M., Ingham, D., Ma, L., Pourkashanian, M.: Techno-economic assessment of vertical axis wind turbine driven RO desalination with compressed air energy storage for remote communities. Desalination **592**, 118094 (2024). ISSN 0011–9164. https://doi.org/10.1016/j.desal.2024.118094
11. Clemons, S.K.C., Salloum, C.R., Herdegen, K.G., Kamens, R.M., Gheewala, S.H.: Life cycle assessment of a floating photovoltaic system and feasibility for application in Thailand. Renew. Energy **168**, 448–462 (2021). ISSN 0960–1481. https://doi.org/10.1016/j.renene.2020.12.082
12. Ellis, G., Schneider, N., Wüstenhagen, R.: Dynamics of social acceptance of renewable energy: an introduction to the concept. Energy Policy **181**, 113706 (2023). ISSN 0301-4215. https://doi.org/10.1016/j.enpol.2023.113706
13. Ma, B., Wang, A.: Exploring the role of renewable energy in green job creation and sustainable economic development: an empirical approach. Energy Strategy Rev. **58**, 101642 (2025). ISSN 2211-467X. https://doi.org/10.1016/j.esr.2025.101642
14. Corona, B., Escudero, L., Quéméré, G., Luque-Heredia, I., Miguel, G.S.: Energy and environmental life cycle assessment of a high concentration photovoltaic power plant in Morocco. Int. J. Life Cycle Assessm. **22**(3), 364–373 (2017). ISSN 1614-7502. https://doi.org/10.1007/s11367-016-1157-y
15. Beniaich, R., En-Nouaary, B., Hmioui, A.: A comprehensive overview of green supply chain management practices in renewable energy systems. In: El Bhiri, B., Hussain, A., Maleh, Y. (eds.) AI and Digital Transformation: Innovations in Supply Chain, Education, and Energy Systems, pp. 111–117. Springer Nature Switzerland, Cham (2025). ISBN 978-3-031-86837-5. https://doi.org/10.1007/978-3-031-86837-5_14

16. Ghosh, A.: A comprehensive review of water based PV: Flotavoltaics, under water, offshore & canal top. Ocean Eng. **281**, 115044 (2023). ISSN 0029-8018. https://doi.org/10.1016/j.oceaneng.2023.115044
17. Riahi, S., Mckenzie, J.A., Sandhu, S., Majewski, P.: Towards net zero emissions, recovered silicon from recycling PV waste panels for silicon carbide crystal production. Sustain. Mater. Technol. **36**, e00646 (2023). ISSN 2214-9937. https://doi.org/10.1016/j.susmat.2023.e00646
18. Niccolai, A., et al.: A review of floating PV systems with a techno-economic analysis. IEEE J. Photovoltaics **14**(1), 23–34 (2024). ISSN 2156-3403. https://doi.org/10.1109/JPHOTOV.2023.3319601
19. Micheli, L., Talavera, D.L., Tina, G.M., Almonacid, F., Fernandez, E.F.: Techno-economic potential and perspectives of floating photovoltaics in Europe. Solar Energy **243**, 203–214 (2022). ISSN 0038-092X. https://doi.org/10.1016/j.solener.2022.07.042
20. Florio, P., Peronato, G., Perera, A.T.D., Di Blasi, A., Poon, K.H., Kämpf, J.H.: Designing and assessing solar energy neighborhoods from visual impact. Sustain. Cities Soc. **71**, 102959 (2021). ISSN 2210-6707. https://doi.org/10.1016/j.scs.2021.102959
21. Almadhhachi, M., Seres, I., Farkas, I.: Sunflower solar tree vs. flat PV module: a comprehensive analysis of performance, efficiency, and land savings in urban solar integration. Results Eng. **21**, 101742 (2024). ISSN 2590-1230. https://doi.org/10.1016/j.rineng.2023.101742
22. Bošnjaković, M., Santa, R., Božič, J.T., Muhič, S.: The future of vertical-axis wind turbines: opportunities, challenges, and sustainability perspectives. Energies **18**(23), (2025). ISSN 1996–1073. https://doi.org/10.3390/en18236369. Company: Multidisciplinary Digital Publishing Institute Distributor: Multidisciplinary Digital Publishing Institute Institution: Multidisciplinary Digital Publishing Institute Label: Multidisciplinary Digital Publishing Institute Publisher: publisher
23. Maghraby, Y.R., Ibrahim, A.H., Tayel, A., Azzazy, H.M.E.S., Shoeib, T.: Towards sustainability via recycling solar photovoltaic panels, a review. Solar Energy **285**, 113085 (2025). ISSN 0038-092X. https://doi.org/10.1016/j.solener.2024.113085
24. Sarkis, J., Dou, Y.: Green Supply Chain Management: A Concise Introduction. October 2017. ISBN 978-1-315-23300-0. https://doi.org/10.4324/9781315233000
25. Li, J., Shao, J., Yao, X., Li, J.: Life cycle analysis of the economic costs and environmental benefits of photovoltaic module waste recycling in China. Resourc. Conserv. Recycl. **196**, 107027 (2023). ISSN 0921–3449. https://doi.org/10.1016/j.resconrec.2023.107027
26. IRENA. Renewable capacity statistics 2025, March 2025. https://www.irena.org/Publications/2025/Mar/Renewable-capacity-statistics-2025
27. Balinski, M., Laraki, R.: Election by majority judgment: experimental evidence. In: Dolez, B., Grofman, B., Laurent, A. (eds.) In Situ and Laboratory Experiments on Electoral Law Reform: French Presidential Elections, pp. 13–54. Springer, New York, NY (2011). ISBN 978-1-4419-7539-3. https://doi.org/10.1007/978-1-4419-7539-3_2

Decision Support Systems and Multicriteria Decision Making

Normalization Tools Effect on Combined Compromise for Ideal Solution (CoCoFISo) Method

Morteza Yazdani[1(✉)], Rôlin Gabriel Rasoanaivo[2,3], and Pascale Zarate[2]

[1] SCODeM Research Group, Universidad Internacional de Valencia – VIU, Valencia, Spain
morteza.yazdani@professor.universidadviu.com
[2] IRIT, Université Toulouse Capitole, Toulouse, France
{rolin-gabriel.rasoanaivo,Pascale.Zarate}@ut-capitole.fr
[3] Ecole Normale Supérieure (ENS), Université de Toamasina, Toamasina, Madagascar

Abstract. Each multi-criteria decision-making method (MCDM) method presents a different equation for normalising the performance matrix. However, the normalisation technique has an impact on the final ranking of alternatives and this has been researched by several studies. We demonstrated the effect of the normalisation tools on the Combined Compromise For Ideal Solution (CoCoFISo) method. Five different normalisation tools were applied and the results showed differences in the rankings compared to the original CoCoFISo. Nevertheless, there are alternatives that have consistent rankings in some tools. Also, the Kendall rank correlation coefficient of the CoCoFISo method compared to the five tools had a maximum rate of 0.62. It was noted that all five tools include two ways of normalising performance matrices, including those for maximising and minimising criteria. CoCoFISo, on the other hand, has only one matrix normalisation technique. This study could pave the way for the addition of another normalisation technique to the CoCoFISo method in order to consider the different directions of the criteria. It also advises MCDM researchers to accurately analyse normalisation tools before choosing one.

Keywords: Multiple criteria decision making · combined compromise for ideal solution (CoCoFISo) · normalization · ranking

1 Introduction

Multiple criteria decision-making methods (MCDM) contain several components like main matrix, weighting process, aggregation process and final ranking index. One of the essential components of MCDM methods is normalization process. An issue in MCDM problems is to deal with independency of decision variables (criteria). Each criterion is representative of a different unit like degree, quality, space etc. and aggregation of such measures in a unified matrix must be determined to make a dimensionless matrix which is equal for all [1, 2].

The vital contribution of decision-making tools in sustainability has become widespread during years. Sustainability in supply chain, water resource management,

A. P. Cabral Seixas Costa et al. (Eds.): ICDSST 2026, LNBIP 582, pp. 193–205, 2026.
https://doi.org/10.1007/978-3-032-25146-6_13

urban development, construction and energy sector, etc. delivered positive consequences and still research projects are developed to reach effective global sustainable objectives. For instance, [3] proposed a combined decision-making approach for Selection of the final urban drainage plan using sustainable and resilient variables. The authors used AHP, TOPSIS and Entropy methods. In another work, [4] addressed MABAC (Multi Attribute Border Approximation area Comparison) multi criteria decision making method method based on Trapezoidal Neutrosophic set to rank resilient supplier performance.

Normalization is introduced as a scaling process to deliver the criteria comparable by eliminating the optimization orientation (benefit, cost), homogenizing and unifying the measurement, and the variation range [5, 6]. Through normalization rule, the data within the decision matrix is transformed to a particular format or standard scale. The role of normalization techniques in MCDM algorithms is to ease decision process, reduce negative effect of each variable and generate the same normalized value for different units of measure. In this way, if for instance we have a decision variable with measure of Kg or Meter or Liter, the normalization process will act as an equilibrium function to generate same concept and eliminate incompatibility [7]

The role of normalization tool is fundamental to shape a homogenized and unified decision matrix. The normalization tool allows decision makers to promote a dimensionless, and unique matrix which suitable for compensatory processing MCDM problems. Normalization tools can develop new version of MCDM as many literatures reflects evidences. Experts choose the normalization technique while they need a special algorithm or follow a concrete objective. Normally it depends on the MCDM structure and preposition [8]. There are plenty of tools that configure a normalized matrix and we try to test them in this study by a specific decision method.

2 Literature Review

At the organizational levels, decision making problems have become complicated by an increasing level of uncertainty and complexity which force policy makers to look for new strategies or disciplines that facilitate the decision processes and deliver a reliable solution. Regarding all the recent technological and social challenges around organization and industries, recent decision-support systems have played a constant and inevitable role in analyses and the identification of critical conditions. MCDM techniques are appreciated tools for configuration of an evaluation system that is build based on the expert's experience and knowledge [9]. The rapid progress of MCDM tools in the last ten years was a milestone in operations research and decision support system studies. Beside the development of several tools like MABAC (Multi-Attributive Border Approximation area Comparison), MARCOS (Measurement of alternatives and ranking according to COmpromise solution), EDAS (Evaluation based on Distance from Average Solution) and CODAS (combinative distance-based assessment), a novel technique that attracted extra attention is called combined compromised solution which presents a mixed approach for alternative ranking system [10, 11].

The idea behind combined compromised solution (CoCoSo) method is attractive for many experts because of delivering an optimal solution easier under expert agreement and consensus [12]. Decision makers who use CoCoSo can rely on solving uncertain decision-making problems effectively. Moreover, the ranking results provided by

CoCoSo are robust in sensitivity analysis and consequently, alternative ranking alters slightly [13]. CoCoSo reflects an innovative MCDM strategy that has priority over methods such as Technique for Order of Preference by Similarity to the Ideal Solution (TOPSIS) [14].

CoCoSo method developed by [15] is a reliable method for the calculation of the optimal consensus score using an integrated framework. Its unique structure allows to construct a more robust model and reach more accurate decisions. Various research studies dedicated to the CoCoSo extension and improvement, while, this study examines the effect of different normalization tools on the improved version of the CoCoSo called Combined Compromise For Ideal Solution (CoCoFISo) developed by [16]. Recently CoCoFISo method is used in literature such as in the risk assessment [17], the analysis of resource availability [18], and the optimization of wire electrical discharge machining [19].

Actually, there are studies extended and developed fuzzy version of CoCoFISo method [20]. integrated the CoCoFISo method with grey theory as CoCoFISo-G. The research explored the green supplier selection problem within the cosmetics industry where the perfume bottles are supplied. Authors considered environmental and social sustainability factors and offered CoCoFISo-G method to undertake uncertainty in expert evaluations. The results showed that it enables a more robust and flexible framework for sustainable supplier evaluation. This extension significantly improves the applicability of the method in real-world, ambiguous decision environments like those found in the green supply chain management of the cosmetics sector [20].

The PF-CoCoFISo (Picture Fuzzy Combined Compromise for Ideal Solution) method represents an innovative extension of CoCoFISo model by integrating picture fuzzy sets (PFSs) to handle a different type of uncertainty. The method evaluated various digital archiving strategies across conflicting criteria such as technological adaptability, preservation standards, accessibility, cultural authenticity, data security, and sustainability. In summary, PF-CoCoFISo extends CoCoFISo to enable experts handling higher degrees of uncertainty and hesitation in expert evaluations, making it particularly effective for particular circumstances [21].

In terms of developing normalization tools on different MCDM tools, [22] conducted a special study and explored the impact of 5 widely used normalization methods on the best alternative determination accuracy and SAW method. The Monte Carlo simulation procedure was designed both optimization directions [8]. applied MCRAT multiple criteria ranking by alternative trace method with implementation of several normalization tools and the results were implicated through case studies [23]. in addition, used five normalization methods combining with the MARCOS method through various scenarios. Finally, author suggested two more normalization tools for MARCOS method [24]. proposed new normalization tool for TOPSIS method. Authors showed that the nature of the transformed normalized data will reflect the similarity of each alternative with the reference solution becoming thus, the decision matrix of homogeneous nature. Rather than notable research works in applying CoCoFISo method, testing normalization tools on that method still was a gap found in the MCDM domain. The advantages and contributions of the proposed decision-making approach are as follows:

- It delivers a new perspective on CoCoFISo and its capacity in handling normalization methods;
- The study promotes a new way for future researchers who are willing to explore such type of sensitivity analysis.

3 Mathematical Tools and Formulas

First, we present the CoCoFISo method [16] as its algorithm is seen in steps below:

Step 1. Formulating the performance matrix composed by the alternatives and the various criteria as shown in Formula 1:

$$x_{\mathrm{ij}} = \begin{bmatrix} x_{11} & x_{12} & \ldots & x_{1n} \\ x_{21} & x_{22} & \ldots & x_{2n} \\ \ldots & \ldots & \ldots & \ldots \\ x_{m1} & x_{m2} & \ldots & x_{\mathrm{mn}} \end{bmatrix}; i = 1, 2 \ldots, m; j = 1, 2 \ldots, n. \tag{1}$$

Step 2. Normalization process in original method is derived by the formulas below:

$$r_{ij} = \frac{x_{ij}}{\sqrt{\sum_{i=1}^{m} \left(x_{ij}\right)^2}} \tag{2}$$

Step 3. Adding the weights of decision criteria through two vectors as seen in Si and Pi formulas:

$$S_i = \sum_{j=1}^{n} \left(w_j r_{ij}\right). \tag{3}$$

$$P_i = \sum_{j=1}^{n} \left(r_{ij}\right)^{w_j} \tag{4}$$

Step 4. Obtaining three aggregation strategies k_{ia}, k_{ib} and k_{ic} according to formulas 5, 6 and 7.

$$k_{ia} = \frac{P_i + S_i}{\sum_{i=1}^{m} (P_i + S_i)} \tag{5}$$

$$k_{ib} = \left(\frac{S_i + P_i}{1 + \frac{S_i}{1+S_i} + \frac{P_i}{1+P_i}}\right) \tag{6}$$

$$k_{ic} = \frac{\lambda(S_i) + (1-\lambda)(P_i)}{\left(\lambda \max_i S_i + (1-\lambda) \max_i P_i\right)}; 0 \leq \lambda \leq 1. \tag{7}$$

Step 5. Generating the ranking of the alternatives according to the value of k_i obtained by formula 8.

$$k_i = (k_{ia} k_{ib} k_{ic})^{\frac{1}{3}} + \frac{1}{3}(k_{ia} + k_{ib} + k_{ic}) \tag{8}$$

The purpose of this work is to apply several normalization methods and observe the behavior and reaction of CoCoFISo algorithm. It presents kind of sensitivity analysis.

In the initial stage of applying the Improved CoCoSo method, we suppose that weights are already given (Table 1). Then we use below normalization tools as, linear sum-based, linear type I, Markovic, linear type II and linear max-min. In each norm we determine formulas for benefit and non-benefit variables (criteria). It should be noticed that we call benefit values when a higher value of the criterion is preferable and in other side, a cost or nonbeneficial criteria is that one a lower value of the criterion is preferable.

Norm 1: LINEAR TYPE I

$$N_{ij} = \frac{x_{ij}}{\text{Max } x_{ij}} \tag{9}$$

$$N_{ij} = 1 - \frac{x_{ij}}{\text{Max } x_{ij}} \tag{10}$$

Norm 2: LINEAR MAX-MIN

$$N_{ij} = \frac{x_{ij} - \text{Min } x_{ij}}{\text{Max } x_{ij} - \text{Min } x_{ij}} \tag{11}$$

$$N_{ij} = \frac{\text{Max } x_{ij} - x_{ij}}{\text{Max } x_{ij} - \text{Min } x_{ij}} \tag{12}$$

Norm 3: LINEAR SUM-BASED

$$N_{ij} = \frac{x_{ij}}{\sum_{i=1}^{m} x_{ij}} \tag{13}$$

$$N_{ij} = \frac{1/x_{ij}}{\sum_{i=1}^{m} 1/x_{ij}} \tag{14}$$

Norm 4: MARKOVIC

$$N_{ij} = \frac{x_{ij}}{\text{Max } x_{ij}} \tag{15}$$

$$N_{ij} = 1 - \frac{x_{ij} - \text{Min } x_{ij}}{\text{Max } x_{ij}} \tag{16}$$

Norm 5: LINEAR TYPE II

$$N_{ij} = \frac{x_{ij}}{\text{Max } x_{ij}} \tag{17}$$

$$N_{ij} = \frac{\text{Min } x_{ij}}{x_{ij}} \tag{18}$$

Table 1. the initial decision matrix

Alternatives	Criteria				
	C_1	C_2	C_3	C_4	C_5
Weights	0.25	0.3	0.1	0.15	0.2
Optimal value	max	Min	max	max	max
A_1	60	0.4	2 540	500	990
A_2	6.35	0.15	1016	3000	1041
A_3	6.8	0.1	1727.2	1500	1676
A_4	10	0.2	1000	2000	965
A_5	2.5	0.1	560	500	915
A_6	4.5	0.08	1016	350	508
A_7	3	0.1	1778	1000	920

Table 2 CoCoFISo method normalized matrix

	C_1	C_2	C_3	C_4	C_5
$A1$	0,97	0,79	0,64	0,12	0,36
A_2	0,10	0,29	0,25	0,73	0,37
A_3	0,11	0,20	0,43	0,37	0,60
A_4	0,16	0,39	0,25	0,49	0,35
A_5	0,04	0,20	0,14	0,12	0,33
A_6	0,07	0,16	0,25	0,09	0,18
A_7	0,05	0,20	0,44	0,24	0,33

4 Results and Discussion

In this section we explore the application of an already decision problem solved in [15] section 3a with a modification on weights. We have seven alternatives and 5 criteria. The weights are also listed in Table 1.

Below, we present the various normalized matrices. We apply the CoCoFISo method, followed by the implementation of various normalization tools. Table 2 presents the matrix normalized according to the CoCoFISo method technique.

Now Tables 3, 4, 5, 6 and 7 each present the normalized matrices by the five different tools.

Taking each of these normalized matrices into account, we applied Eqs. 3, 4, 5, 6 and 7 of the CoCoFISo method. The ranking results are detailed in Table 8.

From this result, we observed that the rankings obtained are different. This means that each normalization tool has its own specific characteristics and affects the final ranking of the alternatives in the case of the CoCoFISo method. In order to determine

Table 3. Normalization by LINEAR TYPE I (Norm 1)

	C_1	C_2	C_3	C_4	C_5
A_1	1,00	0	1,00	0,17	0,59
A_2	0,11	0,63	0,40	1,00	0,62
A_3	0,11	0,75	0,68	0,50	1,00
A_4	0,17	0,50	0,39	0,67	0,58
A_5	0,04	0,75	0,22	0,17	0,55
A_6	0,08	0,80	0,40	0,12	0,30
A_7	0,05	0,75	0,70	0,33	0,55

Table 4. Normalization by LINEAR MAX-MIN (Norm 2)

	C_1	C_2	C_3	C_4	C_5
A_1	1,00	0	1,00	0,06	0,41
A_2	0,07	0,78	0,23	1,00	0,46
A_3	0,07	0,94	0,59	0,43	1,00
A_4	0,13	0,63	0,22	0,62	0,39
A_5	0	0,94	0	0,06	0,35
A_6	0,03	1,00	0,23	0	0
A_7	0,01	0,94	0,62	0,25	0,35

Table 5. Normalization by SUM-BASED (Norm 3)

	C_1	C_2	C_3	C_4	C_5
A_1	0,64	0,04	0,26	0,06	0,14
A_2	0,07	0,12	0,11	0,34	0,15
A_3	0,07	0,18	0,18	0,17	0,24
A_4	0,11	0,09	0,10	0,23	0,14
A_5	0,03	0,18	0,06	0,06	0,13
A_6	0,05	0,22	0,11	0,04	0,07
A_7	0,03	0,18	0,18	0,11	0,13

Table 6. Normalization by MARKOVIC (Norm 4)

	C_1	C_2	C_3	C_4	C_5
A_1	1,00	0,20	1,00	0,17	0,59
A_2	0,11	0,83	0,40	1,00	0,62
A_3	0,11	0,95	0,68	0,50	1,00
A_4	0,17	0,70	0,39	0,67	0,58
A_5	0,04	0,95	0,22	0,17	0,55
A_6	0,08	1,00	0,40	0,12	0,30
A_7	0,05	0,95	0,70	0,33	0,55

Table 7. Normalization by LINEAR TYPE II (Norm 5)

	C_1	C_2	C_3	C_4	C_5
A_1	1,00	0,20	1,00	0,17	0,59
A_2	0,11	0,53	0,40	1,00	0,62
A_3	0,11	0,80	0,68	0,50	1,00
A_4	0,17	0,40	0,39	0,67	0,58
A_5	0,04	0,80	0,22	0,17	0,55
A_6	0,08	1,00	0,40	0,12	0,30
A_7	0,05	0,80	0,70	0,33	0,55

Table 8. Rank by Applying normalization tools in CoCoFISo

	CoCoFISo	Norm 1	Norm 2	Norm 3	Norm 4	Norm 5
A1	1	7	5	1	3	2
A2	3	2	2	3	2	3
A3	4	1	1	2	1	1
A4	2	3	3	4	4	5
A5	6	5	6	7	6	7
A6	7	6	7	6	7	6
A7	5	4	4	5	5	4

the correlation between the CoCoFISo method rankings and normalization tools, we present Kendall's correlation in the Table 9. These coefficients range from 0.14 to 0.62. If the coefficient is close to 1, there is a strong correlation between the CoCoFISo method and the normalization tool.

Table 9. Kendall's rank correlation coefficient

CoCoFISo				
Norm 3	Norm 4	Norm 5	Norm 2	Norm 1
0,62	0,52	0,43	0,33	0,14

Overall result can be analyzed in three ways. First, by comparing the CoCoFISo method's ranking with those of all the normalization tools. Next, by comparing the CoCoFISo method's ranking with each of the normalization tools. Finally, by comparing the results of each normalization tool.

Let us begin with the first type of analysis. Here, no alternative has been able to maintain its position across all tools. Nevertheless, four out of seven alternatives have consistent rankings on two different normalization tools compared by CoCoFISo. These include:

- A2, which ranks third in Norm 3 and Norm 5;
- A5 and A6, which rank sixth and seventh respectively in Norm 2 and Norm 4;
- A7, which retains its fifth position in Norm 3 and Norm 4.

Whereas alternatives A3 and A4, which are ranked fourth and second respectively by the CoCoFISo method, were unable to maintain these positions when all normalization tools were used. Figure 1 details this situation.

The second case study shows us that the number of alternatives with constant ranks based on CoCoFISo method varies from one normalization tool to another. The Norm 3 and Norm 4 normalization tools have a maximum of three alternatives with constant ranks compared by the CoCoFISo. For Norm 3, these are A1, A2 and A7, and for Norm

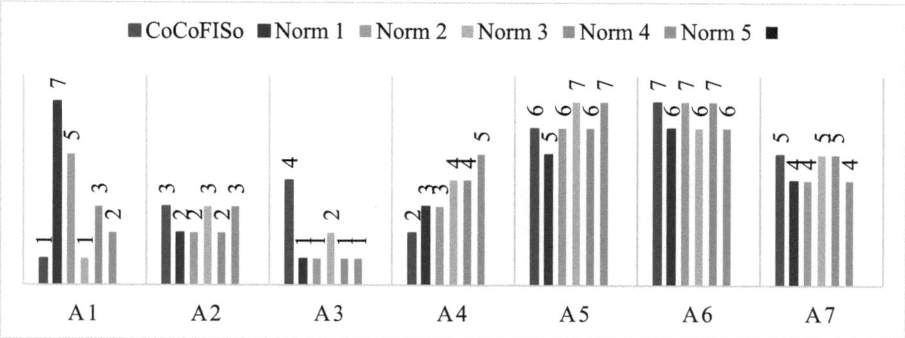

Fig. 1. Stability of CoCoFISo rankings compared to all normalization tools

4, they are A5, A6 and A7. Next, Norm 2 has two alternatives with constant ranks compared to the CoCoFISo method, of which A5 and A6. Norm 5 has only one constant rank alternative, which is A2. However, all ranks are different between the CoCoFISo method and Norm 1. Figure 2 shows this situation.

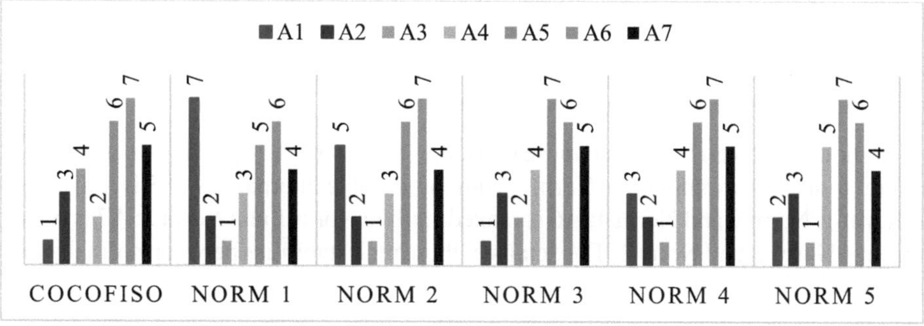

Fig. 2. Stability of CoCoFISo rankings compared to each of the normalization tools

Now, by focusing solely on the results obtained by the five normalization tools without CoCoFISo method, we can draw two situations. One is the comparison of the ranks within the set as a whole, and the other is the comparison of each of them individually.

Among all these normalization tools, it is noted that alternative A3 remains in first place in four normalization tools except for Norm 3. Next, there are three alternatives that have stable rankings on three normalization tools, including:

- A2 ranked second in Norm 1, Norm 2 and Norm 4;
- A7 ranked fourth in Norm 1, Norm 2 and Norm 5;
- A6 ranked sixth Norm 1, Norm 3 and Norm 5.

Otherwise, alternative A5, which ranks sixth in Norm 2, remains constant in Norm 4. However, alternative A1 has different rankings in all five normalization tools. This situation is shown in the Fig. 3.

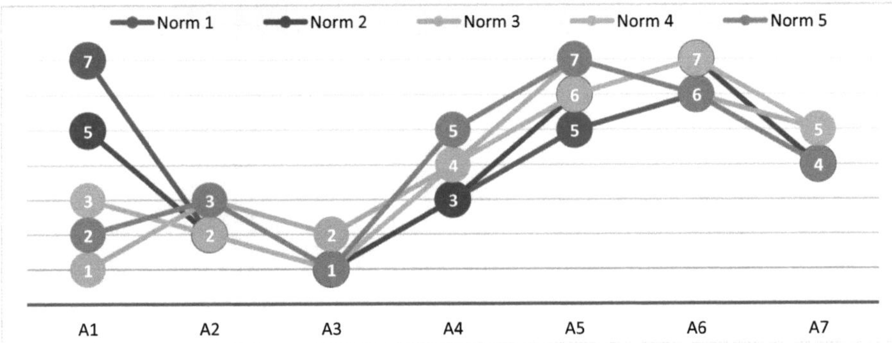

Fig. 3. Ranks stability of all normalization tools

In terms of the stability of the rankings of one normalization tool compared to another, it has been found that three tools present five alternatives with consistent rankings. These are:

- Norm 1, with stable ranking alternatives: A2, A3, A4, A6 and A7;
- Nom 2 and Norm 4 present the same alternatives, including A2, A3, A4, A5 and A7;

The other normalization tools offer two to four stable rank alternatives, including:

- Norm 3 with four alternatives A1, A4, A6 and A7;
- Norm 5 with three alternatives A3, A6 and A7;

It should be noted that for all these tools, the stability of the ranks is dispersed across several tools. For example, for Norm 5 with three stable alternatives, A3 is ranked first in Norm 1, Norm 2 and Norm 4; A6 is ranked sixth in Norm 1 and Norm 3; A7 is ranked fourth in Norm 1, and Norm 2. The details of these rank stabilities between normalization tools can be seen in the Fig. 4.

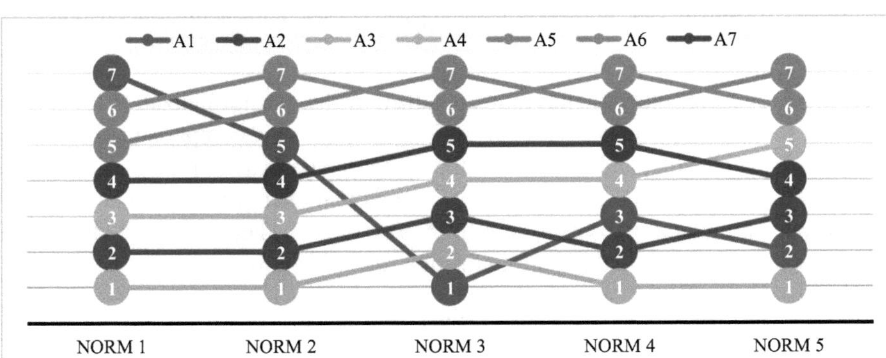

Fig. 4. Stability of ranks between normalization tools

5 Conclusion and Implication

The multi-criteria decision-making is a management discipline that allows alternatives to be ranked and one best option chosen among several options when faced with conflicting criteria. As a result, data normalization is a crucial step in the process of implementing these methods. This research aims to study the impact of different normalization tools applied to the CoCoFISo method. five different tools were selected and using a performance matrix, we first applied the original CoCoFISo method. Thereafter, five normalization were integrated to the method and results were extracted.

The result obtained shows different rankings, but the Kendall rank correlation coefficient calculated indicates that CoCoFISo has the same correlation as these tools. We then analyzed the results obtained in various ways, attempting to compare the rankings produced by the CoCoFISo method with those produced by the normalization tools, and to study the rankings produced by the normalization tools themselves. Overall, from the CoCoFISo method to normalization tools, it was found that some of the alternatives had consistent rankings, but we found some irregular variations also. Through the analysis of the most fitted normalization tools on a MCDM method, we expect optimal results in sustainable and resilient studies [25]. concluded in their work that simulations and validation of various normalization tools within specific case studies, we are able to construct robust resilient or sustainable systems and improve their performance. A double normalization structured for MCDM methods was proposed by [26] in a decision support system for sustainable transportation planning. The proposed double normalization framework that combines linear and nonlinear normalization can act as an instrument to reduce scale distortion and improve fairness. This study implemented new normalization perspective on TOPSIS, MOORA, COPRAS, WSM, and WASPAS model and reflected a robust, interpretable decision environment in a sustainable context. To select the best green energy sources in India, [27] examined the effect of various normalization tools on the assessment of performance ranking of alternative energy sources. The research output can aid in identifying the possible global impact on the society and demonstrate how a supportive decision-making environment can maintain the sustainable and resilient transition in energy sector research.

This study allowed us to conclude that the choice of a normalization method is very important when developing an algorithm for a multi-criteria decision problem. This is due to the normalization tool chosen has an impact on the ranking of alternatives and directly affects the decision made by the decision-maker. The proof is evident in this study, given the different results obtained depending on the normalization tools used. On the other hand, this study is limited to testing the CoCoFISo method with a small size performance matrix and fixed weights. This will make it possible to do the test with a large dataset, use weight variations, apply other multi-criteria methods (like MABAC, EDAS, CODAS, etc) and compare the results with those of CoCoFISo. It opens another avenue of research. Compared to the normalization tools chosen for this study, normalization using the CoCoFISo method is unique. However, it will therefore be interesting to propose two normalization techniques for this method in order to specify the criteria to be maximized and minimized (cost and benefit criteria direction).

References

1. Do, D.T., Nguyen, N.T.: Investigation of the appropriate data normalization method for combination with preference selection index method in MCDM. Oper. Res. Eng. Sci. Theory Appl. **6**(1), 1–22 (2023)
2. Panchal, D., Chatterjee, P., Pamucar, D., Yazdani, M.: A novel fuzzy-based structured framework for sustainable operation and environmental friendly production in coal-fired power industry. Int. J. Intell. Syst. **37**(4), 2706–2738 (2022)
3. Tahmasebi Birgani, Y., Yazdandoost, F.: An integrated framework to evaluate resilient-sustainable urban drainage management plans using a combined-adaptive MCDM technique. Water Resour. Manag. **32**(8), 2817–2835 (2018)
4. Pamucar, D., Yazdani, M., Obradovic, R., Kumar, A., Torres-Jiménez, M.: A novel fuzzy hybrid neutrosophic decision-making approach for the resilient supplier selection problem. Int. J. Intell. Syst. **35**(12), 1934–1986 (2020)
5. Yazdani, M., Jahan, A., Zavadskas, E.K.: Analysis in material selection: influence of normalization tools on COPRAS-G. Econ. Comput. Econ. Cybern. Stud. Res. **51**(1), 59–74 (2017)
6. Zolfani, S., Yazdani, M., Pamucar, D., Zarate, P.: A VIKOR and TOPSIS focused reanalysis of the MADM methods based on logarithmic normalization. arXiv preprint (2020). https://arxiv.org/abs/2006.08150
7. Aytekin, A.: Comparative analysis of the normalization techniques in the context of MCDM problems. Decis. Mak. Appl. Manag. Eng. **4**(2), 1–25 (2021)
8. Ersoy, N., Van Dua, T., Thinh, H.X.: A comparative evaluation of data normalization techniques using different metrics: practical application to a MCDM method. Manuf. Rev. **12**, 19 (2025)
9. Ferreira, F.A., Ilander, G.O.P.B., Ferreira, J.J.: MCDM/A in practice: methodological developments and real-world applications. Manag. Decis. **57**(2), 295–299 (2019)
10. Pamučar, D., Yazdani, M., Tarighi Shaayesteh, M., Pavur, R., Simic, V.: Sustainable performance analysis of remanufacturing suppliers in supply chains: circular economy perspective. In: Supply Chain Forum: an International Journal, pp. 1–30. Taylor & Francis (2025, June)
11. Mohammadpour, M., Afrasiabi, A., Yazdani, M.: Identifying and prioritizing the barriers to TQM implementation in food industries using group best-worst method (a real-world case study). Int. J. Product. Perform. Manag. **73**(10), 3335–3362 (2024)
12. Yazdani, M., Ye, C., Shaayesteh, M.T., Zaraté, P.: Decision support system for waste management: fuzzy group AHP-CoCoSo. Int. J. Prod. Manag. Eng. **13**(1), 77–92 (2025)
13. Ecer, F., Saha, A., Dinçer, H., Yüksel, S.: Evaluating energy management investments for sustainable transition of smart cities through a new intuitionistic fuzzy consensus-based CoCoSo methodology. Renew. Energy. **244**, 122655 (2025)
14. Wittig Vianna, V., dos Santos, M., Gomes, F.S., C., & Lauro, A.: Application of the MPSI-CoCoSo method to rank OECD member countries towards the energy transition. Comput. Econ. **67**, 1–30 (2025)
15. Yazdani, M., Zarate, P., Kazimieras Zavadskas, E., Turskis, Z.: A combined compromise solution (CoCoSo) method for multi-criteria decision-making problems. Manag. Decis. **57**(9), 2501–2519 (2019)
16. Rasoanaivo, R.G., Yazdani, M., Zaraté, P., Fateh, A.: Combined compromise for ideal solution (CoCoFISo): A multi-criteria decision-making based on the CoCoSo method algorithm. Expert Syst. Appl. **251**, 124079 (2024)
17. Yu, Y. et al.: An improved hybrid FMEA method based on spherical fuzzy sets for risk assessment of offshore anchor pile installation. Ocean Eng. **339**, 122180 (2025)

18. Rasoanaivo, R.G.: Decision support system using Centroidous and CoCoFISo methods for analyzing resource availability in high schools. J. Oper. Intell. **3**(1), 197–226 (2025)
19. Sah, S., Sardar, S., Das, D.: Optimization of W-EDM process for superalloy by different objective weight integrated MCDM-a comparative study of CoCoSo and CoCoFISo methods. Int. J. Interact. Des. Manuf. (IJIDeM), 1–26 (2025)
20. Sen, H., Toksoy, M.S.: Green supplier selection with CoCoFISo-G. The Eurasia Proc. Sci. Technol. Eng. Math. **32**, 257–265 (2024)
21. Chang, H.: Decision algorithm for digital media and intangible-heritage digitalization using picture fuzzy combined compromise for ideal solution in uncertain environments. Symmetry. **17**(3), 443 (2025)
22. Kosareva, N., Krylovas, A., Zavadskas, E.K.: Statistical analysis of MCDM data normalization methods using Monte Carlo approach. The case of ternary estimates matrix. Econ. Comput. Econ. Cybern. Stud. Res. **52**(4), 159–175 (2018)
23. Trung, D.D.: Development of data normalization methods for multi-criteria decision making: applying for MARCOS method. Manuf. Rev. **9**, 22 (2022)
24. Acuña-Soto, C., Liern, V., Pérez-Gladish, B.: Normalization in TOPSIS-based approaches with data of different nature: application to the ranking of mathematical videos. Ann. Oper. Res. **296**(1), 541–569 (2021)
25. Vafaei, N., Ribeiro, R.A., Camarinha-Matos, L.M.: Selection of normalization technique for weighted average multi-criteria decision making. In: Doctoral Conference on Computing, Electrical and Industrial Systems, pp. 43–52. Springer, Cham (2018, March)
26. Gopisetty, Y.B., Sama, H.R., Padi, T.R., Patibandla, L.: A double normalization framework for sustainable electric vehicle selection: integrating LOPCOW and RAM in multi-criteria decision-making. J. Oper. Res. Soc. China, 1–46 (2025)
27. Bhowmik, C., Bhowmik, S., Ray, A.: The effect of normalization tools on green energy sources selection using multi-criteria decision-making approach: A case study in India. J. Renew. Sustain. Energy. **10**(6) (2018)

Analyzing User Behavior in the Interpretation of Horizontal Bar Charts for Decision Support Systems

Lorena Vieira Santos Rodrigues[1], Lucia Reis Peixoto Roselli[1,2(✉)], and Ana Paula Cabral Seixas Costa[1,2]

[1] Laboratory for Neuroscience and Behavioral Studies in Decision (NSID), Universidade Federal de Pernambuco, Recife, PE 50630-970, Brazil
[2] Center for Decision Systems and Information Development (CDSID), Universidade Federal de Pernambuco, Recife, PE 50630-970, Brazil
lorena.srodrigues@ufpe.br, {lrpr,apcabral}@cdsid.org.br

Abstract. Graphical visualization assists the exploration of the problem and is a rich tool for decision-making with multiple attributes. Its main role is to enable the decision-maker (DM) to gain a progressive understanding of the problem and holistically evaluate the alternatives clearly in the comparative process before making the final choice. Therefore, the objective and differential of this article is to investigate user performance in tasks of selecting and eliminating alternatives, by considering multiple attributes with similar and different weights in horizontal bar charts, and comparing them to vertical bar charts and tables by conducting a behavioral experiment to improve the decision-making process and usability of Decision Support Systems (DSSs). As a result, the analyses show that the combination of task, weight type, and visualization influences performance. The task of eliminating alternatives took less time for DMs, who performed better when the attributes had similar weights. On the other hand, DMs performed better in the task of selecting alternatives when the weights of the attributes were different. In addition, they performed better and preferred vertical bar charts to tables and horizontal bar charts. Since the results indicate that visualization with horizontal bar charts is associated with poorer performance by DMs in completing tasks, possibly associated with difficulties in interpretation, it is recommended that future studies investigate this format more thoroughly, by incorporating aspects of neuroscience. Such research can provide valuable guidelines for the design and selection of more effective visualizations in DSSs.

Keywords: Decision Support Systems (DSSs) · Holistic Evaluation · Graphical Visualization

1 Introduction

In a decision-making process that involves comparing multiple alternatives while simultaneously considering multiple attributes, graphical visualization plays a fundamental role. It provides accessible visual representations of complex data to facilitate understanding and memorization and to help DMs obtain richer, more intuitive insights into

A. P. Cabral Seixas Costa et al. (Eds.): ICDSST 2026, LNBIP 582, pp. 206–216, 2026.
https://doi.org/10.1007/978-3-032-25146-6_14

the problem. Therefore, clear, informative visualizations need to be integrated into the design of interfaces or DSSs that support this kind of problem [1].

In this context, DSSs have been developed to enable DMs to interact with the system and solve various multi-attribute problems using Multi-Criteria Decision Making/Multi-Criteria Decision Support (MCDM/A) methods. These methods capture DMs' preferences and incorporate them into a mathematical model to generate solutions to their problems; an example is the DSS FITradeoff [2].

This method, from the context of Multi-Attribute Value Theory (MAVT), uses two paradigms: elicitation by decomposition [3], in which preferences are declared by comparing consequences, and holistic evaluation [4] which allows the DM to declare their preferences by comparing alternatives and making a systematic evaluation of the alternatives.

Holistic evaluation can be conducted using different visualizations to facilitate a better understanding of the problem [5]. Given this, behavioral studies have been conducted since 2017 to help refine these methods and improve the usability of DSS by using generic problems with various visualization tools, such as bar, spider, and bubble charts, as well as tables. These analyses and improvements were presented in [6–8], including comparisons of selecting the best alternative and eliminating the worst alternative.

Therefore, by conducting a behavioral study focused on graphical visualizations in the holistic evaluation of FITradeoff DSS, this article seeks to investigate user performance using horizontal bar charts, compare them with vertical bar charts and tables, and provide analyses focused on selection and elimination, and the use of multiple attributes with similar and different weights, with the aim of contributing to the facilitation of decision-making processes in complex contexts.

The article is divided as follows: Section 2 presents background information on graphical formats; Section 3 describes the experiment conducted; Section 4 presents the results; Section 5 discusses the results and presents the conclusions and proposals for future studies.

2 Background

How information presented in graphical formats is comprehended has been widely investigated across diverse areas, from cognitive psychology to applied studies in visual communication and decision-making.

The literature shows that the way charts are structured can significantly influence both how accurately and how quickly users interpret them. Previous research indicates that vertical bar charts tend to be more intuitive and faster to process than horizontal bar charts [9, 10] and tables [11].

For example, in [12], vertical bar charts tend to favor more accurate judgments about quantities when compared to horizontal formats. In addition, [9] observed that response time is shorter when participants analyze vertical bar charts, especially in tasks involving comparing survival and mortality rates between two treatments.

[13] argue that this occurs, in part, because Western readers, accustomed to processing information from left to right, find greater interpretive fluidity in representations that

require this reading pattern, while horizontal charts require vertical observation, which can distract users and cause them to be confused about how to interpret them.

In addition, horizontal bar charts change expectations about how axes are organized, since it is generally expected that the independent variable will be on the x-axis and the dependent variable on the y-axis. This exchange of axes tends to impact the speed and accuracy of processing information [14].

With regard to comparing vertical bar charts and tables, while some studies indicate that both formats can generate similar levels of recall and comprehension [11], preference tends to lean toward vertical bar charts when participants are exposed to more than one format.

In these cases, the familiarity and simplicity of the vertical layout are decisive [15], mainly because vertical bar charts are more typical or common than horizontal bar charts and others [16].

Given these studies, it is clear that graphical choices can significantly alter how people understand and use information. This is even more true in complex decision-making contexts involving multiple attributes. Thus, it must be ensured that decisions supported by DSS are based on visual representations that truly facilitate, and do not hinder, interpretation.

3 Description of the Experiment

This article drew up a structured experiment to investigate user experience and performance during the decision-making process, using charts and tables to identify improvements in the FITradeoff method DSS, specifically in the holistic evaluation stage.

The configuration of the experiment was designed based on a combination of three fundamental elements: type of chart, criteria weights, and task type. The first element consists of three types of graphical display: vertical, horizontal bar charts and tables. For each of these types, the DM must analyze and respond to the task of selecting the best alternative or eliminating the worst alternative. The problems are divided into some visualizations with similar weights for the criteria and others with different weights, obtained via an arithmetic progression, with criterion 1 having the highest weight and criterion 5 the lowest.

The combination of the three elements yielded 12 visualizations, which were presented to the participants. For this purpose, a generic multi-attribute problem was used, without a specific context, consisting of three alternatives and five criteria (as shown in Fig. 1).

Thus, the configuration of these elements, as explored in the experiment, is presented in Table 1.

For example, the notation "B S Sel" indicates an evaluation pattern involving a vertical bar chart, criteria with similar weights and the task of selecting the best alternative.

It should be noted that the sequence in which the visualizations were presented was previously defined by the authors. Randomization was used to determine the sequence of the visualizations and prevent participant bias [17]. Moreover, participants could not

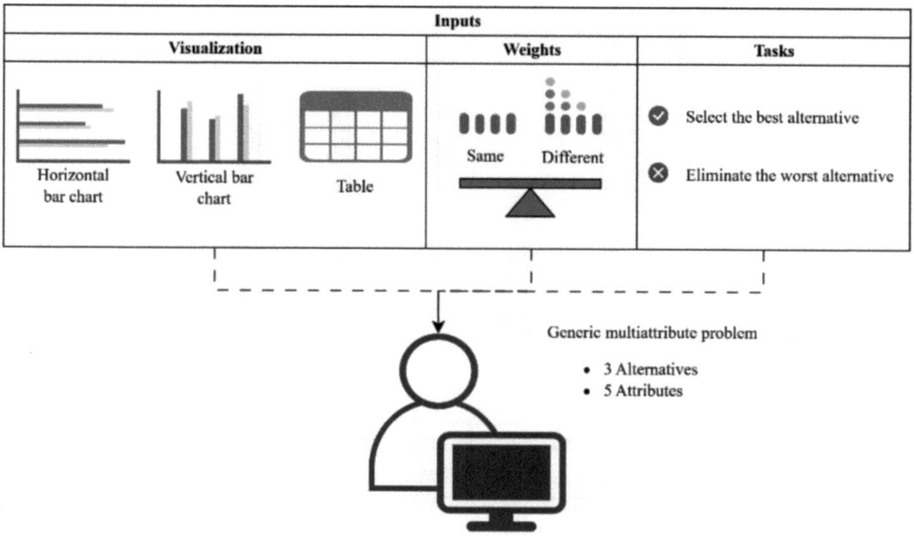

Fig. 1. Structure of the experiment

Table 1. Configuration of the elements

Element	Identifier	Description
Chart types	B	Vertical bar chart
	H	Horizontal bar chart
	T	Tables
Weights	S	Similar Weights
	D	Different Weights
Tasks	Sel	Select the best alternative
	Eli	Eliminate the worst alternative

return to a previous visualization and modify their response after selecting or eliminating the alternative.

Also, Fig. 2 shows examples of some visualizations presented to participants during the experiment.

The survey was conducted at the Federal University of Pernambuco (UFPE) at the NSID (Laboratory for Neuroscience and Behavioral Studies in Decision). The sample consisted of 48 participants (30 men and 18 women), most of whom had knowledge of multi-criteria decision analysis, were aged 18 to 50, and all of whom held a bachelor's, a master's, or a doctoral degree.

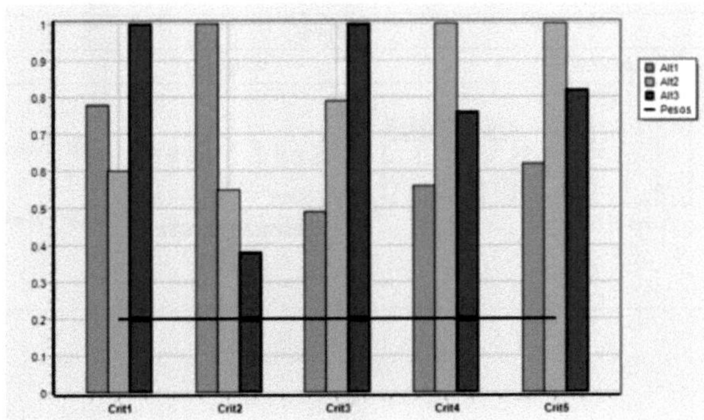

a)

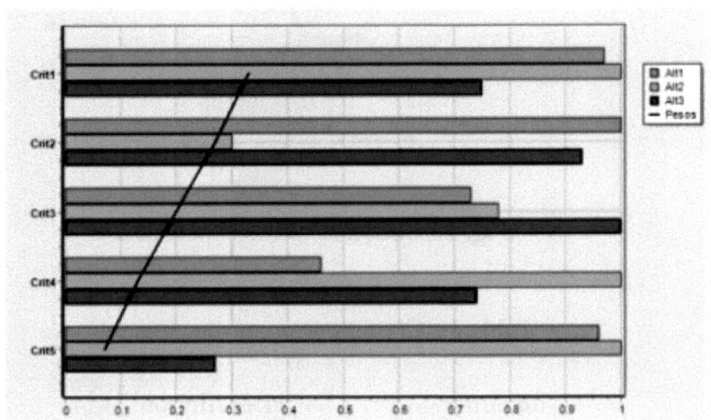

b)

Weights	0.2	0.2	0.2	0.2	0.2
Alt/Crit	**Crit1**	**Crit2**	**Crit3**	**Crit4**	**Crit5**
Alt1	0.6	0.55	0.79	1	1
Alt2	1	0.38	1	0.76	0.82
Alt3	0.78	1	0.49	0.56	0.62

c)

Fig. 2. Visualizations of the experiment a) Vertical bar chart; b) Horizontal bar chart; c) Table

The same conditions were applied to the entire sample of participants: everyone received the same instructions for conducting the experiment, and the same visualizations were presented in the order determined by random selection as stated above. Each participant took fifteen minutes to complete the experiment.

4 Results

4.1 Hit Rate Analysis

The experiment asked participants to analyze each visualization and perform the requested task of selecting the best alternative or eliminating the worst. One result from the data gathered from the experiment was the Hit Rate (HR), which represents the percentage of participants who correctly selected the best alternative or eliminated the worst alternative in each of the twelve visualizations, as previously defined by the authors of the experiment. The correct answer was calculated in advance following the MAVT [3].

Table 2. HR for each visualization

Chart type	Weights	Tasks	HR
B	S	Sel	56.25%
	S	Eli	81.25%
	D	Sel	60.42%
	D	Eli	41.67%
H	S	Sel	39.58%
	S	Eli	81.25%
	D	Sel	47.92%
	D	Eli	27.08%
T	S	Sel	41.67%
	S	Eli	77.08%
	D	Sel	43.75%
	D	Eli	31.25%

Table 2 shows that participants performed better when eliminating the worst alternative in visualizations with criteria of similar weights (i.e., B S Eli with 81.25% against B S Sel with 56.25%). On the other hand, when selecting the best alternative, the best performance was achieved with visualizations with criteria of different weights (i.e., H D Sel with 47.92% against H D Eli with 27.08%), across all chart types.

When analyzing the weights, Table 2 shows that with criteria of similar weights, participants performed slightly better on vertical bar and horizontal charts than on tables (i.e., B S Eli e H S Sel with 81.25% against T S Sel with 77.08%). As to the analysis of criteria of different weights, the same holds true: performance on vertical bar chart is much better than on the other types of chart (i.e., B D Sel with 60.42% against T D Sel with 43.25%).

At the end of the experiment, participants stated their preferences for which type of chart in the holistic evaluation they found best for visualization i.e., vertical and horizontal bar charts, or tables. Their most emphatic response, shown by receiving 62.5% of the votes, was the vertical bar chart, followed by Tables (29.2%) and, lastly, horizontal bar chart (8.3%).

In fact, when analyzing the data, the hit rate (HR) was quite high for the vertical bar chart compared to the horizontal bar charts and table. This finding aligns with the participants' reported preferences in the post-experiment questionnaire.

To certify these results, the McNemar test was conducted. It is a nonparametric test (data do not follow a normal distribution) which is used to analyze paired categorical data to verify whether there is a significant difference in accuracy rates between the types of charts (see Tables 3 and 4).

Table 3. McNemar test for HR comparing weights

Chart types	Number of answers "incorrect to correct"	Number of answers "correct to incorrect"	Result	p-value
B S Eli × B D Eli	4	23	Performance decreased with different weights	p = 0.000
H S Eli × H D Eli	3	29	Performance decreased with different weights	p = 0.001
T S Eli × T D Eli	4	26	Performance decreased with different weights	p = 0.000

Table 3 shows that in all types of charts analyzed and in the elimination task, the number of participants who got the right answer with similar weights and got it wrong with different weights in the elimination task was much higher than the opposite number (incorrect to correct), resulting in significant values in the McNemar test (p-value <0.05). Thus, it can be seen that different weights significantly reduced the number of correct answers in the elimination task.

Another analysis (as shown in Table 4) is that in all comparisons in the three types of charts and involving similar weights, the McNemar test was significant (p-value <0.05), indicating that the elimination task led to a higher number of correct answers than the selection task, as participants changed their answers from wrong to right when they started to eliminate the alternative.

Table 4. McNemar test for HR comparing tasks

Chart types	Number of answers "incorrect to correct"	Number of answers "correct to incorrect"	Result	p-value
B S Sel × B S Eli	20	8	Performance increased with elimination task	p = 0.038
H S Sel × H S Eli	24	4	Performance increased with elimination task	p = 0.000
T S Eli × T S Eli	21	4	Performance increased with elimination task	p = 0.001

This suggests that, cognitively, eliminating incorrect alternatives was a more efficient process than directly selecting the correct one, regardless of the type of chart analyzed.

4.2 Response Time Analysis

Another analysis performed was the participants' response time (RT) in seconds, i.e., how long each took to evaluate the visualization and select the best alternative or eliminate the worst one.

Table 5 shows the relationship between weights and tasks. When considering criteria with similar weights, participants took longer to select the best alternative, whereas when the weights criteria were different, they took longer to eliminate the worst alternative.

Table 5. RT for each visualization

Chart type	Weights	Tasks	Average Response Time (sec)
B	S	Sel	31.25
	S	Eli	21.88
	D	Sel	22.46
	D	Eli	26.92
H	S	Sel	31.35
	S	Eli	22.71
	D	Sel	23.92
	D	Eli	34.21
T	S	Sel	34.98
	S	Eli	24.63
	D	Sel	29.63
	D	Eli	33.90

These results were confirmed using the Wilcoxon statistical test, since the data are continuous, the samples are paired, and the RT values for all views do not follow a normal distribution. The test indicated a significant difference between the tasks and weights. Eliminating is faster when the criteria weights are similar (i.e., $Z = -4.123$ and $p = 0.000$) and selecting is faster when the criteria weights are different (i.e., $Z = -2.519$ and $p = 0.012$), with p-value <0.05.

According to Table 5, comparing the types of charts, it can be seen that the most time was spent analyzing the tables, followed by the horizontal bar charts, and finally the vertical bar charts, which the participants were quickest to evaluate.

The Wilcoxon test also confirmed participants' behavior regarding the charts, indicating that they responded more quickly to vertical bar charts than to horizontal charts, with $Z = -2.832$, $p = 0.005$ (p-value <0.05), after accounting for different criteria weights and the task of eliminating the worst alternative. When the task changed to selecting with different criteria weights, the test was not significant.

On the other hand, participants responded more quickly to horizontal charts than to tables, $Z = -2.27$, $p = 0.023$ (p-value <0.05). In this case, with different criteria weights, the task is to select the best alternative. When the task changed to elimination with different criteria weights, the test was not significant. Similar criteria weights did not yield significant results between the charts.

5 Discussion and Conclusions

To advance behavioral studies on graphical visualizations in holistic assessment, this article investigated the use of horizontal bar charts to improve both decision-making and the design of the DSS, which had not been explored in previous studies [6].

To this end, an experiment was conducted to compare vertical bar charts, horizontal bars, and tables showing variations in criteria with similar and different weights, and to complete tasks involving selecting the best and eliminating the worst alternative.

The findings indicate that participants preferred vertical bar charts, followed by tables and, lastly, horizontal bar charts. Performance results partially aligned with these preferences. Accuracy was higher when participants used vertical bar charts while performance with tables and horizontal bar charts was lower. Notably, vertical bar charts were not only the most preferred but also the format associated with the best overall performance, indicating alignment between decisor preference and obtained performance, as reported in other studies [15, 16].

The results also demonstrate that both task type and criteria weighting significantly influenced performance, affecting accuracy and execution time. When criteria weights were similar, eliminating alternatives was performed more quickly and decisively. In contrast, under differing weights, selecting the best alternative proved more efficient.

This suggests that the cognitive processing strategy adopted by the decision-maker depends on the different elements present in the visualization. This behavior indicates that the complexity associated with criteria weights affects how the DM processes information. Similar weights facilitate direct comparisons to rule out alternative, while different weights may favor strategies that are more focused on identifying the best alternative.

Importantly, the interaction among visualization type, task and weight configuration reveals that performance in decision-making is multidimensional. Vertical bar charts appear to reduce cognitive effort and support clearer comparisons, while horizontal bar charts, possibly due to their structural layout, may increase interpretative difficulty, as suggested in previous literature [9, 10]. Moreover, tables required longer response times, indicating a more analytical processing approach by decision-makers, possibly because they involve the direct interpretation of numerical values.

These findings contribute to the improvement of Decision Support Systems (DSS), particularly those that rely on holistic evaluation approaches, such as the FITradeoff DSS. The results suggest that the choice of visualization format can directly impact both performance and user experience in complex decision-making situations, reinforcing the importance of interface design in DSS development.

Despite the contributions, the study presents limitations. The sample size was limited, and only specific levels of complexity were tested. Therefore, for future work, the sample size could be expanded, neuroscience tools could be used, and the number of alternatives and criteria presented in the graphs could be varied in order to verify whether the negative effects observed remain, intensify, or attenuate at different levels of complexity. At the same time, analyses across different user profiles could be conducted, accounting for factors such as experience, familiarity with visualizations, and domain knowledge further enhance the robustness of these findings.

Acknowledgments. This work had partial support from the Coordination Unit for Improving the Qualifications of Higher Education Personnel – Brazil (CAPES), the Brazilian Research Council (CNPq), and the Foundation for Supporting Science and Technology of the State of Pernambuco (FACEPE), and the National Institute of Science and Technology. The study received funding from CNPq [grant 408976/2024-1; 309237/2023-8; 408976/2024-1] and FACEPE [grant APQ-0484-3.08/17].

Disclosure of Interests Not applicable.

References

1. Miettinen, K.: Survey of methods to visualize alternatives in multiple criteria decision making problems. OR Spectr. **36**, 3–37 (2014). https://doi.org/10.1007/s00291-012-0297-0
2. de Almeida, A.T., de Almeida, J.A., Costa, A.P.C.S., de Almeida-Filho, A.T.: A new method for elicitation of criteria weights in additive models: flexible and interactive tradeoff. Eur. J. Oper. Res. **250**, 179–191 (2016). https://doi.org/10.1016/j.ejor.2015.08.058
3. Keeney, R.L., Raiffa, H.: Decision Making with Multiple Objectives, Preferences, and Value Tradeoffs. Wiley, New York (1976). https://doi.org/10.1017/CBO9781139174084
4. Belton, V., Stewart, T.J.: Multiple Criteria Decision Analysis: An Integrated Approach. Springer US, Boston, MA (2002). https://doi.org/10.1007/978-1-4615-1495-4
5. de Almeida, A.T., Frej, E.A., Roselli, L.R.P.: Combining holistic and decomposition paradigms in preference modeling with the flexibility of FITradeoff. Cent. Eur. J. Oper. Res. **29**, 7–47 (2021). https://doi.org/10.1007/s10100-020-00728-z
6. Ferreira, E.B., de Vasconcelos, T.R.S., Roselli, L.R.P., de Almeida, A.T.: Rebuilding a DSS with feedback from behavioral and neuroscience studies. Int. Trans. Oper. Res. **33**(4), 2759–2780 (2025). https://doi.org/10.1111/itor.13614
7. de Vasconcelos, T.R.S., Roselli, L.R.P., de Almeida, A.T.: Performance of holistic evaluation for multi-criteria decisions comparing selection or elimination of alternatives. In: Decision Support Systems XIV. Human-Centric Group Decision, Negotiation and Decision Support Systems for Societal Transitions, pp. 85–97. Springer, Cham (2024). https://doi.org/10.1007/978-3-031-59376-5_7
8. Ferreira, E.B., Roselli, L.R.P., de Almeida, A.T.: Examining Interface preferences in a decision support system using eye-tracking. In: Decision Support Systems XIV. Human-Centric Group Decision, Negotiation and Decision Support Systems for Societal Transitions, pp. 59–75. Springer, Cham (2025). https://doi.org/10.1007/978-3-031-90863-7_5
9. McCaffery, K.J., Dixon, A., Hayen, A., Jansen, J., Smith, S., Simpson, J.M.: The influence of graphic display format on the interpretations of quantitative risk information among adults with lower education and literacy. Med. Decis. Mak. **32**, 532–544 (2012). https://doi.org/10.1177/0272989X11424926
10. Fischer, M.H., Dewulf, N., Hill, R.L.: Designing bar graphs: orientation matters. Appl. Cogn. Psychol. **19**, 953–962 (2005). https://doi.org/10.1002/acp.1105

11. Brewer, N.T., Gilkey, M.B., Lillie, S.E., Hesse, B.W., Sheridan, S.L.: Tables or bar graphs? Presenting test results in electronic medical records. Med. Decis. Mak. **32**, 545–553 (2012). https://doi.org/10.1177/0272989X12441395
12. Feldman-Stewart, D., Brundage, M.D., Zotov, V.: Further insight into the perception of quantitative information: judgments of gist in treatment decisions. Med. Decis. Mak. **27**, 34–43 (2007). https://doi.org/10.1177/0272989X06297101
13. Pylar, J., Wills, C.E., Lillie, J., Rovner, D.R., Kelly-Blake, K., Holmes-Rovner, M.: Men's interpretations of graphical information in a videotape decision aid. Health Expect. **10**, 184–193 (2007). https://doi.org/10.1111/j.1369-7625.2007.00443.x
14. Kim, S., Lombardino, L.J., Cowles, W., Altmann, L.J.: Investigating graph comprehension in students with dyslexia: an eye tracking study. Res. Dev. Disabil. **35**, 1609–1622 (2014). https://doi.org/10.1016/j.ridd.2014.03.043
15. Kim, S., Trinidad, B., Mikesell, L., Aakhus, M.: Improving prognosis communication for patients facing complex medical treatment: a user-centered design approach. Int. J. Med. Inform. **141**, 104147 (2020). https://doi.org/10.1016/j.ijmedinf.2020.104147
16. Reimann, D., Struwe, M., Ram, N., Gaschler, R.: Typicality effect in data graphs. Vis. Commun. **24**, 114–128 (2025). https://doi.org/10.1177/14703572221130445
17. Roselli, L.R.P., de Almeida, A.T., Frej, E.A.: Decision neuroscience for improving data visualization of decision support in the FITradeoff method. Oper. Res. **19**, 933–953 (2019). https://doi.org/10.1007/s12351-018-00445-1

Theoretical Perspectives on Decision Quality to Inform the Design and Evaluation of Decision-Support Tools

Christina Carrozzo Hellevik[1(✉)], Dina Margrethe Aspen[1], Erica Löfström[2], Ramzi Hassan[3], Christian Klöckner[2], and Ricardo da Silva Torres [4]

[1] School of International Business, Norwegian University of Science and Technology, Trondheim, Norway
{christina.hellevik,dina.aspen}@ntnu.no
[2] Department of Psychology, Norwegian University of Science and Technology, Trondheim, Norway
{erica.lofstrom,christian.klockner}@ntnu.no
[3] School of Landscape Architecture, Norwegian University of Life Sciences, Ås, Norway
ramzi.hassan@nmbu.no
[4] Artificial Intelligence Group, Wageningen University and Research, Wageningen, Netherlands
ricardo.dasilvatorres@wur.nl

Abstract. Decision-support tools and systems have the potential to support higher quality decisions, as they can improve our understanding of complex information, which is the first stage of decision-making. However, a use gap has been identified and there is limited empirical evidence of higher quality decisions resulting from the use of such tools in real decision-making contexts such as sustainability. In this article, we bring together theoretical foundations from Economics, Management, Psychology and Sustainability to not only decompose decision-making processes but, more importantly, the ways in which decision quality can be addressed and assessed. This can form the basis to design and evaluate the effects of decision-support tools on decisions within societal transitions to sustainability. We discuss decision quality following the logic of consequence, appropriateness and meaningfulness in individual and collective decisions, with and without risk. We then review existing threats to decision quality as information-related and cognition-related. Finally, we propose a model to include decision quality considerations in decision-support tool design and evaluation.

Keywords: decision-making · decision quality · decision-support · DSS · DST · sustainability transition

1 Introduction

Decision-makers are increasingly faced with complex and wicked problems characterised by high levels of uncertainty, high stakes, and the absence of single solutions. The Planetary Boundaries framework [36] defines nine socio-environmental problem

A. P. Cabral Seixas Costa et al. (Eds.): ICDSST 2026, LNBIP 582, pp. 217–237, 2026.
https://doi.org/10.1007/978-3-032-25146-6_15

domains on which the future of humanity relies. It can be argued that these challenges we face stem from a succession of poor decisions at all levels of decision-making, from individual to organisational, to local, regional, national and international. In parallel, information technology and decision-support tools are becoming more efficient and sophisticated, and have the potential to help decision-makers navigate current and future challenges through informed and better decisions. Decision-support tools are defined as software-based tools to assist sound decision-making, including the management of environmental risks and planning for impacts in different sectors and regions [46], p. 52, by illustrating possible outcomes [11]. But in all fields providing decision aids and expert systems, a use gap has been identified [49, 14]. This gap consists in a barrier to the operationalisation of the tools in practice and to the integration of the information provided in the decision.

It has been stressed by several scholars that better models are not sufficient to bridge this gap, but that instead, it is necessary to investigate the users and the decision context [1, 14]. The rich field of operations research uses rational decision theory to develop decision-support tools and systems and assist decision-makers in structuring and tackling complex problems. Rational decision-making can be summarised as decision-making based on the prediction of consequences from different alternative scenarios, and the choice of the scenario that best reflects the initially stated goals [42]. The advantages offered by this approach include increased clarity in the problem definition, and transparency in decision-making through goal formulation and traceable methods in problem-solving [30].

Additionally, the mathematical models applied allow the theoretical reconciliation of multiple decision-makers with diverging goals through principles of optimisation [27]. However, there are some problems when applying theory to practice, as Hansson [33] pointed out with his concept of the "tuxedo fallacy". He used this visual metaphor to highlight the fact that much of decision theory and, by extension the design of decision-support systems, is based on the principles of gambling, and throwing dice. These can be far removed from the complex and uncertain real-world decisions we navigate today in the context of societal transition towards sustainable development. The mismatches include the long time horizon considered [50] open-ended nature of the set of alternatives [16], the high levels of uncertainty, the often unstable preferences and multiple value systems of the decision-makers [27], the different types of logic applied in those decision domains due to risks [33], which in turn is dependent on the decision-makers themselves and the context in which the decision takes place [14]. This dispute was already highlighted by Simon [39] and is still unresolved today. It manifests itself in the growing divide between the normative and descriptive strands of decision analysis described by French [14] and others.

Moreover, this model is further complicated by both contextual and bahavioural factors. French [14] makes a case for bridging this divide by recognising and investigating the influence of both context and behavioural issues on decisions, using qualitative insights in addition to the modelling part. Indeed, he argues, decisions are only defined by their context. He usefully makes a comprehensive overview of the major gaps in current MCDM literature and sketches a research agenda for current and coming generations [14].

In addition to the need for research on the contextual and behavioural aspects of decisions, French [14] also highlights the need to investigate the role of uncertainty and risk communication, and how these affect the decision-makers' understanding of cause and effect. In this case, he remarks that, although a large body of literature has highlighted fundamental flaws in human risk perception and understanding of uncertainty, e.g., [41, 43], little research exists on the decision-makers themselves, which is, interestingly, where the fields of decision analysis and risk analysis could connect [14].

The contributions of this paper are threefold. Firstly, we synthesis decision tasks, phases and quality criteria from a trans-disciplinary perspective. Secondly, we propose a conceptual model for the integration of decision theory and decision quality concepts into the design of decision-support tools. Finally, the model is applied to a concrete example of decision-support tool within the context of societal transitions to sustainability. In order to bridge the highlighted gaps and improve the relevance of decision-support tools and systems to real decisions within sustainability transitions, we need to question some of the basic assumptions embedded in decision-support tools (DST) and systems (DSS) design and turn to the theory.

2 Methodology

This paper offers a conceptual synthesis of key literature in the domains of Decision Theory, from the fields of Economics and Management, Psychology and Sustainability. The goal is not to provide a systematic literature review, but rather to compare different perspectives on decision quality, and propose ways to conceptualise it. Furthermore, providing a cross-disciplinary perspective enables us to draw broader recommendations on how to better design and evaluate DSTs and DSSs with the goal of improving decision quality.

3 Decision Theory

3.1 What Is Decision-Making?

Hansson [24] defines decision-making as goal-directed behaviour in the presence of options. This highlights the concept of a goal or vision that drives the design and selection of options or alternatives based on information about the possible courses of action and their potential outcomes. On the basis of these five components of decision-making, namely information, alternative courses of action, choice, goal, and outcomes, one can distinguish different phases in decision-making and their associated tasks.

3.2 Decision-Making Phases

In decision theory, a decision-making process is usually structured into three to six phases. Simon [40], as one of the founders of decision theory, established three phases in decision-making, namely, (1) intelligence gathering, (2) alternative design, and (3) choice (Fig. 1). This is a helpful distinction in theory, although the phases often occur iteratively and, as Witte [60] argues, in parallel rather than sequentially. Nevertheless, it allows us to structure decision problems and determine how a tool can potentially support the various tasks involved. Simon is regarded as the father of behavioural economics as he defined the core concepts in decision theory. More recently, other scholars have built on his work and nuanced the theory on decision-making phases based on observation (i.e., descriptive decision theory), adding insights that may be useful in tailoring decision-support tools.

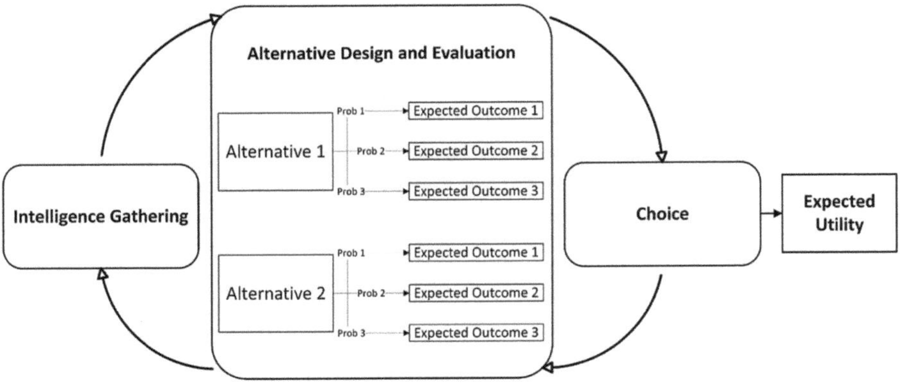

Fig. 1. Main elements of decision-making arranged in three phases based on theory by Herbert Simon, with the ultimate goal of satisficing expected utility

For instance, Brim et al. [4] highlight six phases: (1) identifying the problem, (2) obtaining necessary information, (3) producing possible solutions, (4) evaluating such solutions, (5) selecting a strategy for performance, and (6) implementing the decision. This approach sees decisions as problem-solving tasks Fig. 2. In a similar vein, Spetzler et al. [8] also describe decision-making in six phases: (1) structuring the problem, (2) defining objectives and attributes, (3) developing alternatives, (4) estimating consequences of alternatives, (5) evaluating trade-offs and selecting alternatives, and (6) implementing, monitoring, and reviewing the decision Fig. 2. Risk and uncertainty form an integral part of decision-making [8]. The risks may be of different nature, e.g., financial, environmental, safety, etc. For decisions involving risks for human lives, such as in disaster response, Lindell and Perry [25] distinguish decision-making as comprising five stages: (1) attention, (2) comprehension, (3) acceptance, (4) retention, and (5) action. This approach implies that it takes some effort to accept the risk analysis provided in the information and act upon it in the form of a decision. These four different approaches to structuring decision-making processes are useful when aiming to connect decision-making to the design of decision-support tools as they offer more granularity. Moreover, the concept of risk and uncertainty is a recurrent theme in all the fields explored, and decision-makers' accurate understanding of risk and uncertainty can affect decision quality, as we will see in the next section.

Simon (1960)	Gather Intelligence		Design Alternatives		Choose	
(Brim 1962)	Identify problem	Obtain Information	Produce solutions	Evaluate solutions	Select strategy	Implement decision
Spetzler et al. (2016)	Structure problem	Define objectives & attributes	Develop alternatives	Estimate consequences of alternatives	Evaluate trade-offs & select alternative	Implement, monitor and review decision
Lindell (1992)	Attention	Comprehension	Acceptance	Retention	Action	
Quality criteria Spetzler et al. (2016)	Appropriate frame	Relevant and reliable information	Rich alternatives	Clear values and trade-offs of DM	Sound reasoning in analysis of trade-offs	Commitment to action

Fig. 2. Decision Phases and Quality Criteria Synthesised from the Literature Given in the Left Column

3.3 Decision Quality: Process-Based Definitions

Wong-Parodi et al. [46] argue that decision-support tools are often not developed with the goal of high-quality decision-making in mind and that this should be rectified. But to unpick the effects of DSTs on decisions, we need to clarify the concept of decision quality, which differs depending on the field of study. According to March [27], the intelligence of a decision or action can only be determined in terms of its outcome, which makes it challenging to support decisions due to factors outwith our control. However, Spetzler et al. [8] argue that to discuss decision quality, one has to disentangle decision from outcome. Decision quality concepts, such as rationality and stakeholder engagement, are process-based approaches, whilst impact assessment is outcome-based. We will now go through each of these approaches.

3.3.1 Logic of Consequences: Decision Quality as Internal Consistency

One of the basic and recurring assumptions to explore in judging the quality of the overall decision, is rationality. In Economics, rationality implies that the decision-maker will strive to maximise their expected utility from the decision [42] based on their initial preferences or values. Utility captures a wide array of monetary and non-monetary benefits. So, in practice, the decision-maker attempts to predict the decision outcome, and selects the alternative that will increase their utility or that of their organisation the most Fig. 3. The ability of the decision-maker to predict the outcome of the decision also assumes that the necessary information allows this prediction of outcomes and their likelihood or probabilities [42]. This is where decision-support tools can contribute the most Fig. 3, either by structuring and conveying information in an intuitive way, or by helping predict the outcome of each alternative, using scenarios [38].

However, there are three main criticisms of this approach to decision-making theory. Firstly, the goals of decision-making as utility maximisation is far from accepted as a rule of rational decision-making [37, 39]. Secondly, rationality as an underlying principle in decision-making was disproved through empirical research in behavioural economics highlighting the use of heuristics, or cognitive shortcuts based on past experience rather than the analysis of evidence and consequences of the decision, e.g., [35, 43]. Thirdly,

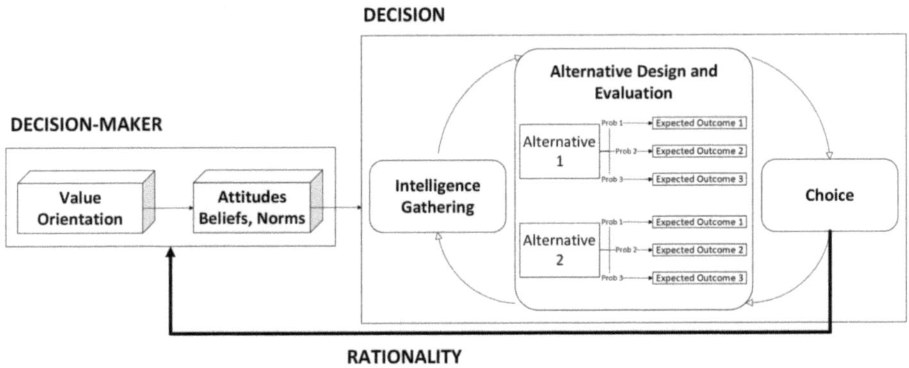

Fig. 3. Model of individual rational decision-making according to the principles of Psychology where rationality consists in an overall consistency between an individual's system of cognitions characterised by their value orientation, attitudes, beliefs and norms, and the choices made. This model builds on Stern's Value-Belief-Norm behaviour model and is otherwise original

the limits of information [39] and the decision-makers' ability to process it [22] is also an important shortcoming. Nevertheless, the latter point has been the subject of an important body of literature looking at how information systems and the way this information is visualised can help get closer to rational decisions.

These criticisms have been partly addressed to allow for the continued study of human decisions. For instance, in Psychology, rationality has been defined as subjective and relative to each particular frame of reference placed within the individual [21]. This frame of reference is made up of a person's system of cognition, including the subject's values, perceived social norms, perceived feasibility of options, and the choices made [21]. In other words, a decision can be very rational within the interpretation frame one decision-maker has, but it may appear irrational for an observer with a different frame of reference. This again implies the ability to anticipate the outcomes of each possible alternative choice. In this line, the tools can aid the users in making an informed decision that is in line with their values, attitudes, and intentions [45]. Empirical research has, here again, demonstrated that humans rarely base their decisions on rationality in a narrow sense alone [48]. However, in spite of inconsistencies observed from an outside perspective between values or utility goals and decisions, a broader understanding of rationality should also take emotional reactions to different choice options and other 'irrational' factors into account.

Minimizing the emotional or social costs of a choice or the wish to minimize the cognitive burden of evaluating many alternatives, for example, can be a rational decision within the frame of reference a person has [21]. Decision-support tools can help push some of these decisions closer to rationality, as proposed in Fig. 4 in a narrower sense by helping to reduce the cognitive and emotional costs of decision-making and help address cognitive biases [48]. Indeed, if we follow the logic of Fig. 4, we see that by providing relevant information about the possible outcomes or consequences of each alternative choice and their probabilities, we can provide decision-support. Additionally, we can help design alternative courses of action and anticipate each alternative outcome. Finally, the decision-maker will attempt to make the choice match original values, beliefs, and norms Fig. 4.

Once a more flexible and comprehensive concept of rationality is in place and the limitations of its assumptions are recognised, we can turn to the way the field of Economics, Management, and Operations Research model decisions. In this case, a high-quality or rational decision has a mathematical turn and is taken to be the optimal choice when balancing criteria, such as multidimensional cost-benefit and mapping them onto the decision-maker's preferences or values, which are synonyms in this field [3] as illustrated in Fig. 5. This can be done by assigning weights to the criteria that reflect those preferences and calculating a resulting value for each alternative Fig. 5. Multi-Criteria Decision Analysis is a commonly used approach, and a good practical example is a web-based tool for choosing holiday packages, e.g., [12], where criteria, such as cost, proximity to destination, carbon footprint, and comfort, are typically considered. The assumption is that utility should be maximised through the experience while balancing the criteria that are important for the decision-maker. The focus is on the choice phase, and mathematical models are developed to rank choices in order of optimality.

This model also implies that there is information about the possible outcomes or consequences of each alternative choice and their probabilities, which is where decision-support tools can play a part. Additionally, such tools can help devise richer alternatives and provide an arithmetic calculation of the optimal outcome based on initial preferences. To conclude this subsection, rationality has been the object of study and debate but is largely inclined towards a subjective interpretation of decision quality, which focuses on the process of matching the ultimate choice with an initial idea by the decision-maker of what would be a good outcome. So, this is a process-based approach using an internal and subjective frame of reference. It does not prescribe what a good decision is in objective terms. However, if we introduce the idea of risk, as will be done in the next subsection, there may be a more objective way of looking at decision quality.

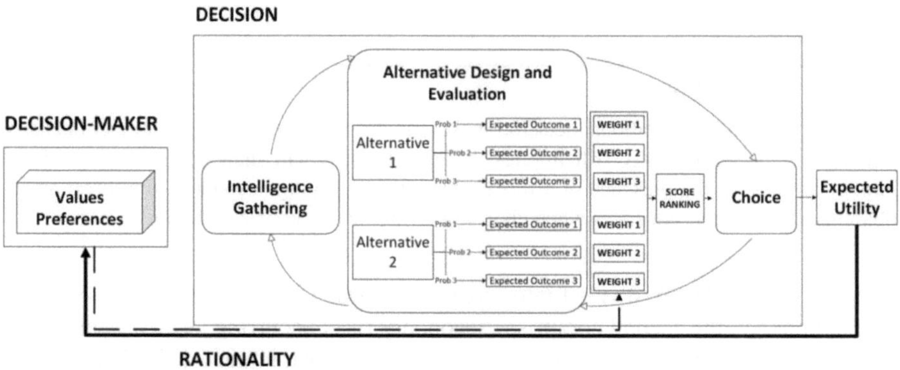

Fig. 4. Proposed model of rational decision-making according to principles of Economics and decision theory, where rationality implies that the decision will increase expected utility. Possible outcomes based on alternative scenarios are assigned weights according to initial preferences, and a final score or ranking is calculated

3.3.2 Decision Quality as Informed Decision in the Presence of Risk

When risk is involved, it is considered rational to take into account information about the nature of the risk and its level, as risk may significantly undermine the expected utility of the decision-maker(s). 'Risk' is defined in decision theory as a quantifiable exposure to an event, or a set of possible outcomes associated with a probability rate [16]. Risk has many definitions; one of the most widely used ones is by Kaplan and Garrick [20] and is composed of a threat, the probability of its occurrence, and the consequences. The definitions above highlight the fact that risk is not an object in itself but rather a notion we associate with a potential threat and an object or system that we consider of value. This leaves room for a subjective and potentially biased [60] evaluation, which will be influenced by the type of information available (evidence) and the way in which it is delivered (framing) but also by the individual's cognitive frame including risk attitudes (Fig. 4). In addition, choices based on the presence of risk can be influenced by the time horizon considered when the decision is made [16] In Fig. 5. we see that the threat can affect the potential outcomes or consequences of each alternative and the ultimate expected utility. Decision-support tools have been used in disaster response research.

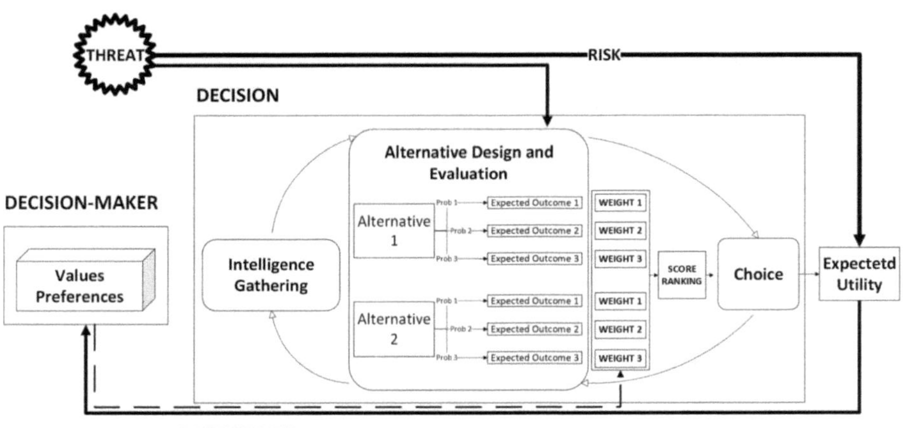

Fig. 5. Proposed model of rational decision-making in the presence of a threat or risk of negative event. In this case, rationality still implies that the decision will increase expected utility or minimise its loss. In the presence of a threat, this involves taking account of the threat and its probability rate and consequences when choosing an alternative course of action

In this case, risks are divided into three categories. The first is the primary risk of imminent danger, the second comprises secondary risks occurring as a consequence of the event, and finally, tertiary risks are defined as long-term impacts [47]. For instance, Wong-Parodi and Small [47] studied decisions that involve risks of property damage and financial losses from climate-change induced natural hazards. Their empirical studies on decisions using visual decision-support tools involve rating decisions as high or low quality, which is a normative approach. Indeed, the assumption is that the participants, once provided with the available information and prediction models about the natural hazard risks they are exposed to, will make the decision that fits with their values, but also the rational decision to limit their losses, as illustrated in Fig. 5. Moreover, their study assesses and aims to change not only knowledge but also beliefs and preferences after exposure to the decision-support tool [47]. Rating decisions in terms of the integration of information in a decision relates more closely to the way decision quality is judged in the field of science and environmental communication through, for instance geographic visualization. In this case, the DST can play a major role in raising awareness around an existing problem or threat, and can also elicit behaviour change by highlighting the need for more pro-environmental behaviour to match the stated values, or even by changing the subjects' priorities [35]. Furthermore, a decision-support tool can help by simulating consequences and probabilities, uncertainty levels, and feedback mechanisms of the different scenarios, as well as by monitoring the state of the system, including vulnerability and resilience [15].

3.3.3 Decision Quality as Consistency with Overarching Plans or Vision

When it comes to collective decisions that regard wider society, such as planning and policy-making, single decisions are often judged based on the extent to which they fit within larger plans and laws, including municipal, regional, national, or international. This can be true when applying both a logic of consequence and a logic of appropriateness, or rule-following [27]. In the latter, decision-makers act according to what is expected of them, looking at past decisions or at other similar organisations and cases [7]. However, when evaluating DSTs, this is rarely considered, and few tools provide this sort of information. Some municipalities have used digital twins and the UN framework of Sustainable Development Goals (SDGs) to attempt visualising the status of the municipality on each SDG [26, 24]. However, there is little research on how this is used in practice and potentially affects decisions.

To a certain extent, this is consistent with the rational model of decision making as, here again, we consider how a decision fits with the original preferences and intentions of the collective as illustrated in Fig. 6. Here, each participant in the decision-making process has their own cognitive frame, including values, beliefs, and norms, and has to, in addition, fit the decision within a feasibility space determined by existing laws, guidelines, and plans.

For collective decisions, the decision analytical models explored in Fig. 4 may also be used to support decision-making. Models to handle group decision-making differ with respect to how divergent goals, preferences, and trade-offs are handled and negotiated. Negotiation models aim to support the group in converging towards common goals and preferences. On the other hand, stakeholder evaluation models keep diverging goals and preferences intact, but assign relative importance to groups to produce aggregated preference information. Lastly, multi-stakeholder models keep diverging preference information intact, but focus on identifying alternatives that produce acceptable outcomes across stakeholder group preferences. However, as we will see in the next section, other approaches are often preferred, which step away from the concept of rationality in an attempt to better reflect the reality of collective decision-making.

3.3.4 Decision Logic of Meaningfulness and Knowledge Co-production

We have hinted at the fact that there are other types of decision logic than that of consequences. These are the logic of appropriateness and the logic of meaningfulness. Dewulf et al. [10] argue that in this third logic, the emphasis is on the interpretation of a decision situation, namely sensemaking. They stress that decision-makers may pick on different elements of the body of evidence to support their frame of interpretation of a particular situation. A remedy they propose is for scientists to interact with decision-makers and co-produce knowledge [10]. This also fits with the stakeholder involvement practices advocated by the field of Science of Actionable Knowledge (e.g., [19]).

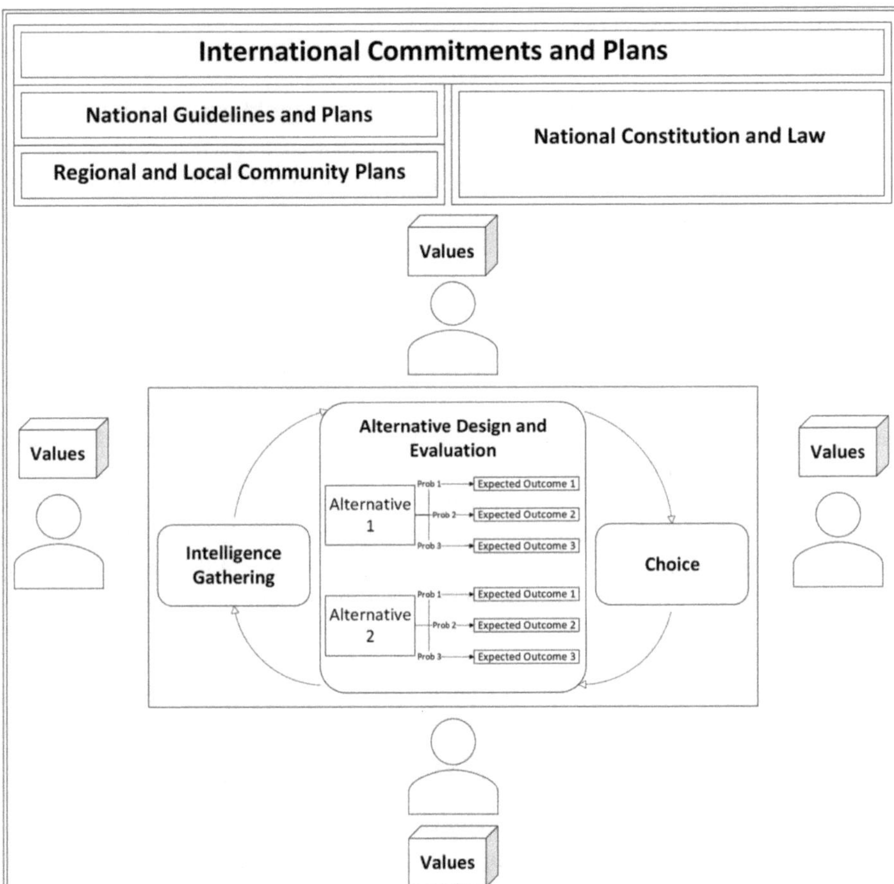

Fig. 6. Proposed model of collective rational decision-making where each participant is characterised by their cognitive system. After deliberation and negotiation, a choice is made among a set of alternatives of alternatives is made which needs to be consistent with local, regional, national and international plans as well as laws and guidelines

3.3.5 Decision Quality as Stakeholder Engagement and Communication

Following the study of collective decisions and the logic of meaningfulness, such as in planning and policy, the concept of stakeholder engagement is crucial when discussing decision quality. The term stakeholder also originates from the field of Management. Stakeholder theory highlights the importance of widening the circle of influence to people beyond the decision-maker or even the organisation making the decision, to those that hold a stake in the decision and may be affected by it [13]. In planning and policy-making decisions, stakeholder engagement and participation replaced rationality theory to more accurately reflect real-world processes, where information and rationality are limited, or bounded, and where discussion and negotiation are frequently used to arrive at a decision [31].

When it comes to decision-support tools in planning and policy decisions, stakeholder engagement and their level of participation have been widely used by the planning-support system community, especially as a proxy for decision quality and planning support effectiveness [18]. Indeed, with decisions involving communities, the assumption is that, if a democratic and participatory process is followed, a higher quality decision will ensue [34]. This view is, of course, highly debatable but is rarely debated as it has its roots in the concepts of democracy and in the deliberative and communicative nature of planning [18]. Going further, Bailey and Grossardt define a sustainable decision as 'a decision that satisfies the maximum proportion of stakeholders and therefore leads to more robust, legitimate, and hence durable decisions' [2]. This view is shared by others in the field of participatory geovisual decision-support systems. The collaborative process can refer both to the dialogue and iterations between the initiative-takers and the stakeholders, or it can refer to the stakeholders discussing a particular issue based on the tool as a reference point (e.g., [5]).

Consequently, the potential of a decision-support tool to stimulate discussions and collaboratively explore different scenarios through its interface and the information it provides will be advantageous. In addition, as communication between the different stakeholders is a crucial part of the decision-making process, DSTs can support this by visualizing the information and foci brought to the discussion by the various stakeholders [17]. One of the advantages of looking at decision quality in terms of stakeholder engagement, in addition to satisfying democratic duties, is the measurability of participation. However, a disadvantage is that it is, again, entirely process-based, and a good process does not guarantee a good outcome [27].

3.4 Decision Quality: Outcome-Based Definitions

As explained in the introduction, to talk about decision quality, one needs to disentangle a decision from its outcome, as the latter is fraught with uncertainty. However, this does not mean that the quality of a decision is only determined by the soundness of its process. On the contrary, some elements of the outcomes can be predicted with relative certainty, and their desirability can be determined, independently from the subjective preferences of the decision-maker.

Taking the example again of spatial planning from an environmental perspective, there is a tradition of carrying out Environmental Impact Assessments (EIA) by considering the environment as a 'common good' [32] that, if damaged, has negative consequences for most and is, therefore, undesirable for society as a whole. EIA is defined by Morgan [31] as the assessment of proposed actions or decisions for their likely implications for all aspects of the environment, from social to biophysical. The goal of such an assessment is to put mitigation measures in place or adapt the original decision to limit the anticipated negative impacts. The term EIA originated in the US but has, over the years, spread to be implemented across the world and is a widely understood term in academia as well as policy-making, planning, and industry. It was developed as a tool to minimise the impact that development projects can have on the environment and has extended to include local societies.

This view of decision quality as having as small a negative impact as possible on the local and global context is well embedded in the fields of sustainability and environmental science communities and is based on the knowledge of long-term geophysical and ecological processes and the extension of ethical responsibility to non-human species and systems [23]. The philosophical stance is that humans are not above other species or nature and the Earth as a whole, but rather form an integral part of it and that they should not damage the integrity of existing natural systems. This is also closely linked to the concept of risk, as the destabilisation of ecosystems leads to rapid runaway changes that ultimately pose a threat to human survival [36].

In summary, decision quality is defined in different ways in the literature according to the discipline and its underlying philosophy. Some disciplines assess the decision-makers for their performance level in achieving a right or optimal answer based on subjective criteria; other disciplines assume the process of stakeholder involvement to be more important than the decision itself. Others still judge the outcome of the decision for the potential impacts on the environment and society, i.e., the decision context. These differences in philosophy, time and geographic frames, and not least process vs. outcome can often result in a lack of understanding among scientists from different disciplines and between scientists and decision-makers or local communities [16]. Indeed, a decision that can seem baffling to some may be perfectly rational to others and vice versa. The study of environmental natural and anthropogenic processes at global scales may predispose scientists to consider global long-term risks of higher priority than short-term local risks or gains. However, the opposite will be true for most people, due to the cognitive bias of proximity [35]. The long time horizon is also consistent with decision-making in the context of societal transitions to sustainable development as highlighted in the Introduction section. This brings us to the definition of potential threats to decision quality, which we will discuss in terms of incomplete information and limited cognition.

3.5 Threats to Decision Quality

Going back to the initial structure of decisions as (1) information or intelligence gathering and processing, (2) alternative design and evaluation, and (3) choice and implementation (Fig. 1), we can identify two main threats to decision quality. Simon coined these as limited information and bounded rationality or limited cognition [39]. The first threat relates to the context, the phenomenon, and information that can and cannot be obtained about it, which we explore through the concept of uncertainty. The second one relates to the decision-makers' own limited cognition, or ability to relate to the evidence available, which we discuss as cognitive biases.

3.5.1 Limits to Information Gathering: Uncertainty

One aspect that is common to all these forms of decision and which can affect decision quality is uncertainty. Indeed, uncertainty levels can be impacted by data availability, data reliability, or by the complexity and unknowns influencing the system. As mentioned in the subsection on rationality in the face of risk, uncertainty differs from, but is one aspect of risk, as the latter consists of consequences associated with probabilities. Probabilities are a measure of what is known, or the evidence available, and a degree of belief about the likelihood that an event will occur [20]. As such, probabilities differ from frequencies, which are based on observed frequency of occurrence and are considered a more reliable form of risk assessment than probabilities [41].

Uncertainty can span over different levels, from low-level to deep uncertainty, where the possible outcomes themselves are unknown. Uncertainty affects the information available at the time of making the decision. Moreover, uncertainty and its interpretation can affect risk perception, which has been shown to be largely subjective [41]. Three forms of uncertainty are distinguished [6], namely epistemic uncertainty, ontological uncertainty, and ambiguity.

Firstly, epistemic uncertainty stems from a lack of knowledge about decision problem elements. As an example, estimating global greenhouse gas emissions may be subject to epistemic uncertainty due to insufficient data coverage and quality. This kind of uncertainty may be reduced through more, and more precise data collection. Reducing and communicating uncertainty effectively has been the focus of countless articles in all the domains of science. Hierarchies of evidence and frameworks are created, and more and more precise measuring tools are developed to address this issue [9]. The IPCC (Intergovernmental Panel on Climate Change) AR5 group proposes a framework to combine multiple types of uncertainty such as confidence level and likelihood of occurrence [29]. This sort of approach can be used to understand and manage uncertainty and risk in decision-making.

Secondly, ontological uncertainty, or aleatory uncertainty, originates from the randomness in nature of a phenomenon [44]. This is typical for information pertinent to system behaviour, such as e.g., predicting global climate change effects and weather extremes. Aleatory events can be a huge determining factor in decision-making as they can focus attention on the need for an immediate response as opposed to long-term strategies [9].

Lastly, there is ambiguity, where information may be subject to multiple interpretations that are simultaneously valid [6]. This could be due to linguistic vagueness, such as considering sea level rise as partly 'moderate' and partly 'extreme' for levels close to the border between the two categories. Ambiguity can also be deliberate to allow more flexibility or wiggle room, especially in the decision-making phase where goals and values are defined [7]. Ambiguity may also originate from variability in subjective values, judgements, and frames of knowledge, which leads to multiple, potentially contradictory, interpretations of the same information. In the context of societal decisions involving risks, Cairney et al. [9] advise the scientific community to address this issue and reduce the ambiguity by collaborating on establishing a dominant way to frame the problems and communicate them as simple stories.

3.5.2 Limits to Information Processing: Cognitive Biases

As highlighted in the section on rationality in the face of risk, risk perception is subjective and closely related to the way information, such as uncertainty, is processed and judgements are made. It is known in decision theory that limited cognition or 'bounded rationality' [39], and limited information drive decision-makers to apply some coping strategies, known as heuristics. In Psychology, heuristics, along with other coping cognitive processes such as editing, decomposition, and framing [28], can lead to cognitive biases.

Indeed, heuristics involve the use of memories and previous experience of a problem rather than the calculation of future probabilities prescribed by the rationality principle. As decision-makers are unable to understand, retain and consider all the evidence relevant to each individual decision, they can adopt two strategies, one characterised as rational, as it consists in a goal-directed prioritization of certain kinds of information, and the other as irrational, as it involves emotions, intuition and beliefs based on past experience [9].

The resulting cognitive biases are defined as systematic errors that can affect judgement [3], especially in the presence of uncertainty. This implies that even if all the knowledge about each alternative and the consequences of electing these were available in a format that limited the cognitive load, decision-makers might still use these coping techniques to choose alternatives efficiently and resulting in biased, irrational decisions. Cognitive biases can intervene at any point in a decision-making process and are not restricted to one particular phase.

4 Implications for the Design and Evaluation of Decision-Support Tools

We have now reviewed and discussed the concept of decision quality viewed from a multitude of fields studying decisions and working to improve their quality. The synthesis provided highlights that there are various approaches to measuring decision quality that go beyond mere internal consistency between goals and choices, as this method is highly subjective. Indeed, in the context of societal transitions to sustainable development, other factors intervene, such as the involvement of relevant stakeholders in a meaningful way, the consistency with stated national and international goals and laws, the integration of risk information in the decision, and the impacts of the decision on the environment, i.e., our life-support system. The quality of decisions from those viewpoints can also be established in a normative way based on available knowledge and predictive models.

The goal of improving decision quality in each phase should motivate the design of decision-support tools, and greater awareness of decision quality metrics should be acknowledged and specific goals defined at the onset of tool development. In order to facilitate this, we provide a detailed model of decision phases and their associated quality criteria based on the theory provided in the previous sections (Figs. 3, 4 and 5). This should enable a more thorough and iterative evaluation of decision-support tools at each stage of decision-making. The model should also inform the design of decision-support tools keeping in mind decision quality as a main goal [46].

Through more systematic evaluation of tool and decision quality, the field can build robust theory on what elements of the decision-support tools are more or less successful. This can inform the design of DSTs to improve their relevance and effectiveness in facilitating societal transitions to sustainable development, thereby reducing the use gap. Furthermore, through confrontation with real decision contexts, we can aim to ultimately reconcile normative and descriptive theories in an attempt to achieve transitions to sustainability at all levels.

In problem structuring, the boundaries of the problem should be defined, an appropriate frame applied (i.e., scale and time horizon), and knowledge gaps identified to proceed with information gathering. As decisions do not happen in a void, we add context mapping, in our model Fig. 7, which involves looking at potential repercussions of a decision and mapping relevant stakeholders and frameworks. Moreover, the fundamental principles of rational decision-making consist of designing rich alternatives and predicting their potential consequences Fig. 3. This is because choice usually consists of the selection of one of the alternatives defined; so, to avoid being constrained by limited alternatives, Spetzler et al. [8] recommend 'thinking outside the box'. Therefore, in the Alternatives part of our model Fig. 7, we add Design of rich and creative alternatives and Prediction of consequences of alternatives, including risk, probabilities and uncertainty.

Exploring these predictions at various levels of precision, including local scale and wider scale (hence the concept of stakeholders and decision context) involves zooming in and out of a particular decision. Furthermore, as decisions are rarely taken individually, there is a need for tools that encourage discussion and help reach an informed consensus (Fig. 7). As to addressing threats to rationality, a decision-support tool should help reduce cognitive cost and minimise cognitive and emotional biases, providing relevant and accurate information on risks and helping overcome cognitive limitations when envisaging the future. Therefore, another addition to this section is Reasoning and Discussion, based on [8] quality criteria 'Sound reasoning'.

Secondly, emotional factors, in an era of global environmental challenges that are often man-made through a multitude of individual and collective poor decisions, and whose risks are difficult to fathom in a long-term perspective, the need for tools that help transcend the limited nature of our cognitive abilities to shape a sustainable future is more obvious than ever [48]. Finally, in the choice part of our model Fig. 7, we add the aspects related to the implementation of the choice, as part of the decision. In this section, we add Planning, including commitment to action, which is a quality criterion highlighted by Spetzler et al. [8]. We also add Implementation and Revision, which reflects the iterative nature of decision-making [44].

We now exemplify the model by looking at a specific problem domain. Using the theory discussed and conceptualised in Sect. 3 as a backdrop, we follow our model step by step to design and evaluate a decision-support tool in the domain of spatial planning for societal transition to sustainable development. Firstly, a spatial decision-support tool, such as an urban digital twin, needs to have functionalities that help structure the

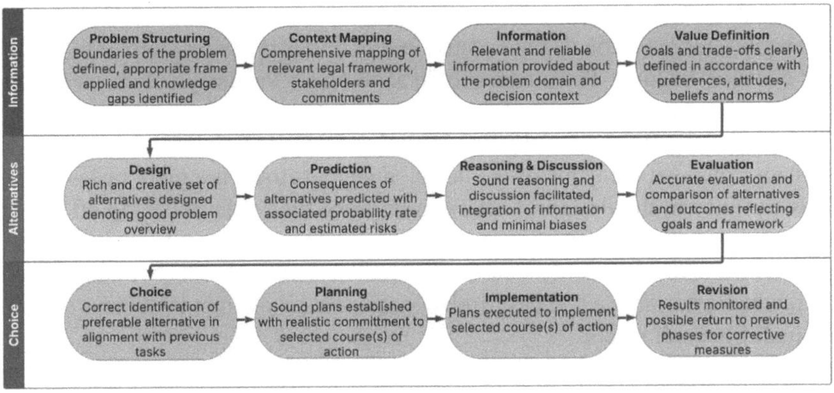

Fig. 7. Conceptual model highlighting the high-level tasks involved in decision-making and their associated quality criteria, based on Simon's three decision-making phases.

problem. By defining spatial boundaries around an area of interest and allowing for zooming in and out and adding different spatially-explicit information layers, one can see what issues may arise with a particular plan. An example can be that a desired plan encroaches on a protected area or an area of cultural interest, or that there may be a need to fill out part of the sea to expand the area. In the spatial decision-support tool, layers of information on municipal land and their designation are added and can be turned on to observe any overlaps with a particular area, such as with an area planned for recreation. This initial Problem Structuring can naturally lead to Context Mapping by highlighting stakeholders, i.e., people that may be affected by the plan, and legal frameworks relevant for the case. The role of a decision-support tool here could be to link up the layers of information on the area in question to the relevant laws that protect those interests. Although some of the laws can be vague and leave room for ambiguous interpretation, some laws are more specific; such as the law against filling in sea areas, and a red flag could be raised when planning with a digital tool linking relevant laws to plans. Additionally, being able to draw a radius around the area in question to determine direct stakeholders can be useful and allow for zooming in and out to increase that radius according to the impact of the plan.

Then, Information can consist of links to repositories of relevant information such as species present or contamination levels of the soil or water, or demographic information in a certain area. In Value Definition, a decision-support tool should enable the decision-makers to define the goal of a plan, such as 'making an area more attractive' or 'providing housing for a particular demographic group'. These can be more or less precise, but the more precise they are, the more accurate one can be in matching a decision with the stated goals. In the case of an urban digital twin, this capability is not directly present. The tool can serve as a support for discussion on these goals, but there is room for making such tools more comprehensive to enable goals definition and finding optimal solutions. This would also make spatial planning decisions more transparent and focus on decision quality in terms of rationality, or internal consistency between goals and choice, which is often lacking in this sector [27].

In the second phase of a decision-making process, namely Alternative Design, a decision-support tool should facilitate the creation and exploration of different scenarios. Here, one can embed models, such as demographic changes, mobility predictions, or climate change impacts. The goal of this phase is to design different courses of action based on predicted outcomes, and the latter is where a decision-support tool can best aid decisions. We mentioned cognitive and emotional biases and here, by playing with the time horizon and visualizing futures [35], decision-support tools have the potential to help users overcome the proximity bias and optimism bias, for instance [48]. Going back to the urban digital twin example, it is possible to simulate sea-level rise and show what areas would be most affected by flooding. As demographic models are embedded in the tool, one can show predictions of housing demand in the near future. And to incorporate other aspects of decision quality, the environmental impact assessment of the different scenarios should also be included.

While the influence of emotions on the general population in spatial planning and policy-making has received attention, the impact of emotions on decision-makers remains notably neglected. Traditionally, it has been assumed that planners and policy-makers are not swayed by emotions, a standpoint supported by earlier studies, such as Vining & Schroeder [51]. Yet, it is essential to recognise that decision-makers are not immune to emotions. The incorporation of decision-support tools that are visually explicit and have the power to involve the senses, and evoke emotional responses through future scenarios, for instance [22, 48], can indeed shape a decision-maker's attitudes and preferences, both broadly and in relation to specific projects. Research by Klöckner & Löfström [22] highlights that such tools can serve as potent agents in steering stakeholders to rethink their preferences and attitudes through the elicitation of emotional responses.

As to Reasoning and Discussion, the literature shows that certain interfaces are more effective in stimulating discussions, such as maptables [52]. This can be taken into account when designing decision-support tools for multiple decision-makers.

Perhaps the aim, rather than finding an optimal compromise among the different goals and values, could be for stakeholders and decision-makers to reach consensus on the best alternative in the light of the stated goals, legal frameworks and available predictions. Decision-support tools have the potential to foster community-centric approaches, serving as a bridge between decision-makers and communities [53], encouraging dialogue and participatory decision-making processes [54]. This is not always provided by decision-support tools used in spatial planning, but opening the tools to citizens for annotation and comments could foster a citizen science approach and open the dialogue beyond decision-makers. Ideally, they could stimulate adaptive planning wherein strategies evolve based on the feedback and emotional responses garnered, giving rise to initiatives that are environmentally-sustainable and in harmony with community sentiments. Sound reasoning was mentioned above, and the possibility to quickly turn on and off simulated future scenarios is important in reminding the decision-makers of the consequences of different alternatives. In the Evaluation part, quantitative methods of scoring the different alternatives based on stated goals can be visualised in an intuitive way, such as a heatmap, or bar graph, or other type of graphic, and can help orient decision-makers in evaluating the best alternative for their goals.

This evaluation should ideally inform the Choice part, and a feedback mechanism that helps decision-makers understand whether their choice is the best alternative based on the information available and goals stated can be useful. Finally, the tasks relating to the implementation of the decision, namely, Planning, Implementation, and Revision, could be included in the form of an integrated planning tool.

5 Conclusion

We have reviewed existing theories about decision-making phases and ways of evaluating decision quality. These include process-based approaches, such as rationality, and stakeholder engagement for collective decisions. And outcome-based approaches in the form of impact assessment, especially in problem domains within sustainability and societal transitions. Secondly, we have explored potential threats to decision quality in the form of limited information, i.e., uncertainty, and limited cognition, i.e., cognitive biases. Thirdly, we have provided a conceptual model to relate detailed decision phases with quality criteria. Future research should focus on applying this theory and further bridging the gaps among the different research fields explored with the goal of high decision quality in mind. This involves reporting not only the successes but also the failures of DSTs in practical decision cases, to better support our societies and decision-makers towards a sustainable future.

References

1. Adelman, L.: Evaluating Decision Support and Expert Systems. Wiley-Interscience, New York (1992)
2. Bailey, K., Grossardt, T.: Toward structured public involvement: justice, geography and collaborative geospatial/geovisual decision support systems. Ann. Assoc. Am. Geogr. **100**(1), 57–86 (2010). https://doi.org/10.1080/00045600903364259. ISSN: 0004-5608,1467-8306. URL: http://www.tandfonline.com/doi/abs/10.1080/00045600903364259 (visited on 01/16/2024)
3. Belton, V., Stewart, T.: Multiple Criteria Decision Analysis: An Integrated Approach. Springer Science & Business Media, New York (2002)
4. Brim, O., et al.: Personality and Decision Processes: Studies in the Social Psychology of Thinking. Stanford University Press, Stanford (1962)
5. Brömmelstroet, M.T., Schrijnen, P.M.: From planning support systems to mediated planning support: a structured dialogue to overcome the implementation gap. Environ. Plan. B Plan. Des. **37**(1), 3–20 (2010). https://doi.org/10.1068/b35019. ISSN: 0265-8135, 1472-3417. URL: http://journals.sagepub.com/doi/10.1068/ b35019 (visited on 12/18/2023)
6. Brugnach, M., et al.: Toward a relational concept of uncertainty: about knowing too little, knowing too differently, and accepting not to know. Ecol. Soc. **13**(2), 16 (2008)
7. Brunsson, N.: The Consequences of Decision-Making. Oxford University Press, Oxford (2007)
8. Spetzler, C., Winter, H., Meyer, J.: Decision Quality: Value Creation from Better Business Decisions. Wiley, Hoboken (2016)
9. Cairney, P., Oliver, C., Wellstead, A.: To bridge the divide between evidence and policy: reduce ambiguity as much as uncertainty. Public Adm. Rev. **76**, 399–402 (2016). https://doi.org/10.1111/puar

10. Dewulf, A., et al.: Usable environmental knowledge from the perspective of decision-making: the logics of consequentiality, appropriateness, and meaningfulness. Curr. Opin. Environ. Sustain. **42**, 1–6 (2020)
11. Dicks, V.L., Walsh, J.C., Sutherland, W.J.: Organising evidence for environmental management decisions: a '4S' hierarchy. Trends Ecol. Evol. **29**(11), 607–613 (2014)
12. Dimara, E., Bezerianos, A., Dragicevic, P.: Conceptual and methodological issues in evaluating multidimensional visualizations for decision support. IEEE Trans. Vis. Comput. Graph. **24**(1), 749–759 (2018). https://doi.org/10.1109/TVCG.2745138. ISSN: 1077-2626. URL: http://ieeexplore.ieee.org/document/8019855/ (visited on 12/18/2023)
13. Freeman, R.: Strategic Management: a Stakeholder Approach. Pitman, Boston (1984)
14. French, S.: Reflections on 50 years of MCDM: issues and future research needs. EURO J. Decis. Process. **11**, 100030 (2023). https://doi.org/10.1016/j.ejdp.2023.100030. ISSN: 21939438. URL: https://linkinghub.elsevier.com/retrieve/pii/S2193943823000031 (visited on 01/30/2026)
15. Haimes, Y.Y.: On the complex definition of risk: a systems-based approach. Risk Anal. **29**(12), 1647–1654 (2009). https://doi.org/10.1111/j.1539-6924.2009.01310.x. ISSN: 0272–4332, 1539–6924. URL: https://onlinelibrary.wiley.com/doi/10.1111/j.1539-6924.2009.01310.x (visited on 01/16/2024)
16. Hansson, S.: Decision Theory: A Brief Introduction. KTH Royal Institute of Technology, Stockholm (2005)
17. Hassan, R.: Computer visualizations in planning: computer techniques for visualization of development scenarios for historically important landscapes and urban spaces. A case study: Nablus. Ås (2002) ISBN: 82–575–0516-1
18. Healey, P.: Planning through debate: the communicative turn in planning theory. Town Plan. Rev. **63**, 143–162 (1992)
19. Jagannathan, K., et al.: A research agenda for the science of actionable knowledge. Environ. Sci. Policy. **144**, 174–186 (2023). https://doi.org/10.1016/j.envsci.2023.03.004. ISSN: 14629011. URL: https://linkinghub.elsevier.com/retrieve/pii/S1462901123000680 (visited on 12/18/2023)
20. Kaplan, S., Garrick, B.J.: On the quantitative definition of risk. Risk Anal. **1**(1), 11–27 (1981). https://doi.org/10.1111/j.1539-6924.1981.tb01350.x. ISSN: 0272–4332, 1539–6924. URL: https://onlinelibrary.wiley.com/doi/10.1111/j.1539-6924.1981.tb01350.x (visited on 01/16/2024)
21. Klöckner, C.: The Psychology of pro-Environmental Communication - Going beyond Standard Information Strategies. Palgrave Macmillan, London (2015)
22. Klöckner, C., Löfström, E.: Disruptive Environmental Communication. Springer Nature, Cham (2022)
23. Leopold, A.: A Sand County Almanac. Ballantine, New York (1970)
24. Leplat, L., et al.: VIZPLAN: a visual analytics platform for the assessment of multidimensional indicators over time. ISPRS Ann. Photogramm. Remote Sens. Spat. Inf. Sci. **X-4/W3–2022**, 127–134 (2022). https://doi.org/10.5194/isprs-annals-X-4-W3-2022-127-2022. ISSN: 2194–9050. URL: https://isprsannals.copernicus.org/articles/X-4-W3-2022/127/2022/ (visited on 12/18/2023)
25. Lindell, M., Perry, R.: Behavioral Foundations of Community Emergency Planning. Hemisphere Publishing Corp, Washington (1992)
26. Major, P., et al.: The use of a data-driven digital twin of a smart city: a case study of Ålesund, Norway. IEEE Instrum. Meas. Mag. **24**(7), 39–49 (2021). https://doi.org/10.1109/MIM.2021.9549127
27. March, J.G.: A Primer on Decision-Making: How Decisions Happen. Simon and Schuster, New York (1994)

28. March, J.: Decisions and Organizations. Blackwell, New York (1988)
29. Mastrandrea, M.D., et al.: The IPCC AR5 guidance note on consistent treatment of uncertainties: a common approach across the working groups. Clim. Change. **108**(4), 675–691 (2011). https://doi.org/10.1007/s10584-011-0178-6. ISSN: 0165–0009-1480. URL: http://link.springer.com/10.1007/s10584-011-0178-6 (visited on 02/13/2024)
30. McIntosh, B., et al.: Environmental decision support systems (EDSS) development – challenges and best practices. Environ. Model. Softw. **26**(12), 1389–1402 (2011). https://doi.org/10.1016/j.envsoft.2011.09.009. ISSN: 13648152. URL: https://linkinghub.elsevier.com/retrieve/pii/S1364815211002076 (visited on 01/30/2026)
31. Morgan, R.: Environmental Impact Assessment: A Methodological Approach. Springer Science & Business Media, New York (1999)
32. Ostrom, E.: Coping with tragedies of the commons. Annu. Rev. Polit. Sci. **2**(1), 493–535 (1999)
33. Ove, H.S.: From the casino to the jungle: dealing with uncertainty in technological risk management. Synthese. **168**, 423–432 (2009). https://doi.org/10.1007/s11229-008-9444-1
34. Pelzer, P., et al.: Planning support systems and task-technology fit: a comparative case study. Appl. Spat. Anal. Policy. **8**(2), 155–175 (2015). https://doi.org/10.1007/s12061-015-9135-5. ISSN: 1874-463X, 1874–4621. URL: http://link.springer.com/10.1007/s12061-015-9135-5 (visited on 12/19/2023)
35. Richter, I., et al.: Advancing the potential impact of future scenarios by integrating psychological principles. Environ. Sci. Policy. **140**, 68–79 (2023). https://doi.org/10.1016/j.envsci.11.015. ISSN: 14629011. URL: https://linkinghub.elsevier.com/retrieve/pii/S1462901122003616 (visited on 12/18/2023)
36. Rockström, J., et al.: A safe operating space for humanity. Nature. **461**, 472–475 (2009). https://doi.org/10.1038/461472a. ISSN: 0028–0836, 1476–4687. URL: https://www.nature.com/articles/461472a (visited on 01/30/2026)
37. Sen, A.: Rational fools: a critique of the behavioral foundations of economic theory. Philos Public Aff. **6**, 317–344 (1977)
38. Shaw, A., et al.: Making local futures tangible—synthesizing, downscaling, and visualizing climate change scenarios for participatory capacity building. Glob. Environ. Change. **19**(4), 447–463 (2009). https://doi.org/10.1016/j.gloenvcha.2009.04.002. ISSN: 09593780. URL: https://linkinghub.elsevier.com/retrieve/pii/S0959378009000363 (visited on 12/18/2023)
39. Simon, H.: Rational decision making in business organizations. Am. Econ. Rev. **69**, 493–513 (1979)
40. Simon, H.: The New Science of Management Decision. Prentice-Hall, Englewood Cliffs (1960)
41. Slovic, P.: Perception of risk. Science. **236**(4799), 280–285 (1987)
42. Tisdell, C.: Concepts of rationality in economics. Philos. Soc. Sci. **5**, 259–272 (1975)
43. Tversky, A., Kaneman, D.: Judgment under uncertainty - heuristics and biases. Science. **185**(4157), 1124–1131 (1974)
44. Walker, W., et al.: Defining uncertainty: a conceptual basis for uncertainty management in model-based decision support. Integr. Assess. **4**(1), 5–17 (2003). https://doi.org/10.1076/iaij.4.1.5.16466. ISSN: 1389–5176. URL: http://www.tandfonline.com/doi/abs/10.1076/iaij.[59]1.5.16466 (visited on 12/19/2023)
45. Witte, E.: Field research on complex decision-making processes - the phase theorem. Int. Stud. Manag. Organ. **2**, 156–182 (1972)
46. Wong-Parodi, G., et al.: A decision science approach for integrating social science in climate and energy solutions. Nat. Clim. Chang. **6**(6), 563–569 (2016). https://doi.org/10.1038/nclimate2917. ISSN: 1758-678X, 1758–6798. URL: https://www.nature.com/articles/nclimate2917 (visited on 12/19/2023)

47. Wong-Parodi, G., Small, M.J.: A decision-centered method to evaluate natural hazards decision aids by interdisciplinary research teams. Risk Anal. **41**(7), 1118–1128 (2021). https://doi.org/10.1111/risa.13261. ISSN: 0272–4332, 1539-URL: https://onlinelibrary.wiley.com/doi/10.1111/risa.13261 (visited on 12/18/2023)
48. Zhao, J., Luo, Y.: A framework to address cognitive biases of climate change. Neuron. **109**(22), 3548–3551 (2021). https://doi.org/10.1016/j.neuron.2021.08.029. ISSN: 08966273. URL: https://linkinghub.elsevier.com/retrieve/pii/S0896627321006267 (visited on 12/18/2023)
49. Ziaja, S.: Role of knowledge networks and boundary organizations in coproduction: a short history of a decision-support tool and model for adapting multiuse reservoir and water-energy governance to climate change in California. Weather Clim. Soc. **11**(4), 823–849 (2019). https://doi.org/10.1175/WCAS-D-19-0007.1. ISSN: 1948–8327, 1948–8335. URL: https://journals.ametsoc.org/doi/10.1175/WCASD-19-0007.1 (visited on 11/05/2025)
50. Frantzeskaki, N., Loorbach, D., Meadowcroft, J.: Governing societal transitions to sustainability. Int. J. Sustain. Dev. **15**(1–2), 19–36 (2012)
51. Vining, J., Schroeder, H.: Emotions in environmental decision making: rational planning versus the passionate public. In: Social Science in Natural Resource Management Systems. Routledge, London (1988)
52. Flacke, J., Shrestha, R., Aguilar, R.: Strengthening participation using interactive planning support systems: a systematic review. ISPRS Int. J. Geo Inf. **9**(1), 49 (2020). https://doi.org/10.3390/ijgi9010049
53. Dembski, F., et al.: Urban digital twins for smart cities and citizens: the case study of Herrenberg. Germany. Sustainability. **12**(6), 2307 (2020). https://doi.org/10.3390/su12062307
54. Gallo, J.A., Lombard, A.T., Cowling, R.M.: Conservation planning for action: end-user engagement in the development and dual-centric weighting of a spatial decision support system. Land. **12**(1), 67 (2022). https://doi.org/10.3390/land12010067

A Framework to Help Adapt Strategic Decision Support Systems, to Assist the Early Detection of Indications of Strategic Drift

Werner Podkowik(✉), Brian Lehaney, and Ria Wild

Worcester Business School, University of Worcester, City Campus, Castle Street, Worcester WR1 3AS, UK
podw1_21@uni.worc.ac.uk

Abstract. This research aims to create a framework to help adapt strategic decision support systems, to assist the early indications of strategic drift (SD). SD describes deterioration of competitiveness, related to changes in the business environment. This may lead to a loss of strategic position in the market, followed by crisis and bankruptcy. Business failures for 2025 were projected to be over 23,900 in Germany and over 26,250 in the UK, the highest levels in a decade. These can inflict significant economic harm on creditors and employees. Early response to SD indicators could make the difference between survival and demise. The costs of strategic drift are not just immediately financial. Its resource waste impacts sustainability and positive transition.

Adapted Decision Support Systems (DSS) should be able to provide managers with early warnings that may help avoid such crises. Desirable and feasible DSS technology is best developed by means of a thorough and iterative problem structuring phase. Issues with this phase are often attributed to be the cause of DSS failures, rather than to the technology itself. That is, good technology requires good problem structuring.

The aim is to be achieved via four stages. These include a comprehensive literature review on strategic frameworks, strategic leadership, and corporate culture in organisations to detect SD, and a critique of existing DSS in relation to SD (Conceptual Model 1). The results are discussed with academic strategy experts and revised to produce Conceptual Model 2. This will be exposed to the critique of practitioners and professional strategists. Adjustments will be made accordingly to produce Conceptual Model 3, from which the 'final framework' will emerge. This will be assessed against historical case studies. To help reduce bias and improve credibility, the research will use data source and method triangulation. Different respondents will be used for each component to help avoid self-fulfilling prophecies.

Keywords: Strategic · Decision Support · Drift · Management · Sustainability

A. P. Cabral Seixas Costa et al. (Eds.): ICDSST 2026, LNBIP 582, pp. 238–253, 2026.
https://doi.org/10.1007/978-3-032-25146-6_16

1 Background and Introduction

In today's world, business dynamics incorporate a high degree of uncertainty [1]. Pressures from ethical consumers and ethical investors (amongst other factors) require organisations to have a sustainable vision [2]. This, in turn, needs strategic decision-making, and therefore suitable strategic decision support systems (DSS). Artificial intelligence (AI) and machine learning (ML) are increasingly important. Nonetheless, creating an appropriate system requires human knowledge and management [3].

This research aims to create a framework to help adapt strategic decision support systems, to assist the early detection of indications of strategic drift (SD). Strategic drift describes a gradual deterioration of competitiveness, often related to changes in the business environment. This may result in strategic misalignment and may lead to a loss of strategic position in the market, followed by organisational crisis, major transformation and possibly bankruptcy [4].

Modern enterprises operate in complex and dynamic conditions with the elements of surprise Environmental scanning is of vital importance. Information about the possibility of a significant loss, such as bankruptcy, liquidation of a large part of the enterprise or the occurrence of long-term, structural crisis is of a particular relevance. Information about the possible threats is of useful warning value only if it is acquired suitably early when it is possible to take preventive measures. The key point of prevention is to avert irreversible destructive processes [5]. Business failures for 2025 were projected to be over 23,900 in Germany and over 26,250 in the United Kingdom, reaching the highest levels in over a decade. These can inflict significant economic harm on creditors, employees, and other stakeholders [6–8].

It is the tendency for business strategies to develop incrementally based on cultural influences. They fail to keep pace with a changing environment and companies tend to allocate the same resources to their functional units year after year. Additionally, the business environment will become more challenging, and business disruptions will become more frequent and impactful due to rapid technological change, globalization, and climate change. A study of 215 major UK firms from 1983 to 2003 identifies only four companies without organizational crisis, major transformation, or bankruptcy. Unfortunately, SD is hard to detect. Several strategic frameworks exist to analyse the company and its external environment, which often lead to ambiguous results, making it challenging to identify the need for change [9]. A 2006 paper states communication with customers is critical for SMEs [10]. It is inferred here that it is similarly important for inter-departmental communication in larger organisations, as well as external communications with customers and suppliers. The instruments of strategic management should be able to help provide solutions by assisting the early identification of SD. Successful development of the organisation needs the right choice of appropriate strategies, and the identification of the need for change, if SD is to be avoided. If it fails, SD will be the consequence [11].

This paper reports on Conceptual Model 1, the development of Conceptual Model 2, and next steps. The model syntheses the findings from the comprehensive literature review in the areas of strategy, strategic leadership, and corporate culture. Figure 1 shows an overview of the development of Conceptual Model 1.

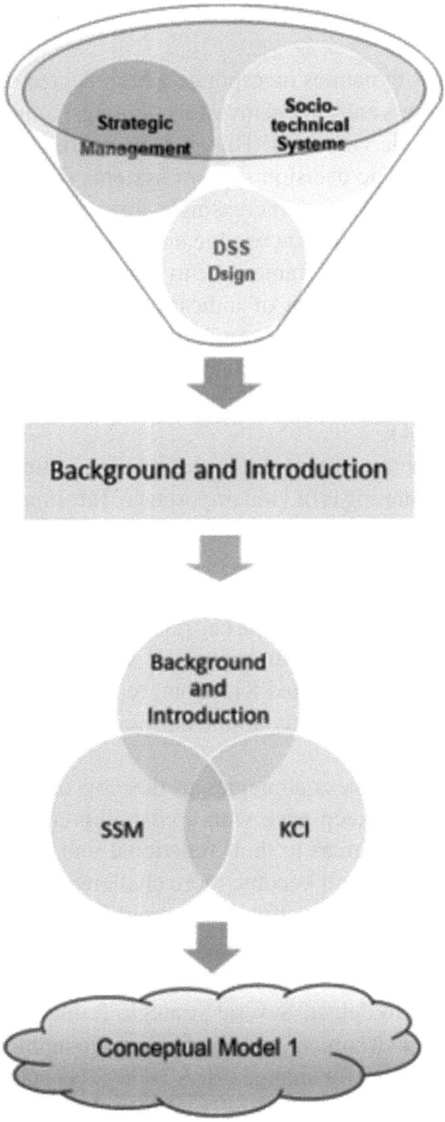

Fig. 1. Overview of the development of Conceptual Model 1.

2 Some Major Issues with Decision Support Frameworks and Models

As stated previously, the framework proposed in this thesis is a type of strategic decision support system (DSS). Issues related to business failures were discussed in the previous section. This section outlines some issues related to decision support systems and related information technology (IT) failures.

2.1 The CAD Failure of the London Ambulance Service [12]

This is a relatively old case, but it remains important for at least three reasons:

(i) It outlines many key failings.
(ii) It explicitly highlights issues when it was attempted to use technical solutions to problem structuring issues that require resolution rather than solution.
(iii) It can serve a as a baseline benchmark to help consider more recent cases.

This case reports on the findings of an independent inquiry into the failed implementation of a Computer Aided Despatch System (CAD) by the London Ambulance Service (LAS) in autumn 1992. The failings led to severe operational disruption, prolonged ambulance response times, and crisis in public confidence.

Technically, the system functioned as designed. However, the inquiry concluded that the system was neither ready, nor trusted at deployment, due to technical, managerial, and human factors. These factors are further addressed in the Theory section.

Failings

- Rushed and overambitious implementation
- Poor procurement and supplier selection
- Inadequate project management
- Technical and communication deficiencies:
- Human and organisational issues
- Governance and accountability gaps:

Recommendations

- Implement future CAD systems in phased stages with thorough testing, quality assurance, and fallback provisions.
- Before deployment, ensure system reliability, resilience, and user buy-in.
- Employ a qualified and experienced IT Director and a dedicated Project Manag
- Revise procurement guidelines to emphasise quality assessment over cost alone.
- Use professional Project Management and independent quality assurance.
- Establish a dedicated LAS board committee for project oversight
- Rebuild trust through consultation with staff and unions.
- Provide consistent, comprehensive training for all control and ambulance staff.
- Develop joint training to bridge the gap between control and field operations.
- Restructure management to empower divisions and improve control.
- Foster industry relations through joint committees and single-table bargaining.
- Provide management training, consistent policies, and stress support.
- Openly publish performance metrics, such as response times.
- Adopt open public relations, engage communities, and improve complaint handling.
- Secure dedicated funding for vehicle replacement and maintenance.

2.2 Other DSS Fails

- Decision Support System Fail in Formula 1 [13]
- The Hidden Risk Of Clinical Decision Support System Malfunctions [14]

- Decision Support Systems Fail - Importance Of Usability [15]

From these, some common issues emerge, including:

- Failure to support critical evaluation
- Erosion of trust and reliance dynamics
- Narrow problem framing
- Strategic distortion
- Design for transparency and explainability
- Support the human as a final decision-maker
- Balance support with appropriate friction
- Manage DSS affordances and pressure
- View DSS as a performative actor
- Establish dedicated oversight creating a formal, multidisciplinary body with clear authority – not just responsibility.

2.3 Summary

Strategic drift is traditionally understood as a failure to adapt to external environmental shifts, often attributed to leadership myopia, cognitive inertia, and organisational rigidity. Conventional detection methods, reliant on periodic reviews, are limited by their reliance on data quality and unbiased interpretation, which are challenging in practice. DSSs appear to offer a solution. Their capacity for continual data processing and sophisticated algorithms helps overcome this human limitation and provides an objective, vigilant sensor for early detection of strategy misalignment.

The cited cases of DSS failures challenge this assumption. The failure of the London Ambulance Service, the performative design failure in Formula 1, the silent technical malfunctions in clinical systems and the human-cognitive failures induced by poor design, demonstrate that DSS are not magic formulas for success. Instead, the failures reveal a critical paradox: The tool deployed to detect strategic drift can become its own enemy. A flawed DSS does not merely fail to see drift. Through its embedded logic, data limitations, and design, it can actively construct a distorted strategic reality that leads to strategic drift itself.

Furthermore, the failures presented in these cases are not isolated. They form a compounding, interconnected cycle where a flaw in the organisational domain (e. g., poor governance) creates conditions for failure in the technical domain (e. g., technical problems), which then leads to issues in the human domain (e. g., user error). Early warning of potential strategic misalignment therefore requires a new detection model that integrates these socio-technical aspects, adopts and adapts the recommendations from these cases, and unifies continual sensing of governance, technical integrity, and human-cognitive interaction in a single framework. It will enable environmental scanning in an improved way and will also encourage managers to question their strategic assumptions to assess whether the company remains on track. In order to incorporate the principles noted here, such a framework should be holistic and iterative, rather than forcibly linear-sequential, Technical proof of concept should not preclude ongoing monitoring and control actions from other knowledge areas.

3 Theory and DSS Failures and Recommendations from Cases

"In recent years *problem structuring* (soft methods) has flourished as an area of academic and practitioner interest. A range of methodologies has been developed including Soft Systems Methodology (Checkland, 1981, Wilson, 1984). SSM draws from a number of principles, including:

- emphasis on problem identification, problem structuring, and problem resolution.
- no automatic acceptance of existing structures.
- involvement of those being researched in the research.
- challenging approaches to 'norms'
- structured approaches to problem identification.
- increased sense of model ownership, with increased likelihood of model confidence and increased probability of implementing recommendations" [16]

As indicated previously, this research is about problem structuring. SSM maps closely to Jürgen Habermas's theory of knowledge-constitutive interests (KCI). This link is explained by Lehaney et al. [17]. This section is based on that source. KCI postulates that human knowledge is shaped by three fundamental interests: technical, practical, and emancipatory, corresponding to different forms of enquiry and understanding.

Technical (technological) interests focus on causal explanations and engineering the environment. This interest aims at controlling and predicting phenomena, often linked to technological advancements and practical applications in various fields. A real-life example would be determining the physical requirements of a bridge. The structure, foundations, and so on.

Practical (social) interests relate to communication, discussion, understanding and interpretation. This relates to culture and social practices and the meanings individuals derive from their interactions within society. For example, where is the best place to put the bridge to address the interests of the populations of two towns on opposite sides of a river? A technical approach is designed to solve. A practical approach is intended to resolve.

Emancipatory (political) interests may be considered as addressing power. It is sometimes thought of as the entity that can start or stop a system. "You have five days to agree where to put the bridge. If you do not reach agreement, I shall decide where to place it".

This theory provides a basis for understanding how different forms of knowledge serve distinct human purposes. From the foregoing, however, it is evident KCI views knowledge as neither neutral nor objective, as it is influenced by the interests that shape its acquisition and deployment. It can, however, be made as transparent as possible. This is achieved in the proposed framework by incorporating KCI.

> "Humanity is about to be handed almost unimaginable power, and it is deeply unclear whether our social, political, and technological systems possess the maturity to wield it." [18]

Table 1 shows how KCI may be applied to the failings and recommendations from the LAS case. This one case table is used in this paper for the sake of brevity. Similar tables have been created for all the cases. In summary:

Table 1. KCI applied to the failings and recommendations from the LAS case

Case	**T**echnical, **P**ractical, **E**mancipatory	DSS Failures and Recommendations
LAS Failures	P, E	Rushed and overambitious implementation
	P	Poor procurement and supplier selection
	T, P	Inadequate project management
	T	Technical and communication deficiencies
	P	Human and organisational issues
	P, E	Governance and accountability gaps
LAS Recommendations	T	Implement future CAD systems in phased stages with thorough testing, quality assurance, and fallback provisions.
	T	Before deployment, ensure system reliability, resilience, and user buy-in.
	P, E	Employ a qualified and experienced IT Director and a dedicated project manager.
	T, P	Revise procurement guidelines to emphasize quality assessment over cost alone.
	P	Use professional project management and independent quality assurance.
	E	Establish a dedicated LAS board committee for project oversight.
	P	Rebuild trust through consultation with staff and unions.
	T, P	Provide consistent, comprehensive training for all control and ambulance staff.
	P	Develop joint training to bridge the gap between control and field operations.

(*continued*)

Table 1. (*continued*)

Case	**T**echnical, **P**ractical, **E**mancipatory	DSS Failures and Recommendations
	P, E	Restructure management to empower divisions and improve control.
	P, E	Foster industry relations through joint committees and single-table bargaining.
	P, E	Provide management training, consistent policies, and stress support.
	T, P, E	Openly publish performance metrics, such as response times.
	P, E	Adopt open public relations, engage communities, and improve complaint handling.
	P, E	Secure dedicated funding for vehicle replacement and maintenance.

- the LAS failings and recommendations are largely practical or emancipatory.
- the F1 failings and recommendations are all practical or emancipatory, with none being technical.
- the clinical DSS failings were largely technical, with the recommendations being a mix of technical and practical.
- the DSS fail case failings were largely technical, with the recommendations being a mix of technical and practical.

What can be drawn from this discussion and how does it relate to this research?

- Many of the failings and recommendations relate to problem structuring, rather than to the technical aspects of decision support systems.
- Addressing an inappropriate KCI at a particular time may result in poor results. For example, providing a technical solution to the bridge issue is likely to result in failure if the location has not been agreed through practical resolution.

Figure 2 shows the foregoing concepts linked as an approach to DSS development

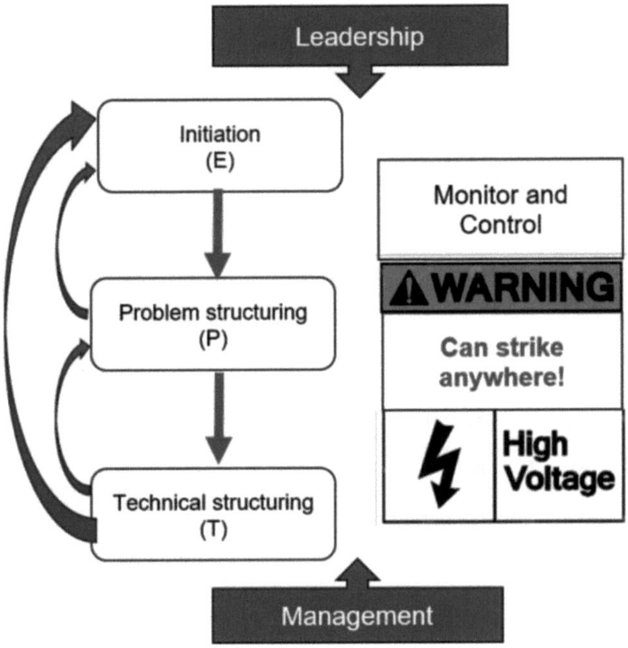

Fig. 2. KCI concepts linked as an approach to DSS development

4 Overview of Methodology

This research is grounded in pragmatism and uses a mixed methods research design, involving qualitative and quantitative data collection and analysis a sequential exploratory research design was chosen.

A literature review on strategic drift and DSS was used to help develop Conceptual Model 1 (CM1). This was revised to create CM2 by using the results of semi-structured interviews with academic strategy experts. CM2 will be exposed to practice by an expert opinion survey with strategic practitioners using the Delphi technique to help find the most appropriate elements and processes for strategic drift detection for the 'final' framework (CM3).

CM3 will be validated by demonstration and evaluated for its detection capabilities through the application to well-documented historical cases of SD. The semi-structured interviews will be analysed using thematic analysis and a self-developed relevance grid, assessing prevalence, DSS recommendations, and anticipatory detection potential. The relevance grid operates on the principle that themes scoring highly across all categories, receive the highest priority for inclusion in CM3.

Strategic principles and elements that achieve consensus will be incorporated into CM3. In the final research stage, the qualitative case study will apply retrospective pattern-matching as primary analysis technique to validate CM3 detection capabilities against historical cases.

Data and method triangulation will be applied to help reduce bias and improve credibility via convergence. In addition, triangulation will help provide some divergence, to help provide a richer picture. Figure 3 illustrates the overview discussed here.

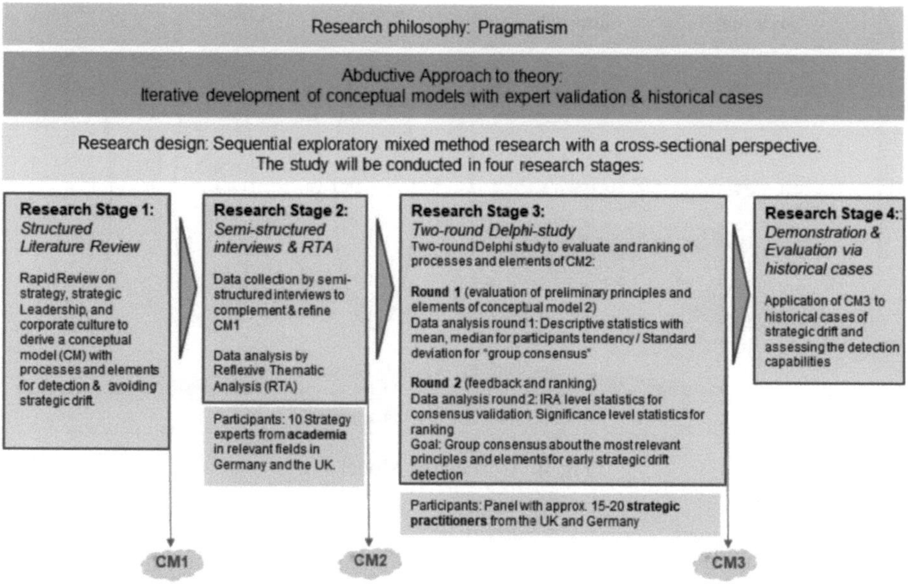

Fig. 3. Methodology

5 Conceptual Model 1

Conceptual Model 1 (CM1) was designed around the strategic decision-making process and consists of two primary constructs: (1) Overarching initiatives that enhance the organisation's resilience to strategic drift, and (2) the concrete strategic planning steps necessary for making informed strategic decisions for the future. These are shown in Fig. 4.

Both constructs aim to detect the need for change at an early stage. To achieve this, a set of strategic principles was selected from the literature, representing frameworks and tools with the best potential to identify strategic drift.

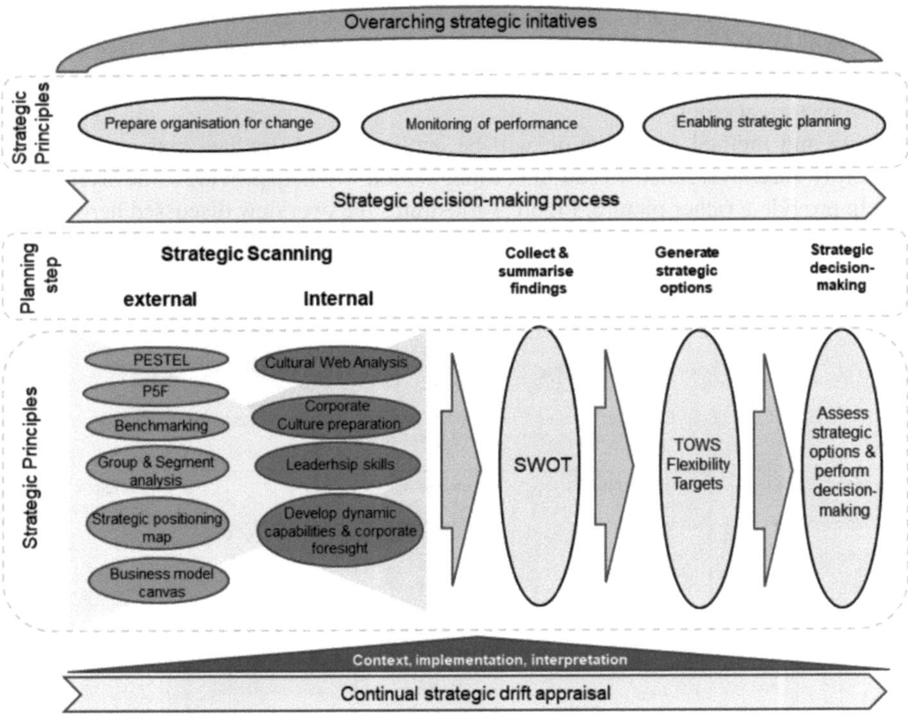

Fig. 4. Conceptual Model 1

For the overarching strategic initiatives, a range of specific strategic principles were identified, including review of strategic KPIs, focus on change management, and monitor organizational commitment. These principles were chosen to assess the current strategic position of the organization and enable initiatives that foster quicker adaption to both, external and internal changes. Strategic principles were selected to align with the steps of the strategic planning process steps. These principles are designed to identify strategic drift and support a systematic approach to strategy decision-making. Key steps in this process include external and internal strategic scanning of the organization, systematic collection and summary of the findings, the generation of strategic options, and their rigorous evaluation to enable a transparent future strategy decision.

The strategic principles for both overarching initiatives and the strategic planning process are further decomposed into strategic elements. These elements represent concrete strategic tasks derived from the literature that must be executed to ensure meaningful outcomes for each strategic principle. For example, the strategic principle leadership skills is broken down in actionable strategic elements, as shown in Fig. 5.

Each strategic principle must be implemented and adapted to the specific context of the organization. The appraisal for strategic drift is a regular activity conducted throughout the strategic decision-making process.

> Information processing is an essential part of any decision-making process … However, business managers are facing unprecedented challenges in dealing with ever growing abundant data and struggling to make sense of it to make well informed decisions. As Artificial Intelligence (AI) is making profound impact on every aspect of business operations and management, it has the great potential in assisting/replacing business managers' information processing activities. [19]

In light of the foregoing, where desirable and feasible, artificial intelligence (AI) will be advocated for both the external and internal strategic scanning components of the framework. In addition to previous arguments in favour of AI, it offers the possibility of continuous, rather than continual, environmental scanning. This is not, however, part of the problem structuring activity as such and would be a choice made by individual organisations. It is therefore not addressed any further in this paper.

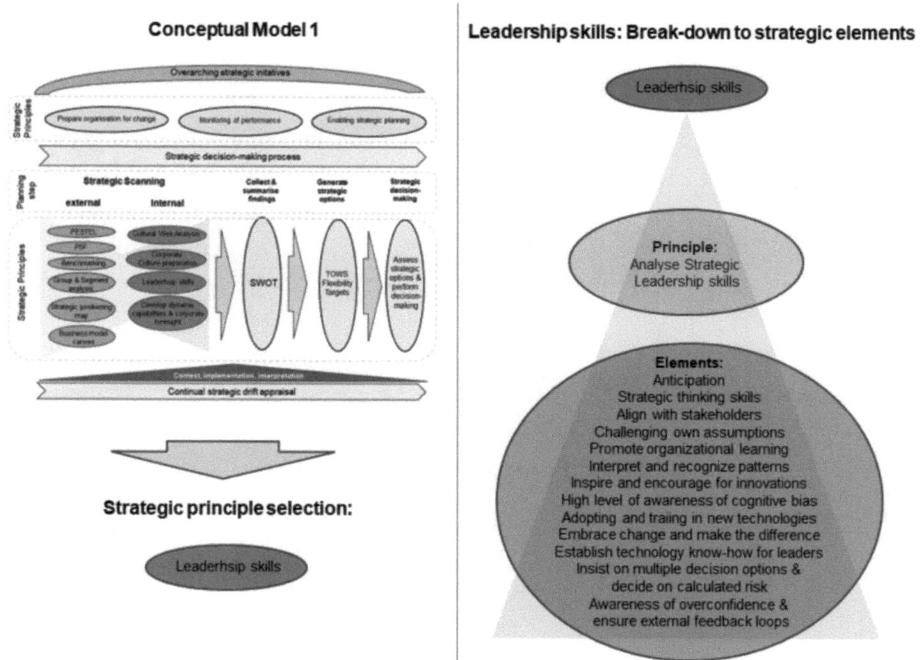

Fig. 5. Break-down of leadership skills into strategic elements

6 Conceptual Model 2

As the next step, CM1 has been further developed through semi-structured interviews with strategy experts from academia to that refine and enhance the model. Initial refinements occurred at the strategic principle level. Experts emphasized the necessity of clear organization-wide communication to detect strategic drift within overarching initiatives. Furthermore, within the generate strategic options principle, the critical integration of formal risk management processes, tools and actions was highlighted.

The current version of CM2 is shown in Fig. 6, with newly integrated strategic principles highlighted in green, and dotted lines denoting the refined strategic elements. For instance, the enhanced strategic principle leadership skills is operationalized into its constituent strategic elements, as shown in Fig. 7. The newly integrated or refined strategic elements are highlighted in blue.

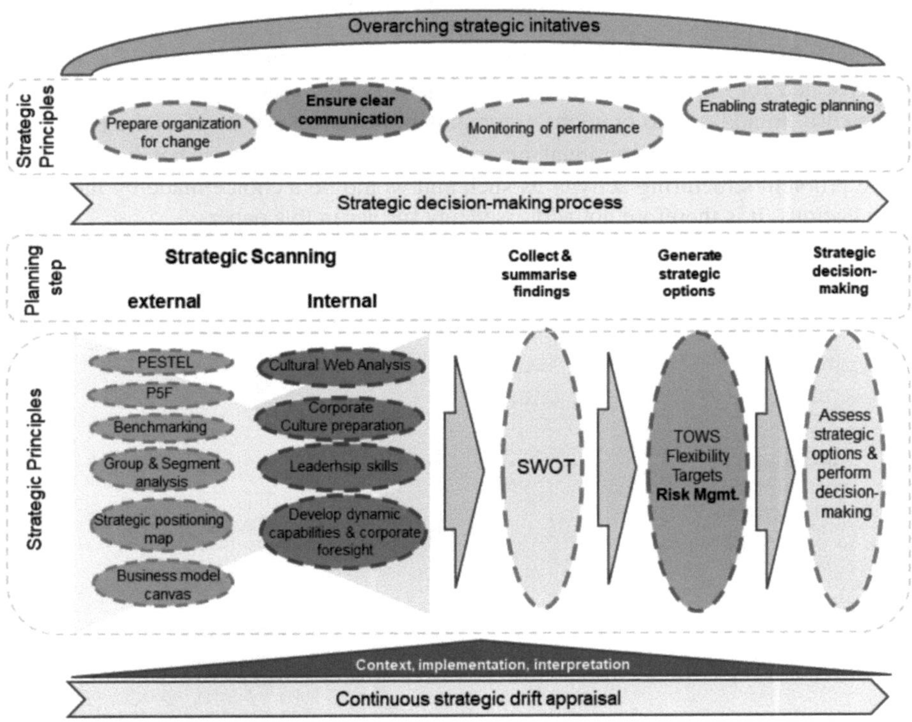

Fig. 6. Projected Conceptual Model 2

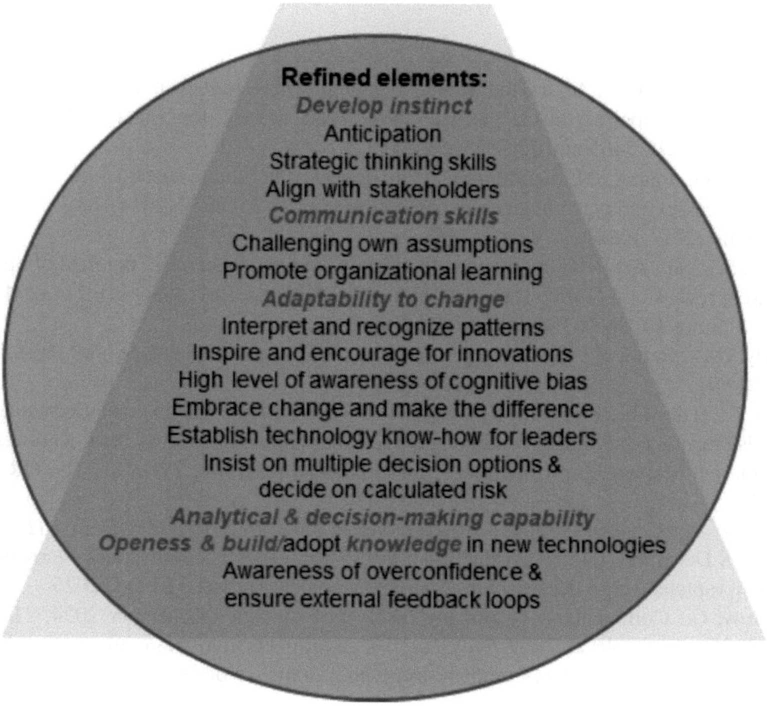

Fig. 7. Break-down of leadership skills into refined strategic elements (CM2)

7 Conclusions

Based on the analysis of Decision Support System failures and the application of the problem structuring theory, this paper concludes that a robust framework for the early detection of strategic drift must be a fundamentally socio-technical system. It cannot rely on technical solutions alone but must integrate a practical (organizational, cultural) and emancipatory (governance, power) dimension for early strategic drift detection. The proposed iterative conceptual model, developed with expert engagement from academia and practice, aims to create a framework that could serve as an input for a strategic Decision Support System. This system would be designed not to replace human judgment but to augment distributed organisational cognition, fostering a resilient partnership between decision-makers and the support system to navigate dynamic environments and avert the significant economic and sustainability costs of strategic drift. It would thereby be mitigating the inherent risk that the support system itself becomes a source of strategic drift.

References

1. Sirmon, D.G., Hitt, M.A., Ireland, R.D.: Managing firm resources in dynamic environments to create value: looking inside the black box. Acad. Manag. Rev. **32**(1), 273–292 (2007). https://doi.org/10.5465/amr.2007.23466005
2. Eccles, R.G., Ioannou, I., Serafeim, G.: The impact of corporate sustainability on organizational processes and performance. Manag. Sci. **60**(11), 2835–2857 (2014). https://doi.org/10.1287/mnsc.2014.1984
3. Gupta, S., et al.: Artificial intelligence for decision support systems in the field of operations research: review and future scope of research. Ann. Operat. Res. **308**(1–2), 215–274 (2022). https://doi.org/10.1007/s10479-020-03856-6
4. Handy, C.B., Bennis, W.G.: The Age of Unreason. Harvard Business School Press, Boston, MA (1989)
5. Dohn, K., et al.: The early warning concept in the system supporting knowledge management in polish machine-building industry enterprises. In: Proceedings - KIM 2013, Knowledge and Information Management Conference: Sustainable Quality, pp. 211–221. The OR Society (2013)
6. Martinez, M.: German corporate bankruptcies to surge to a decade high in 2025, Reuters Media, 8 December (2025). Available at: https://www.reuters.com/business/german-corporate-bankruptcies-surge-decade-high-2025-2025-12-08/. Accessed 21 Dec 2025
7. McElroy, G.: Commentary - Company Insolvency Statistics December 2024, 21 December (2025). Available at: https://www.gov.uk/government/statistics/company-insolvencies-december-2024/commentary-company-insolvency-statistics-december-2024. Accessed: 21 Dec 2025
8. Hawkins, S.: The UK Insolvency Market: Predictions for 2025 | 16 Dec 2024 | Insights | Bishop Fleming, 21 December (2024). Available at: https://www.bishopfleming.co.uk/insights/uk-insolvency-market-predictions-2025. Accessed 21 Dec 2025
9. Johnson, G., Whittington, R.: Exploring Strategy: Text and Cases. Pearson, Harlow, UK (2023)
10. Chen, S., et al.: Toward understanding inter-organizational knowledge transfer needs in SMEs: insight from a UK investigation. J. Knowled. Manag. **10**(3), 6–23 (2006). https://doi.org/10.1108/13673270610670821
11. Zafirova, T.: Strategic drift and strategic crisis management of organization. China-USA Busi. Rev. **13**(7) (2014). https://doi.org/10.17265/1537-1514/2014.07.006
12. Finkelstein, A.: Report of the inquiry into the London ambulance service: Report to The Communications Directorate, South West Thames Regional Health Authority (1993). Available at: http://www0.cs.ucl.ac.uk/staff/A.Finkelstein/las/lascase0.9.pdf. Accessed: 21 Dec 2025
13. Aversa, P., Cabantous, L., Haefliger, S.: When decision support systems fail: insights for strategic information systems from Formula 1. J. Strat. Inform. Syst. **27**(3), 221–236 (2018). https://doi.org/10.1016/j.jsis.2018.03.002
14. Wright, A., et al.: Analysis of clinical decision support system malfunctions: a case series and survey. J. Am. Med. Inform. Assoc. JAMIA. **23**(6), 1068–1076 (2016). https://doi.org/10.1093/jamia/ocw005
15. Brauner, P., et al.: What happens when decision support systems fail? — the importance of usability on performance in erroneous systems. Behav. Inform. Technol. **38**(12), 1225–1242 (2019). https://doi.org/10.1080/0144929X.2019.1581258
16. Lehaney, B., Paul, R.J.: Soft systems methodology and simulation modeling. In: Proceedings of the 28th conference on Winter simulation – WSC'96, the 28th conference, Coronado, CA, United States, 12/8/1996–12/11/1996, pp. 695–700. ACM Press, New York (1996). https://doi.org/10.1145/256562.256788

17. Lehaney, B., et al.: Beyond Knowledge Management. IGI Global, Hershey, PA (2004) Available at: http://services.igi-global.com/resolvedoi/resolve.aspx?doi=10.4018/978-1-59140-180-3
18. Amodei, D.A.: The Adolescence of Technology: Confronting and Overcoming the Risks of Powerful AI, 27 January (2026). Available at: https://www.darioamodei.com/essay/the-adolescence-of-technology. Accessed 31 Jan 2026
19. Duan, Y., Cao, G., Ong, V., Xu, M., Dietzmann, C.: AI in information processing for decision making - perceptions and use intention of UK managers. In: Nunes, M.B., Isaías, P., Powell, P., Kommers, P. (eds.) Proceedings of the IS 2025 + ES 2025, pp. 423–427. IADIS (2025) https://www.iadisportal.org/is-es-2025-proceedings

Large Language Models and Data-Driven Decision Support

Large Language Models in MCDA Weight Elicitation: A Systematic Review

Michail Rafail Ouzounis(✉), Jason Papathanasiou, Georgios Tsaples, and Georgios Drogalas

University of Macedonia, Egnatias 156, 54636 Thessaloniki, Greece
m.ouzounis@uom.edu.gr

Abstract. The rapid advancement of Large Language Models has sparked growing interest in their integration into Multi-Criteria Decision Analysis, particularly in weight extraction methodologies. However, despite the increased research activity of recent years, a unified methodological framework that clearly guides the use of LLMs in the weighting process is still missing, along with consensus on how to ensure transparency and reliability in the generated results. The aim of this study is to systematically review the literature from 2020 to 2025 and critically investigate how LLMs have been used in the field of MCDA for weight extraction. The findings reveal significant methodological heterogeneity in terms of both prompt engineering and evaluation protocols, including comparisons with human experts. Despite their potential, current LLM-based weight elicitation approaches often fail to consistently capture human judgment. Finally, the review highlights the lack of comprehensive methodologies, discovering that the field is still in its early stages and requires efforts to standardize and develop transparent, valid, and reproducible methodologies for the use of LLMs in criterion weighting.

Keywords: Large Language Models (LLMs) · Multi-Criteria Decision Analysis (MCDA) · Weight Elicitation · Decision Support Systems · Prompt Engineering · Reliability and Consistency Analysis

1 Introduction

Decision-making is a valuable tool for selecting the best alternative solution, as it often requires the simultaneous evaluation of many and, in many cases, conflicting criteria. Multi-criteria decision analysis (MCDA) offers a structured methodological tool for addressing such complex problems, allowing decision-makers to combine quantitative and qualitative factors in a comparable manner. This technique is used in various areas of everyday life, such as environmental management or strategic planning. Despite the maturity of the methodology, one of the most essential and decisive stages is weight elicitation, i.e., determining the relative importance attributed to each criterion.

Assigning weights is a fundamental step in MCDA, since it specifies the impact of each criterion on the overall decision. To illustrate, when evaluating renewable energy technologies, participants may consider criteria such as cost, sustainability, efficiency,

A. P. Cabral Seixas Costa et al. (Eds.): ICDSST 2026, LNBIP 582, pp. 257–269, 2026.
https://doi.org/10.1007/978-3-032-25146-6_17

and social acceptance, but these criteria may not all have the same weight. Scientific research has identified several classic methods for the weighting process, such as the Analytic Hierarchy Process (AHP), the Simple Multi-Attribute Rating Technique (SMART), and the Best-Worst Method, which rely heavily on human judgment and consistent pairwise comparisons. Although these methods have been scientifically analyzed, the results are often influenced by human bias, and the elicited weights may reflect behavioral inconsistencies rather than mathematical inconsistencies of the aggregation models themselves, particularly as the number of criteria increases.

In recent years, there has been some debate as to whether traditional approaches can truly express the complexity of human judgment under conditions of uncertainty. Factors such as question order and context can significantly influence weights, leading to unstable results. At the same time, there are objective weight extraction methods such as Entropy, CRITIC, and MEREC, which seek to a greater extent to extract impartiality based on the data of the other alternative options. However, these methods often fail to represent the underlying value-based preferences that guide real-world decisions. As a result, there is a conflict between objectivity and representativeness, with the result that neither subjective nor objective methods provide a complete solution.

Recent advances in AI offer novel approaches for improving MCDA, not only by automating procedures but also by analyzing patterns, deriving preferences from past data, and ensuring more consistent expert evaluations. The rapid progress of Large Language Models (LLMs) such as GPT-5, GPT-4, Gemini 3, and Sonar introduces a new dimension to decision-making. These models can process natural language, interpret textual information, and transform qualitative descriptions into quantitative data, such as MCDA weight estimates. Early studies have already experimentally investigated how LLMs can take on the role of experts by producing consistent comparisons of criteria and documenting the reasoning behind the proposed results. The combination of MCDA and AI represents an interaction between human judgment and computational intelligence. Using LLMs for weight elicitation can lower the cognitive demand, while improving the stability and transparency of outcomes. Acting as virtual agents, LLMs can combine multiple approaches, reduce bias, and generate more reliable estimates. Nevertheless, evidence regarding the reliability, consistency, and interpretability of LLM-derived results remains limited.

This review examines trends, approaches, and gaps in integrating AI into decision-making systems. It reviews the current state of research, highlights the possibilities and challenges of this process, and provides a synthetic interpretation of the findings to date. Through this process, the paper aims to understand and highlight the gaps in the research community and to present the usefulness and benefits of artificial intelligence in the MCDA process. Accordingly, the review addresses the question of how existing literature conceptualizes and evaluates the role of large language models in MCDA weight elicitation and where further methodological development is required.

2 Theoretical Background

A prerequisite for appreciating recent developments in Multi-Criteria Decision Analysis (MCDA) is familiarity with the process of weighting criteria, both with and without the use of artificial intelligence. This chapter reviews the conceptual and methodological background of MCDA, with emphasis on the methods and challenges of weight elicitation. The main classical techniques and the central theoretical challenges of consistency and subjectivity are reviewed, followed by a discussion of attempts to connect MCDA with LLMs to establish a coherent framework for evaluating their capabilities, limitations, and future prospects in the weighting and decision-making process.

2.1 Foundations of Multi-Criteria Decision Analysis (MCDA)

MCDA is a tool developed to solve complex, multi-dimensional decision-making problems that cannot be addressed through a one-sided analysis. In the theoretical framework, complex problems are converted into structured analysis models, helping researchers and decision-makers to understand more clearly the relationships between criteria and how they affect the outcome [1, 2].

Any decision problem in MCDA is classified into sets of criteria, solutions, and preferences. Each criterion addresses a particular aspect of the problem, such as economic, social, or environmental, which may conflict with other criteria. The use of weights highlights the relative influence of each criterion on the aggregated assessment.

The overall value of each alternative is calculated as a linear combination of the criteria and their corresponding weights. The mathematical relationship describing the process is given by the formula:

$$V_j = \sum_{i=1}^{n} w_{i.} \cdot s_{ij} \tag{1}$$

where $\sum_{i=1}^{n} w_i = 1$

In the formula above, V_j is the total score of alternative j, with w_i representing the weight of criterion i and s_{ij} measuring the performance of alternative j regarding that criterion. The total number of criteria is represented by n. The weights are normalized to ensure that their sum equals one, allowing results to be compared and importance to be distributed consistently across criteria. Weights are a decisive factor in determining the overall assessment where even small variations can substantially affect the ranking of alternatives, making the weight elicitation process critical for the reliability and stability of decisions [3, 4]. A variety of methods have been developed to calculate weights, categorized by levels of objectivity and subjectivity, with the aim of balancing outcomes [5].

2.2 Challenges and Limitations of Weight Elicitation

Weight elicitation is perhaps one of the most important processes, as it is responsible for the reliability and stability of the results obtained after multi-criteria analysis. Although in theory weights reflect the preferences of the decision maker, in reality difficulties arise as preferences may be vague and contradictory. Furthermore, the complexity of the problems and cognitive fatigue can negatively affect the results, causing uncertainties.

Weighting techniques are scientifically divided into subjective and objective. Subjective methods are based on the judgment, experience, and perception of experts or participants. Some of the methodologies mentioned above are AHP, SMART, and BWM, which are widely accepted and influenced by cognitive biases, the order of questions, and the difficulty of maintaining consistency when there are many criteria. For the above reasons, considerable care is required in the process of collecting preferences and interpreting the weights produced. In contrast, objective methods aim to determine weights based solely on data performance without human opinion. Examples of algorithms are the Entropy Method, CRITIC, and MEREC [6–8]. These methodologies approach the problem in a neutral manner, differentiating the data of the criteria, but often ignoring value preferences, limiting the possibility of real decision-making. The balance between subjectivity and objectivity is a key point in the literature, as absolute dependence on expert opinions can lead to inconsistency, while full automation with objective methods can alienate and distance the process from reality. For these reasons, several recent studies propose hybrid weighting models that combine these two approaches to achieve something realistic. In addition, sensitivity analysis plays an important role in showing us the degree of influence of weights on the final result. The analysis contributes to the stability, evaluation, and reliability of the results in exchange for the high cognitive and computational load required.

Based on the above, the need for new weight extraction approaches that would overcome the weaknesses that exist today is becoming increasingly urgent. The integration of artificial intelligence, and more specifically Large Language Models, which can extract preferences with little cognitive overhead, may open new avenues for addressing these issues.

2.3 Artificial Intelligence and Large Language Models in the Context of MCDA

The integration of Artificial Intelligence and Large Language Models into MCDA is one of the most important and critical developments in contemporary research on decision support systems. The rapid increase in computing power, the availability of data, and the development of language models such as GPT-5, GPT-4, Gemini 3, and Sonar have opened new horizons in preference modeling and weight extraction in a more efficient, explainable, and less error-prone manner [9, 10]. Artificial intelligence has been used in multi-criteria analysis to optimize results since its inception, but recent learning systems, especially LLMs, present a new form of cognitive support where the machine is not limited to calculation but can also understand, interpret, and formulate answers that resemble those of humans. This creates the possibility for a new type of weighting in which there is transparency, as the results are interpreted in natural language and converted into numerical weights.

Recent studies suggest that combining LLMs with established methods like AHP or BWM can help reduce subjectivity and ease the cognitive load by acting as virtual experts. Beyond substituting human judgment, LLMs contribute to expanding domain knowledge and enhancing the consistency of outcomes. The introduction of GPT-5 and Gemini 3 in 2025 represented an important step forward, providing advanced multi-step and long-context reasoning capabilities that are especially useful for complex multi-criteria problems with interdependent criteria. At the same time, the Gemini and Sonar models offer a high degree of adaptability to specific domains, making them suitable for applications such as risk assessment and strategic decision-making. An additional innovation is the development of multi-agent LLM systems in which different models work collaboratively, simulating the opinions and preferences of different experts, thereby improving stability and variety of results. Despite these positive developments, the literature acknowledges that the application of LLMs in MCDA remains in its infancy. Most existing studies are either experimental or focus on limited application contexts, without fully evaluating the reliability, repeatability, and explainability of results across different problem domains. Furthermore, the lack of unified evaluation protocols makes it difficult to compare approaches and assess the actual contribution of LLMs to the decision-making process.

Overall, this integration will lead to fundamental flexibility in decision support, enabling the understanding of natural language, context analysis, and the creation of cognitive assistance. However, in order to fully exploit this potential, further empirical research and comparative evaluation against traditional methods are required to ensure the scientific validity and practical usefulness of these new approaches.

2.4 LLMs in Individual and Group Weight Elicitation Contexts

Weight assessment using LLM may present different approaches depending on whether the decision-making context concerns an individual or group object. In individual decision-making, weights reflect the personal values, attitude towards risk, and priorities of an individual decision-maker. In such cases, the use of LLM presents both opportunities and challenges. Dialogue interaction carries the risk of reducing cognitive load, but it can help clarify preferences and support structured thinking. However, there is a risk that the suggestions generated by the model may inadvertently guide the decision-maker towards generalized or average preference structures, greatly reducing personal judgment.

In contrast, group decision-making frameworks introduce complexity as weighting factors arise from the combination of multiple stakeholders. Under these circumstances, LLMs can offer greater potential value by synthesizing different views and identifying areas of convergence and divergence for the derivation of final weights. The ability to adapt information from different sources and create scenarios facilitates the production of more accurate results.

Therefore, two different categories emerge, which must be examined separately for individual and group environments, as the benefits may not be symmetrical in both cases.

2.5 Prompt Engineering in LLM-Based Weight Elicitation

Prompt engineering is a fundamental element for the use of Large Language Models in weight extraction processes in MCDA. The structure, specialization, and context of prompts directly influence the responses generated by LLMs, shaping both the numerical outputs and the explanatory reasoning provided by the model. In practice, prompts can set constraints and normalizations (for example, requiring the resulting weights to sum to 1) and formulate scenarios and preferences to avoid the risk of misaligned responses. Iterative prompting strategies further enhance the mining process by allowing decision makers to refine and reevaluate weight distributions through successive interactions, thus improving interpretability and alignment with subjective preferences. The methodology concerns the correct formulation of the problem so that the results produced are transparent and consistent in their reproducibility. Consequently, systematic examination of prompt engineering is essential to ensure that weight extraction is aligned with MCDA objectives.

3 Research Methodology & Results

For this paper, a gradual search was conducted in order to progressively improve the search field and ensure as much completeness and relevance as possible. The initial search was conducted in the Scopus database, which contains high-quality publications and is particularly strong in advanced and established research results. This search yielded a very limited number of relevant results on the use of large language models for weight extraction in the context of MCDA. To make the research more comprehensive, an expanded search was conducted via Google Scholar, which includes a wide range of academic materials beyond journal articles. The search covered the period 2020–2025, during which there was growth and increase in the use of LLM applications in various types of tasks, and more specifically in decision-making. Boolean logic was applied in the advanced search (AND, OR) to combine keywords.

Using the keywords "multi-criteria decision analysis" OR "MCDA" AND "weight extraction" OR "criteria weighting," the initial search identified approximately 2,700 records. This search was intended to cover the general research framework concerning MCDA, but it also included many studies outside the specific subject of weight extraction based on LLM. Figure 1 illustrates the annual publication trend for studies combining MCDA and weight elicitation.

Next, publications containing keywords such as "multi-criteria decision analysis" OR "MCDA" AND "large language models" OR "LLM" OR "GPT" were found. This search yielded 759 publications, reflecting the growth of the field, a fact linked to the maturation of models such as GPT-5, GPT-4, Gemini, Claude 3, Llama 3, and Gemma, but also highlighting the still fragmented nature of research integrating artificial intelligence with decision-making methodologies. Figure 2 summarizes the number of publications identified through this keyword combination per year during the period 2020–2025. However, many of these studies concerned general artificial intelligence applications or decision support systems without directly focusing on weight extraction in MCDA.

In the final step, and in order to further narrow down the publications related to the topic of weight extraction using LLM, a final search was performed using a combined query. The search was conducted using the following Boolean query:

```
("multi-criteria decision analysis" OR "MCDA") AND ("weight
extraction" OR "criteria weighting") AND ("large language
models" OR "LLM" OR "GPT")
```

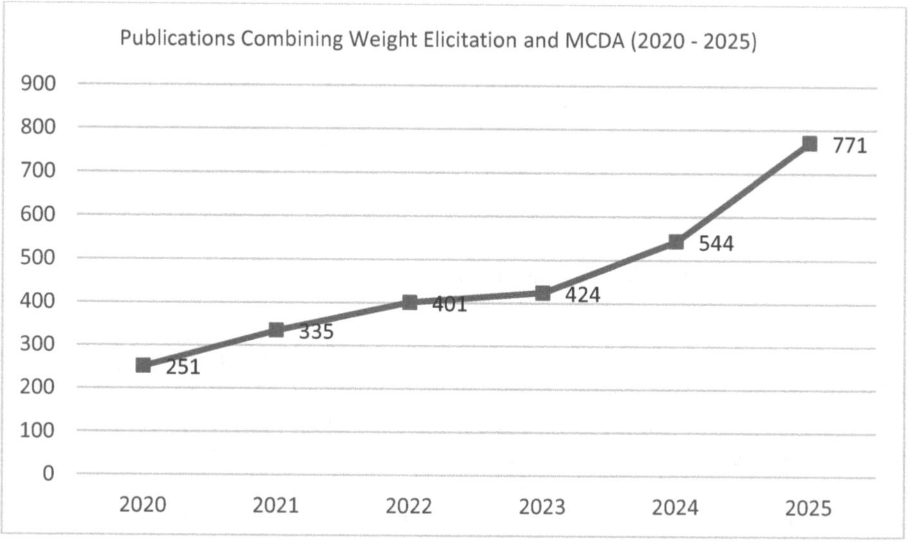

Fig. 1. Publications Combining Weight Elicitation and MCDA

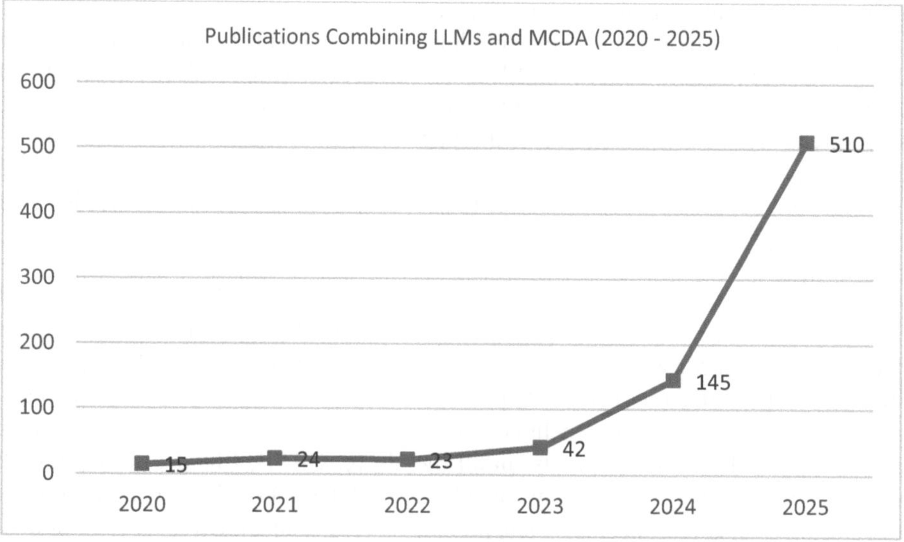

Fig. 2. Publications Combining LLMs and MCDA

Predefined inclusion criteria were applied to the findings to ensure relevance to the search topic. Studies were included only if (i) they explicitly examined the use of LLMs in weight extraction or MCDA applications, (ii) they provided methodological or empirical contributions relevant to decision support, and (iii) they were written in English. Duplicate publications and studies that did not meet these criteria were excluded during the review of titles and abstracts (Fig. 3).

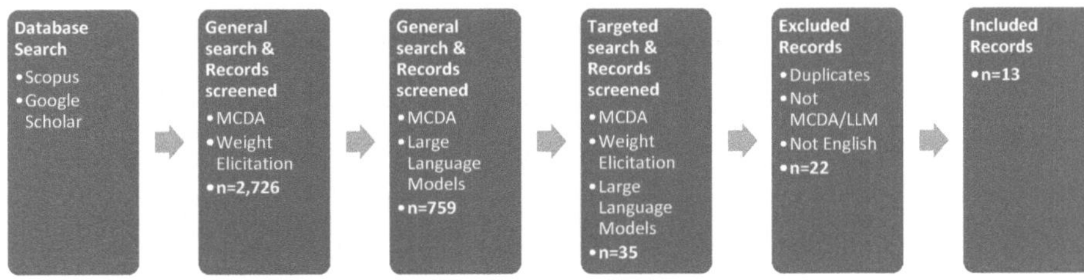

Fig. 3. Systematic Literature Review Flowchart

The final specialized search yielded 35 articles, from which, after applying the inclusion criteria, the final sample consisted of 13 publications Table 1.

The analysis of the thirteen identified works, including peer-reviewed publications and preprints, highlights a research field that is particularly new but at the same time fragmented in terms of methodology, the use of LLMs, and the process of weight extraction in Multi-Criteria Decision Analysis (MCDA). Although most publications were made as recently as 2025, there is already a trend of trust and utilization of LLMs either as virtual experts, as supportive interpretation mechanisms, or as multi-agent automatic grading systems. However, even in the most technically mature papers, there are important caveats that must be mentioned (Fig. 4).

Of the thirteen articles, only twelve implement a real weight elicitation process using LLMs (e.g., Multi-agent LLM MAS, LLMs as virtual AHP experts, RAD, BWM–LLM synergy, GPT-4 AHP Aidm, IVHFF TOPSIS). These constitute the core of the literature that is directly relevant to the subject of this paper, and more specifically, within this category, it is observed that AHP and fuzzy variants have a higher frequency of occurrence for weight extraction. Although these methods show potential, their dominance is based on small samples with no comparative tests between different LLMs or standardized command protocols.

An important finding is that only four papers actually perform benchmarking of LLMs, and even then in a limited problem environment. Paper [11] (Multi-agent LLM MAS) compares LLMs within a multi-agent environment, while Paper [12] compares six different LLM architectures in an AHP problem with verification against human experts. Paper [19] (IVHFF TOPSIS) benchmarks genetic chatbots, but without a comparative consistency check or fixed prompts. In all the above, GPT-4, Claude, or Gemini are mainly used, while no paper uses new models such as GPT-5.1 or Gemini 3. The absence of a controlled benchmarking that would include a fixed prompt, fixed seed, and common decision task is evident.

Equally limited is the comparison of benchmarking and human experts, which appear only in Papers [11, 12, 17]. Although these studies attempt to investigate the reliability of LLM-generated weights, the samples are small, and statistical validation remains at a basic level. There is a lack of more in-depth testing, such as sensitivity analysis on multiple prompts, robustness under model change, cross-model agreement tests, or analysis of systematic biases (LLM bias fingerprints). Therefore, the performance of LLMs as preference extraction mechanisms remains poorly documented to date.

Table 1. Findings of Studies Related to LLM Applications in MCDM

Work	Paper	Weight Elicitation	Benchmarking LLMs	Comparison Expert	Prompt Engineering Used
[11]	Multi-agent LLM MAS	✔	✔	✔	✖
[12]	LLMs as virtual AHP experts	✔	✔	✔	✔
[13]	Trustworthiness Eval via AHP	✔	✔	✔	✖
[14]	IF-PROBID & LLMs	✖	✖	✖	✔
[15]	RAD Framework	✔	✖	✖	✔
[16]	BWM & LLM Synergy	✔	✖	✖	✔
[17]	AHP & GPT-4 Aidm	✔	✖	✔	✔
[18]	IDEMATEL for GenAI selection	✔	✖	✖	✖
[19]	IVHFF TOPSIS GAI chatbots	✔	✔	✖	✖
[20]	Tourism Recommender GPT-3.5	✔	✖	✖	✔
[21]	GPT Agents for Policy Modeling	✔	✖	✖	✔
[22]	AI/ML Approaches for Heterogeneous MCGDM	✔	✖	✖	✔
[23]	AHP & LLM for Cyber Evaluation	✔	✖	✖	✔

On the issue of prompt engineering, the picture is even more unclear. Although Papers [12, 15, 16, 17, 20, 21, 22, 23] use specially designed prompts, only Paper [12] offers complete and systematic documentation of different prompt strategies. The rest provide fragmentary examples without repeatability and without variables that allow the process to be reproduced. With the above research, the literature remains empty in terms of standardization and reproducible prompt protocols for MCDA weight elicitation, which makes it difficult to compare studies and impossible to evaluate LLMs comprehensively. Another important feature is the absence of systematic reliability checks of LLM results. Only two papers explicitly apply control mechanisms such as IQR filtering, Kendall W

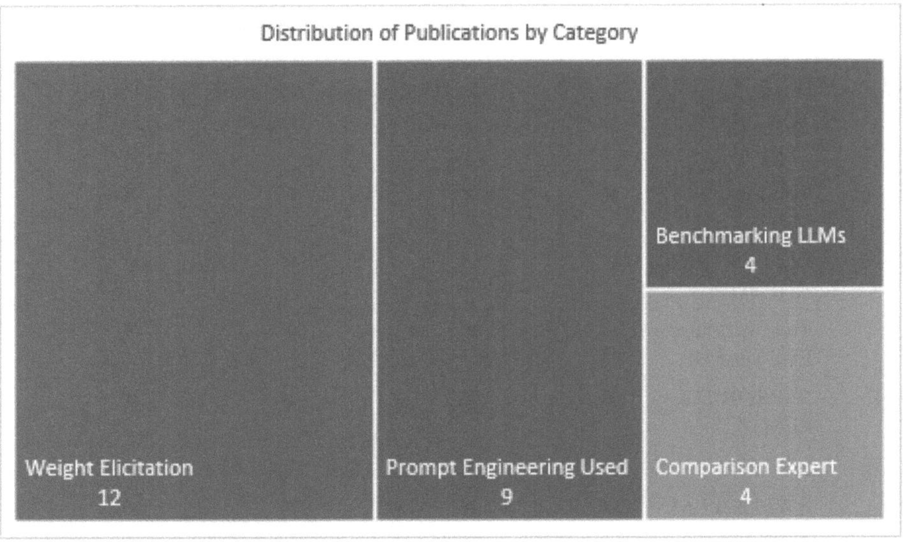

Fig. 4. Distribution of Works by Category

consensus, and sensitivity analysis. The rest of the studies either limit themselves to descriptive comparisons with human experts or do not perform any form of consistency checking. There are no tests such as multiple runs with identical prompts and comparative studies of results.

The gradual refinement process from broad MCDA research to LLM related studies and finally to weight extraction applications, ensured that the selected corpus was both focused and representative of the available evidence.

4 Discussion

The analysis of the studies shows that the use of Large Language Models in weight extraction for MCDA is in its early stages but is developing very rapidly. The fact that few studies combine LLMs and MCDA indicates that this is a new field in which standards and methodologies have not yet been firmly established. The creation of a virtual expert and its ability to compare and justify its results reduces the cognitive load required of the participant, but it cannot be considered a reliable source as there is not yet sufficient evidence that the outputs are more stable or reliable than those of human experts. Similarly, the documentation of the results in relation to the prompts used is incomplete. There is no analytical structure to check the numerical effects that a change in the prompt may have, nor any methodology for monitoring the different results that the use of different LLMs (GPT, Claude, Gemini) may produce. The same problem is observed in the relationship between humans and LLM, where only a few studies compare the results with those of human experts, while statistical verification remains at a basic level.

Assessing the reliability of the weights produced is an even greater challenge. Few studies apply consistency or stability checks, such as Kendall W, IQR filtering, or sensitivity analysis. The rest rely on descriptive comparisons without checking for multiple repetitions or variations between different models. This leaves open the question of how consistently and repeatably LLMs produce valid results, which is critical for any MCDA application.

LLM-based weight extraction introduces a more interactive approach compared to traditional MCDA procedures. Through natural language dialogue, decision-makers can iteratively clarify and refine their preferences, enhancing interpretability and engagement in the process. This feature extends existing weight extraction methods rather than replacing them, although its methodological reliability still requires systematic empirical validation.

Nevertheless, the studies highlight significant potential by focusing on the use of multi-agent approaches where more than one model collaborates or converges on common preferences, offering greater stability. The ability of LLMs to interpret text, extract criteria, and justify their choices in natural language is something that can be studied because it improves the transparency of the process. Overall, LLMs can offer a new form of support in the multi-criteria analysis process by reducing cognitive load. However, scientific documentation remains limited, and more systematic research with controlled experiments is needed to stabilize the methodology of the process.

5 Conclusion

The purpose of this study was to conduct a systematic literature review on how Large Language Models are used in the weight extraction process in the context of Multi-Criteria Decision Analysis (MCDA). The review involved recording existing research to identify current trends and, above all, to investigate whether the absence of a unified methodological framework for LLM-based weight elicitation has affected the quality and comparability of research results. A search of the literature for the period 2020–2025 showed that the field of research is in an early but rapidly developing stage. The findings show that no standardized methodology is being followed, with the result that no reliable picture of the results is being created. There is also a lack of comprehensive comparative benchmarking between different LLMs and comparisons of specific people in different fields with specific prompt text. In combination with benchmarking, it is also necessary to check the reliability that each model can extract.

The above weaknesses have a direct impact on the final results obtained from MCDA. The use of non-standardized and unverified weighting procedures can lead to incorrect estimates, affecting the stability of decisions. Taking the above limitations into account, we believe that future research could focus on developing a unified standardized framework for weight extraction with LLMs, which would include the latest models, stable test scenarios, and comparative reliability and validity studies with the help of human experts.

Future research should therefore go beyond descriptive reviews and focus on empirical validation and comparative evaluation of weight extraction based on LLM. Controlled experimental studies comparing weights derived from LLM with traditional extraction techniques and human expert judgments will provide stronger evidence regarding the stability and reliability of decisions. Comparative evaluation across different models and application domains and proper study of prompt engineering are necessary to assess methodological consistency and reproducibility. Such efforts will contribute to the development of scientifically sound frameworks that enhance the reliability and transparency of MCDA applications involving LLM.

I hope that this review will serve as a useful reference point for the research community and contribute to the development of a more scientifically sound approach to the use of LLMs in the field of MCDA, and in particular in the weight extraction process.

Acknowledgements. This research did not receive any specific grant from funding agencies in the public, commercial or not-for-profit sectors.

Disclosure of Interests The authors declare that they have no known competing financial interests or personal relationships that could have appeared to influence the work reported in this paper.

References

1. Greco, S., Ehrgott, M., Figueira, J.R. (eds.): Multiple Criteria Decision Analysis: State of the Art Surveys, 2nd edn. Springer, Heidelberg (2016)
2. Mardani, A., Jusoh, A., Zavadskas, E.K., et al.: Multiple criteria decision-making techniques and their applications – a review of the literature from 2000 to 2014. Econ. Res. Ekon. Istraž. **28**(1), 516–571 (2015)
3. Belton, V., Stewart, T.J.: Multiple Criteria Decision Analysis: An Integrated Approach, 2nd Print. Kluwer Academic Publishers, Dordrecht (2003)
4. Kou, G., Ergu, D., Peng, Y.: Evaluation of decision-making methods in MCDA: a comparative study. Omega. **103**, 102360 (2021)
5. Zavadskas, E.K., Turskis, Z., Kildiene, S., et al.: State of art surveys of overviews on MCDM/MADM methods. Technol. Econ. Dev. Econ. **20**(1), 165–179 (2014)
6. Wang, X., Luo, Y.: A review of objective weighting methods in multi-criteria decision making. J. Syst. Sci. Syst. Eng. **18**(4), 456–472 (2009)
7. Diakoulaki, D., Mavrotas, G., Papayannakis, L.: Determining objective weights in multiple criteria problems: the critic method. Comput. Oper. Res. **22**(7), 763–770 (1995)
8. Pamučar, D., Stević, Ž., Vuković, K.: Recent developments in objective weighting methods in MCDA. Expert Syst. Appl. **220**, 119736 (2023)
9. Svoboda, P., Lande, D.: Artificial intelligence in decision support systems: emerging trends. J. Decis. Anal. **12**(3), 45–68 (2024)
10. Leon, R.: Large language models for multi-criteria decision analysis: applications and perspectives. Int. J. AI Res. **17**(1), 101–130 (2025)
11. Vahidnia, M.H.: Multi-agent systems of large language models as weight assigners: an approach to collaborative weighting in spatial multi-criteria decision-making. Geomatica. **77**(2), 100071 (2025)
12. Memduhoğlu, A., Fulman, N., Polat, N., Ataş, T.: Large language models as virtual experts? Evaluating AHP-based criteria weighting performance for solar power plant site selection. Expert Syst. Appl. **299**(Part B), 130171 (2026)
13. Aksoy, M., Adem, A., Dağdeviren, M.: Trustworthiness evaluation of large language models using multi-criteria decision making. IEEE Access. **13**, 168183–168201 (2025)

14. Anhao, F., Karbassi Yazdi, A., Tan, Y., Ocampo, L.: Integrating large language models into a novel intuitionistic fuzzy PROBID method for multi-criteria decision-making problems. Mathematics. **13**(17), 2878 (2025)
15. Wu, H., Zhang, H., Chen, W., Xia, J.: Retrieval augmented decision-making: a requirements-driven, multi-criteria framework for structured decision support. arXiv abs/2505.18483 (2025)
16. Briegel, H., Tipdecho, T.: A synergistic integration between large language models and the best-worst method. In: Rezaei, J., Brunelli, M., Mohammadi, M. (eds.) Advances in Best–Worst Method Lecture Notes in Operations Research, pp. 23–39. Springer, Cham (2025)
17. Dehghanimohammadabadi, M., Kabadayı, N.: The ai-driven decision-making (Aidm) framework: integrating AHP and ChatGPT-4 for supplier selection. SSRN (posted 24 Oct 2024) 2024
18. Rădulescu, C.Z., Rădulescu, M.: Criteria analysis for the selection of a generative artificial intelligence tool for academic research based on an improved group DEMATEL method. Appl. Sci. **15**(10), 5416 (2025)
19. Ilieva, G.: Extension of interval-valued hesitant Fermatean fuzzy TOPSIS for evaluating and benchmarking of generative AI chatbots. Preprints.org (posted 7 Jan 2025) 2025
20. Chasanidou, D., Boletsis, C., Krogstie, J.: User experience in sustainable tourism: enhancing transparency in tour guide app recommendations. In: Proceedings of the 3rd International Conference of the ACM Greek SIGCHI Chapter (CHI Greece 2025), pp. 55–60. ACM, New York (2025)
21. Ariaso Sr., D.A., Gorro, K.D., Balijon, D., Balijon, M.: Integrating fine-tuned GPT with agent-based economic Modeling for transparent wage policy decisions. Int. J. Adv. Comput. Sci. Appl. **16**(8) (2025)
22. Abu-Faty, H.G., Kafafy, A., Abdel-Raouf, O., Hadhoud, M.M.: Integrating generative AI and machine learning classifiers for solving heterogenous MCGDM: a case of employee churn prediction. Sci. Rep. **15**, 15645 (2025)
23. Kampourakis, V., Kavallieratos, G., Spathoulas, G., Gkioulos, V., Katsikas, S.: LLM-assisted AHP for explainable cyber range Evaluation arXiv abs/2512.10487 (2025)

An LLM-Augmented System-Theoretic DSS for Resilient Transitions of Smart Ports

Xiaofang Wu[1,2,3](✉) and Yu Cao[1]

[1] Jimei University, Xiamen 361021, China
xiaofang.wu@jmu.edu.cn

[2] Fujian Provincial University Engineering Research Center for Smart Port and Shipping, Xiamen 361021, China

[3] Xiamen Key Laboratory of Green and Smart Coastal Engineering, Xiamen 361021, China

Abstract. As smart ports evolve with digital and intelligent technologies, their growing complexity challenges conventional vulnerability assessment methods in identifying risks from dynamic interactions and control failures. Current approaches largely focus on climate change and natural hazards and heavily rely on expert judgment and predefined indicators. Their assessment processes are also often time-consuming and labor-intensive. These limitations hinder the capture of dynamic risks and the reproducibility of analyses during the transformations. Hence, this study proposes a novel, reproducible Decision Support System (DSS) that integrates a Large Language Model (LLM) with System-Theoretic Process Analysis (STPA) and the CRITIC method for dynamic vulnerability diagnosis in smart ports. Unlike prior applications of LLMs in safety analysis, this work frames LLM-augmented STPA as a coherent and transferable DSS for resilience governance, with a fully documented prompting strategy provided in supplementary materials to ensure reproducibility. A case study of the Hairun automated container terminal demonstrates the system's capability to replicate core UCAs and most expert-identified causal factors and uncovers vulnerabilities related to information system interactions and temporary operational states that are often overlooked by conventional methods. Overall, this study not only offers a systematic and reproducible DSS for port resilience governance but also provides a transferable framework for integrating LLMs into resilience analysis of complex socio-technical systems.

Keywords: Systematic Vulnerability Diagnose · System-Theoretic Process Analysis · Large Language Models · Automated Container Terminal

1 Introduction

Ports facilitate over 80% of global trade by volume and function as critical nexuses within international supply chains. In recent years, the rapid advancement of digital and intelligent technologies such as the Internet of Things, autonomous robotics, cloud computing, and artificial intelligence has fundamentally reshaped port operation paradigms, accelerating the global transition toward "Smart Port" [1, 2]. This transition shows substantial gains in operational efficiency, resource utilization, and environmental sustainability through predictive maintenance, intelligent scheduling, or energy optimization.

A. P. Cabral Seixas Costa et al. (Eds.): ICDSST 2026, LNBIP 582, pp. 270–286, 2026.
https://doi.org/10.1007/978-3-032-25146-6_18

Scholars are also aware of growing vulnerabilities for cyber-physical interdependencies, tighter coupling between digital platforms and physical equipment, evolving organizational roles, and fragmented regulatory frameworks [3–5]. Smart ports are exposed to emergent risks such as cascading failures, amplified operational disruptions, and expanded threat surfaces associated with cyberattacks [6]. There is a need to capture system-level interactions and transition-induced vulnerabilities beyond isolated component failures.

However, existing studies on port vulnerability have predominantly focused on climate change and natural hazards like sea-level rise, extreme weather events, earthquakes, and tsunamis [7–10], and systematic methods for integrating the port digital and intelligent concerns into holistic vulnerability assessments remain underdeveloped [3, 11]. The absence may risk underestimating transition-induced vulnerabilities and ultimately compromise not only port resilience but also broader economic stability and national security. Additionally, current port vulnerability assessments commonly are based on predefined indicators, combining with fuzzy methods, geographic analysis, or simulation models [8, 12–16]. They are limited to represent dynamic interactions. Moreover, most of them heavily rely on expert judgment, which may constrain the assessments' objectivity and reproducibility, thereby affecting the effectiveness and efficiency of decision-making.

Therefore, this study aims to propose a dynamic, systematic decision support system (DSS) for diagnosing vulnerabilities during smart port transitions. It would be grounded in System-Theoretic Process Analysis (STPA), which conceptualizes safety and vulnerability as emergent properties arising from inadequate control and coordination in complex socio-technical systems [17]. In order to support dynamic, reproducible, and scalable, a Large Language Model (LLM) will be embedded across all analytical phases. The research question is how a STPA-based DSS is augmented by LLMs to enable reproducible and scalable vulnerability diagnosis in smart port transitions.

This research contributes to developing a systematic and reproducible decision support methodology. Its primary contribution lies not merely in applying LLMs to STPA, but in formalizing an LLM-augmented STPA workflow into a coherent DSS for resilience governance.

2 Literature Review

2.1 Current Port Vulnerability Assessments

Table 1 reviews the current methods employed for port vulnerability assessments. We can find that they are commonly grounded in a multi-criteria decision-making (MCDM) approach, with the integration of specific expert-driven weighting methods (e.g., Analytic Hierarchy Process (AHP) and Best-Worst Method (BWM)), objective data-driven techniques (e.g., Entropy and Criteria Importance Through Intercriteria Correlation (CRITIC)), and fuzzy or evidential reasoning methods to address uncertainty and complex interdependencies.

2.2 Resilience and STPA in Ports

Increasingly, risks arise not from component failures but from dysfunctional interactions among well-functioning components [22], while resilience relies on effective control

Table 1. Current methods for port vulnerability assessments

Methods	Description	Source
AHP-Entropy-Grey Relational Analysis	It combines the AHP and Entropy methods for objective and subjective weight determination, followed by Grey Relational Analysis to evaluate and rank alternatives based on multi-criteria performance.	[18]
AHP-Weighted Sum Model	It employs the AHP to derive expert-based weights for indicators of seaport exposure and sensitivity, which were then aggregated using a Weighted Sum Model to construct a composite index of climate vulnerability.	[7]
Fuzzy Theory-Evidential Reasoning	It uses fuzzy set theory to handle linguistic and uncertain data and integrates evidence reasoning to aggregate multi-source information under uncertainty, enabling a robust and standardized vulnerability assessment across ports.	[19]
Fuzzy Evidential Reasoning-Fuzzy TOPSIS	It employs Fuzzy Evidential Reasoning to integrate diverse and uncertain expert judgments into multi-level vulnerability estimates, and utilizes Fuzzy TOPSIS to quantify port bottleneck coefficients by evaluating alternatives against ideal solutions under fuzzy environments.	[12]
Fuzzy Cognitive Maps-Sensitivity Model-GIS	It employs Fuzzy Cognitive Maps to model causal interdependencies among vulnerability factors, a Sensitivity Model to identify system-critical variables based on their active and passive influences, and GIS to visualize and analyze spatial-functional relationships within port infrastructure systems.	[20]
DEMATEL-ISM-BWM	It employs an integrated DEMATEL-ISM approach to analyze the interrelationships and hierarchical structure among vulnerability factors, and applies the BWM method to determine their relative weights and identify key vulnerability points.	[21]
BWM-CRITIC-Cloud Model	It integrates the BWM for subjective weight determination, the CRITIC for objective weighting, and the Cloud Model for handling fuzziness and randomness in decision-making processes.	[13]

Note: TOPSIS: Technique for Order Preference by Similarity to Ideal Solution; GIS: Geographic Information Systems; DEMATEL: Decision-Making Trial and Evaluation Laboratory; ISM: Interpretive Structural Modeling

structures, feedback mechanisms, and constraint enforcement across hierarchical levels [17]. To analyze such emergent properties, STPA has proven effective by modeling systems as dynamic socio-technical control systems [23].

In port contexts, STPA has been applied to security governance, interorganizational risk management, and automated terminal operations, demonstrating its effectiveness in identifying vulnerabilities arising from unsafe interactions, coordination failures, and flawed control logic [24–26]. Empirical studies have further revealed critical causal factors, supporting holistic risk diagnosis [25]. However, existing applications remain largely expert-driven and manual, leading to subjective, static, and time-consuming analyses. This limitation motivates the integration of LLMs to enhance analytical efficiency and reproducibility in smart port assessments.

2.3 LLMs for Resilience Decision Support and STPA

LLMs are increasingly recognized for supporting resilience-oriented decision-making in complex socio-technical systems-such as disaster response, cyber resilience, and critical infrastructure management-due to their ability to process disparate data and generate real-time insights amid uncertainty [27–29]. They are often regarded as human-centered cognitive tools that enhance situational awareness, scenario interpretation, and measure prioritization [27, 28].

Some studies have begun to integrate LLMs into STPA-based analyses for identifying unsafe control actions (UCAs) and causal scenarios, demonstrating gains in analytical efficiency and coverage [30, 31]. However, their efforts remain largely task-oriented or safety-focused, without systematically framing LLM-augmented STPA as a coherent DSS for resilience governance [32]. This study directly addresses this gap by embedding the LLM-STPA synergy within a broader, multi-phase DSS architecture that includes objective evidence weighting, thereby shifting the focus from isolated safety analysis to integrated resilience decision-making.

3 Methods

This study employs the formal STPA procedure to identify UCAs and causal factors, and the CRITIC objective weighting method to assess vulnerability levels (see Fig. 1). The resulting DSS integrates LLMs throughout all analytical phases to enable systematic and reasoning-enhanced vulnerability diagnosis.

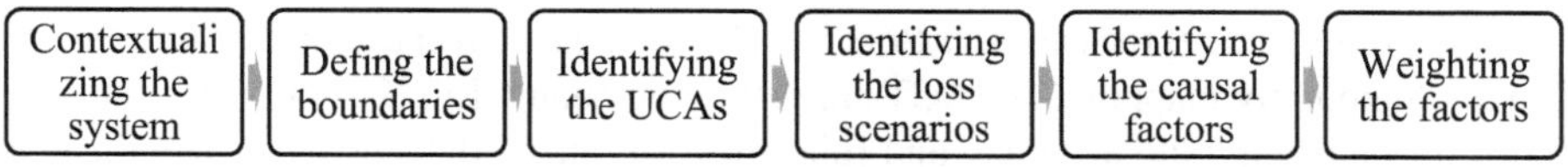

Fig. 1. A LLM-augmented STPA-based DSS for resilient transitions of a smart port.

The proposed DSS framework integrates LLMs across six sequential phases, including system familiarization via textual input and interactive clarification, boundary definition using STPA handbooks [17] and predefined losses and hazards [25], control action-wise identification of UCAs, loss scenario generation based on STPA categories, causal factor identification through causal micro units extraction and taxonomy mapping, and evidence-based factor weighting. To ensure the reproducibility and transparency of the LLM-assisted analysis, a detailed, multi-phase prompting strategy which may include the specific instructions, context-setting prompts, and output format requirements for each of the six phases is needed.

4 Case Study and Results

4.1 Case Context and Data Sources

This study chooses the Hairun Automated Container Terminal (ACT) at the Port of Xiamen as a case. This port is the first in China to carry out an intelligent upgrade, with pilot operations commencing in 2022 and final acceptance successfully completed in December 2023. The specific renovations are demonstrated in terminal operating system (TOS), vessel handling, yard operations, horizontal transportation, tractor guidance, 5G communication, gates, security systems, tallying, industrial control network security, and facility maintenance [33]. We collect all available data related to the case, including academic publications and publicly available technical reports verified by the project team, to ensure the relevance, authority, and comprehensive coverage.

4.2 Vulnerability Diagnosing by Promoting the DSS

Following the structured prompting strategy detailed in the supplementary materials, the DSS which is powered by the Chat Generative Pretrained Transformer (ChatGPT) identified the UCAs for the Hairun ACT (see Table A in the Appendix). The UCAs are identified based on the 23 CAs from supervision, managerial, operational, and physical layers in terms of the Hairun ACT description. The CA items are filtered by ChatGPT from the original 47 CAs, applying a subset analysis based on the "safety-critical first" principle. Those uncovered CAs primarily focus on finer-grained physical execution actions (e.g., hoist, trolley, and gantry) and supporting actions for management, safety, and signaling (e.g., alarm lights, access control, and isolation). This granularity selection can help avoid combinatorial explosion and redundant modeling while prioritizing coverage of the system's most unique and hazardous control loops. Regardless, we can find that most of the UCAs cover the first three categories and the LLM anchors deviations in control actions to specific operational contexts within the ACT. Moreover, the "improper duration" category is mainly empty, with an exception for CAs 21 and 23.

For another, we get the causal factors (see Table 2) through causal micro-unit extraction and taxonomy mapping. An ACT case study of Zhang, Liu [25] provides category codes for enhancing the accuracy and robustness of our model's analysis. We observe that the LLM-generated causal factors achieved a substantial overlap with the expert-derived factors from Zhang, Liu [25]. Of the 23 factors identified in the reference study [25], 20 were replicated by our DSS, yielding an overlap rate of approximately 87%. The three factors not identified (i.e., E2, E3 and H4) represent external, low-frequency, high-impact events, which were outside the scope of the input evidence describing the Hairun ACT's internal operations. Conversely, numerous new factors were identified,

Table 2. Causal factors generated by ChatGPT for Hairun ACT (*Newly identified factors)

Types	Factors	Connotation	Corresponding UCAs
Management	M1: Poor communication and coordination	Low efficiency in information exchange and collaboration among stakeholders in smart ports	3-3, 4-1, 5-1, 5-2, 19-1
	M2: Inadequate safety education and training	Insufficient implementation of safety education, training and assessment for operators	20-2
	M3: The low level of emergency response	Inadequate emergency plans, capabilities and implementation effects	3-3, 5-2, 23-4
	M4: Imperfect safety rules and regulations	Deficiencies in safety guidelines, protocols and legal frameworks	3-3, 4-1, 5-1, 7-2, 19-1
	M5: Poor supervision	Insufficient or ineffective monitoring and guidance by supervisory authorities	3-1, 7-2, 21-4, 23-4
	M6*: Teleoperation multitasking overload	Single remote operator monitors multiple RTGs, causing attention saturation, delayed response and missed prompts	17-1
Environment	E1: Adverse weather	Natural factors such as heavy rain and dense fog affect port operations	3-3, 10-2, 11-1, 11-3, 14-3, 17-2, 18-1, 21-3, 23-2
	E2: Poor vessel stability	Inadequate vessel stability, prone to capsizing and deviation risks	N/A
	E3: Congestion of collection and distribution	Poor logistics environment behind the port, low collection and distribution efficiency	N/A
	E4*: Mixed-traffic intrusion with non-cooperative human-driven vehicles	Manual vehicles intrude IGV corridors, forcing emergency rerouting/braking and increasing collision risks	13-3
Human	H1: Lack of expertise and experience	Insufficient professional skills and knowledge reserve of operators	15-1

(*continued*)

Table 2. (*continued*)

Types	Factors	Connotation	Corresponding UCAs
	H2: Lack of safety awareness	Insufficient understanding and attention to safety procedures and risks	2-2, 14-3, 16-2, 17-2, 20-2
	H3: Inappropriate handling actions	Non-compliant and incorrect operations, failing to follow established processes	2-2, 14-3, 15-1, 15-2, 16-2, 21-2, 22-3
	H4: Terrorist attack	Violent sabotage against ports by individuals or groups	N/A
	H5*: Failure to follow digital procedures	Non-compliance with automated processes, putting the system in an unsafe state	9-3, 20-1, 21-4, 22-3
Information system	S1: Poor control algorithm performance	Insufficient performance of port equipment control algorithms, failing to reach optimal level	1-1, 1-2, 1-3, 2-1, 2-2, 2-3, 3-1, 3-2, 4-1, 4-2, 4-3, 5-1, 5-2, 5-3, 6-1, 8-3, 9-1, 9-2, 9-3, 10-1, 10-2, 11-1, 11-2, 11-3, 15-3, 16-3, 17-3, 18-1, 18-2, 18-3, 19-2, 19-3, 20-1, 20-2, 20-3, 21-2, 21-3, 21-4, 22-1, 22-2, 22-3, 23-1, 23-2, 23-4
	S2: Poor schedule algorithm performance	Insufficient performance of port operation and logistics scheduling algorithms	6-2, 6-3, 7-1, 7-2, 7-3, 8-1, 8-2, 8-3, 12-1, 12-2, 12-3, 13-1, 13-2, 13-3, 23-4

(*continued*)

Table 2. (*continued*)

Types	Factors	Connotation	Corresponding UCAs
	S3: Cyber-attack	Malicious sabotage of port computer systems and digital infrastructure	5-1
	S4: Communication network failure	Terminal communication system interruption, failing to transmit data normally	1-1, 1-3, 2-2, 3-2, 4-2, 4-3, 5-2, 5-3, 6-2, 6-3, 7-1, 7-3, 8-2, 9-2, 9-3, 10-3, 11-2, 12-1, 12-2, 12-3, 13-1, 13-2, 13-3, 14-1, 14-3, 15-1, 15-2, 16-2, 17-1, 17-3, 18-3, 19-2, 20-1, 20-2, 20-3, 21-1, 22-1, 22-3, 23-3
	S5: Insufficient computing power	Computing power unable to meet data processing and task execution needs	18-3, 23-3
	S6*: HMI workflow cue ambiguity for supervisory approvals	Faint and unclear HMI supervisory approval cues, prone to omission and delay	2-1
	S7*: HMI change-notification deficiency for safety thresholds	Unclear HMI notification of safety threshold changes, misleading decisions	10-3
	S8*: Alarm overload obscures safety-critical confirmation prompts	Alarm flooding masks critical confirmation prompts, causing operational risks	14-1
	S9*: Alarm overload suppresses actionable guidance for required operator steps	Lack of step-by-step HMI guidance under alarm overload, leading to missed steps	14-1
	S10*: Readiness-state opacity: prompt shown without exposing failed readiness checks	"Proceed" prompts hide failed readiness checks, misleading unsafe operations	14-2
	S11*: Critical alert suppression/occlusion on HMI	Critical alerts are suppressed/occluded, undetected by operators in time	14-2

(*continued*)

Table 2. (*continued*)

Types	Factors	Connotation	Corresponding UCAs
	S12*: Sensor-consistency non-disclosure to operators	Undisclosed cross-sensor discrepancies, affecting diagnostic accuracy	14-2
	S13*: Teleoperation viewpoint distortion	Camera perspective distortion causes spatial misjudgment in remote operations	15-2
	S14*: Authority/lockout cue deficiency during auto-landing	Unclear HMI authority/lockout status during auto-landing, causing command conflicts	15-3
	S15*: Resume workflow cue deficiency	Unhighlighted manual re-authorization after E-stop, leading to operational hazards	16-1
	S16*: Clearance status not latched/visible	Unlatched and unrecorded clearance status, affecting restart decisions	16-1
	S17*: Post-clear restart readiness not explicitly disclosed	Unclear restart readiness conditions, increasing restart risks	16-1
	S18*: Alarm semantics ambiguity	Alarms fail to distinguish caution from stop, causing improper operator responses	21-2
	S19*: Insufficient mission-resynchronization status visibility	Invisible mission resynchronization status, causing cross-layer desynchronization	19-1
Facility	F1: Sensor malfunction	Abnormal sensor operation or inaccurate data	1-2, 3-3, 6-3, 10-1, 11-1, 15-1, 18-1, 18-2, 19-3, 21-1, 21-2, 22-2, 23-1, 23-2
	F2: Low digital technology application level	Inadequate port digital technology application, failing to meet demands	17-2, 20-1

(*continued*)

Table 2. (*continued*)

Types	Factors	Connotation	Corresponding UCAs
	F3: Inappropriate maintenance and inspection	Inadequate maintenance, inspection and replacement of port equipment	10-1, 13-2
	F4*: Insufficient visual operating conditions for remote verification	Poor lighting and camera conditions, reducing remote visual verification reliability	17-1

especially those related to human-machine interface design and information system feedback (e.g., S6* to S19*). These were not present in the original study [25] as they are specific to the upgraded ACT's digital systems and emergent operational contexts, such as teleoperation and complex alarm management.

4.3 Supporting Resilience Decision

Based on all available evidence associated with the Hairun ACT, we can get vulnerability scores for each indicator using the formula by ChatGPT:

$$Score = \sum(\text{Evidence Strength} \times \text{Evidence Weight}) \div \sum \text{Evidence Weight},$$

where Evidence Weight = (5 − Uncertainty Rating)

Table 3 gives an example of the evidence mining for the vulnerability metrics. Specifically, ChatGPT extracts relevant evidence for each indicator, assigns it an evidence strength score (1-5, with 5 being strongest), and documents associated uncertainties (rated 5-1, with 5 being most uncertain). A composite score is subsequently computed via a weighted formula incorporating both strength and uncertainty ratings.

We also assign the weights by using CRITIC to profile the vulnerability levels (see Table 4). There, M3 and H2 emerged as the most influential vulnerability metrics, attaining the highest objective weights of 0.175 and 0.124, respectively. Meanwhile, a cluster of high-scoring indicators, including M6, E4, H1, S4, S8, S11, S13, S15–17, S19, F1, and F4, consistently contributed a substantial weight of approximately 0.083, reflecting their significant and non-redundant role in differentiating the port's vulnerability profile.

5 Discussions

This study validates that LLMs can effectively identify UCAs, replicate a substantial portion of expert-derived causal factors, and generate consistent vulnerability weightings within a complex ACT system. This supports the growing consensus that well-guided LLMs can facilitate structured vulnerability analysis, extending their applicability from conceptual frameworks to real-world ACT operations.

Table 3. An example of evidence mining results

Indicators	Evidence and Source	Evidence Strength	Remarks/Uncertainty	Score
M1: Poor Communication and Coordination	"The security system of the port is vulnerable to attacks via the information supply chain... Multi-modal transport connection points are at risk." [34]	4	Uncertainty Rating: 2 (Medium), because the exact mechanisms of communication failures are not fully specified. The quote indicates that communication failures can be exploited but doesn't provide granular details about how these vulnerabilities manifest on the ground. The context is a bit general.	4.2

Table 4. CRITIC weights for the vulnerability metrics of Hairun ACT

Indicators	Scores	σ_j	r_{jk}	C_j	Weights	Indicators	Scores	σ_j	r_{jk}	C_j	Weights
M1	4.2	0.19	0.69	12.85	0.025	S7*	4.8	0.44	0.69	12.85	0.057
M2	4.8	0.44	0.69	12.85	0.057	S8*	4.9	0.56	0.63	14.95	0.083
M3	3.2	1.00	0.57	17.48	0.175	S9*	4.8	0.44	0.69	12.85	0.057
M4	4.0	0.44	0.69	12.85	0.057	S10*	4.8	0.44	0.69	12.85	0.057
M5	4.2	0.19	0.69	12.85	0.025	S11*	4.9	0.56	0.63	14.95	0.083
M6*	4.9	0.56	0.63	14.95	0.083	S12*	4.8	0.44	0.69	12.85	0.057
E1	4.0	0.44	0.69	12.85	0.057	S13*	4.9	0.56	0.63	14.95	0.083
E4*	4.9	0.56	0.63	14.95	0.083	S14*	4.8	0.44	0.69	12.85	0.057
H1	4.9	0.56	0.63	14.95	0.083	S15*	4.9	0.56	0.63	14.95	0.083
H2	5.0	0.78	0.61	15.89	0.124	S16*	4.9	0.56	0.63	14.95	0.083
H3	4.1	0.28	0.71	11.68	0.032	S17*	4.9	0.56	0.63	14.95	0.083
H5*	4.5	0.39	0.72	11.36	0.044	S18*	4.8	0.44	0.69	12.85	0.057
S1	4.2	0.19	0.69	12.85	0.025	S19*	4.9	0.56	0.63	14.95	0.083
S2	4.1	0.28	0.71	11.68	0.032	F1	4.9	0.56	0.63	14.95	0.083
S3	4.8	0.44	0.69	12.85	0.057	F2	4.5	0.39	0.72	11.36	0.044
S4	4.9	0.56	0.63	14.95	0.083	F3	4.7	0.39	0.72	11.36	0.044
S5	4.3	0.28	0.71	11.68	0.032	F4*	4.9	0.56	0.63	14.95	0.083
S6*	4.7	0.39	0.72	11.36	0.044						

Note: σ_j: Standard Deviation; r_{jk}: Correlation Coefficient; C_j: Conflict Index

In particular, LLM-generated UCAs demonstrate greater operational specificity by embedding control biases within concrete ACT scenarios, and exhibit sensitivity to temporal persistence, especially regarding prolonged operations in degraded states, a critical precursor to accidents. The use of factor chain decomposition further enhances the reproducibility and traceability of causal analysis, strengthening the credibility of LLM-derived explanations.

However, compared to expert studies such as the Yangshan ACT analysis [25], the LLM failed to identify certain low-frequency, high-impact risk factors (e.g., poor vessel stability, terrorist attacks), indicating that its outputs remain constrained by the scope and quality of input evidence. This limitation also contextualizes the emergence of several new information system-related factors identified in this study. While common factors such as equipment failure showed consistent weighting across cases, discrepancies in factors like inadequate safety training and low digital technology adoption reflect the distinct risk priorities in upgraded versus fully ACTs.

Overall, LLMs prove valuable in elucidating interaction-level vulnerabilities, enhancing temporal reasoning, and translating abstract control theories into operationally relevant insights, though their effectiveness remains bounded by input constraints. We thus advocate for treating LLMs as collaborative supports rather than substitutes in structured resilience analysis.

The DSSs transferability lies in its generic components: STPA logic, prompt template structure, and CRITIC weighting. Application remains context-dependent, requiring tailored inputs (system description, CAs, evidence). To adapt this DSS to another system, analysts should: (1) redefine system boundaries (losses, hazards, control structure); (2) populate initial prompts with new context; (3) curate a relevant evidence corpus. The framework thus provides a repeatable process for surfacing vulnerabilities, while outputs remain specific to the input system description.

6 Conclusions

This study applies LLMs to support resilience decision-making for upgraded ACTs, combining STPA-based vulnerability assessment with CRITIC weighting to evaluate their potential in supporting judgments within complex socio-technical systems. The approach clarifies the role and governance of LLMs in ACT resilience analysis and provides a transferable methodology for broader system applications.

Findings show that with clear system boundaries, structured prompting, and evidence mining, LLMs can reproduce core unsafe control actions and causal mechanisms while delivering preliminary weight analyses. This signifies LLMs' potential to support the ACT system resilience decision-making, particularly in early stages. Notably, the LLM-augmented STPA offers improved operational specificity and temporal sensitivity, contributing to bridging abstract models and automated operations and revealing risks related to prolonged degraded states often missed by traditional methods, respectively. Causal traceability is enhanced through micro-factor decomposition and mapping, though LLM outputs remain limited by input scope, omitting some external and low-frequency high-impact risks.

Given its single-case focus, generalizability of specific findings is limited. Future work should cross-validate this DSS across other ACTs and infrastructures, comparing LLM outputs with context-specific expert assessments. Quantitative validation (e.g., correlating CRITIC weights with expert priorities) and integration of external disturbance mechanisms (e.g., market shocks, extreme weather) into the evidence base are needed to enhance robustness in real-world scenarios.

Acknowledgments. This study was funded by Natural Science Foundation of Fujian Province, China (grant number 2024 J01719).

Disclosure of Interests The authors declare no conflicts of interest.

Supplemental Materials The prompting strategy of the LLM-STPA for Hairun ACT can be found at Zenodo (DOI: 10.5281/zenodo.18720553).

Appendix

Table A. UCAs generated by ChatGPT for Hairun ACT

CAs	UCAs			
	Not providing the CA leads to a hazard	Providing the CA leads to a hazard	Providing a potentially safe CA at inappropriate time	Improper CA duration
(1) Supervision layer				
CA1: Set operating mode (normal / degraded / recovery / maintenance)	1-1: Operations control does not switch to degraded/recovery mode when key sensing (CPS/OCR) is unreliable, so automation continues with unsafe separation and mis-handoffs. [Ha1, Ha5]	1-2: Operations control selects "normal automation" during mixed-traffic conditions without sufficient readiness, increasing proximity violations in shared zones. [Ha1, Ha2]	1-3: Mode is changed to "recovery/maintenance" after automation has already started executing queued moves, creating conflicting instructions across layers. [Ha5]	N/A
CA2: Approve resume after comms interruption / system pause	2-1: Resume approval is not granted after comms restoration, causing ad-hoc manual interventions and uncoordinated moves that degrade instruction correctness across layers. [Ha5]	2-2: Resume is authorized while an obstacle alarm remains unresolved, allowing equipment/IGV to move into an occupied area. [Ha1, Ha2]	2-3: Resume is approved before controllers have re-synchronized states (task/position/lock status), leading to contradictory execution commands. [Ha5]	N/A
CA3: Order system-wide pause/stop	3-1: A site-wide pause is not issued when abnormal states propagate (e.g., repeated E-stops, widespread alarms), allowing unsafe distances to persist during automation. [Ha1]	3-2: A system-wide stop is triggered unnecessarily during active hoisting/landing, increasing the chance of losing stable container control (e.g., abrupt halt mid-landing). [Ha3]	3-3: The stop is initiated too late, after multiple assets have entered a conflict geometry, making separation no longer maintainable. [Ha1, Ha2]	N/A
CA4: Dispatch on-site response (obstacle removal / isolation / emergency handling)	4-1: On-site response is not dispatched after a safety system E-stop, leaving the obstacle/occupied area uncleared and prompting unsafe restart attempts. [Ha1, Ha2]	4-2: On-site personnel are dispatched into an active automated zone without effective area isolation, increasing human exposure near moving equipment. [Ha1]	4-3: Response is dispatched after operations already approved restart, causing mixed human presence during resumed motion. [Ha1]	N/A
CA5: Enforce OT/ICS security response (segment isolation / credential revocation / tighter ACL)	5-1: Security containment is not executed during suspected compromise, allowing inaccurate or manipulated control instructions to propagate. [Ha5]	5-2: Overly aggressive isolation blocks legitimate control/feedback channels, degrading control accuracy and leading to inconsistent automation commands. [Ha5]	5-3: Security controls are applied mid-operation without coordination (e.g., during handoff), causing partial command delivery and inconsistent equipment states. [Ha5]	N/A
(2) Managerial layer				

(continued)

Table A. *(continued)*

CAs	UCAs			
	Not providing the CA leads to a hazard	Providing the CA leads to a hazard	Providing a potentially safe CA at inappropriate time	Improper CA duration
CA6: iTOS generates & releases work orders (quay/transport/yard)	6-1: iTOS fails to issue the required work order update when container identity/position changes, so downstream systems execute against wrong task context. [Ha5]	6-2: iTOS releases a work order with incorrect target bay/yard block, sending equipment into unintended operating areas with higher conflict likelihood. [Ha1]	6-3: iTOS releases transport/yard orders before quay-side handoff readiness is established, causing premature IGV/RTG motions and unsafe proximity. [Ha1, Ha2]	N/A
CA7: iTOS sets work-order priority and time window	7-1: iTOS does not update priority/time window during congestion, leading to prolonged high-density operations that increase exposure to proximity violations. [Ha1]	7-2: iTOS prioritizes throughput by compressing windows beyond safe execution capacity, increasing the likelihood of instruction mismatch across layers. [Ha5]	7-3: Priority/time window updates are issued after equipment has committed to a conflicting sequence, producing wrong-order execution. [Ha5]	N/A
CA8: Middle-layer allocates resources (bind QC/RTG/IGV to tasks; reserve zones)	8-1: Middle-layer does not reserve/lock the required zone for a handoff, allowing another asset to enter and violate safe separation. [Ha1]	8-2: Middle-layer assigns two assets (e.g., two IGVs) to overlapping pickup/drop points without conflict resolution, creating container-transport proximity hazards. [Ha2]	8-3: Resource binding is performed in the wrong order (IGV released before RTG/QC task lock), producing conflicting movements at the handoff. [Ha5]	N/A
CA9: Middle-layer pauses/resumes by task/equipment/zone	9-1: Middle-layer does not pause a specific zone after repeated abnormal deviations, allowing unsafe operations to continue in that zone. [Ha1, Ha5]	9-2: Middle-layer pauses only part of the chain (e.g., IGV paused but QC continues into handoff), creating instruction inconsistency between layers. [Ha5]	9-3: Resume is issued before "reset complete" and state reconciliation, so equipment acts on stale/incorrect control context. [Ha5]	N/A
CA10: Configure execution thresholds (safe height, deviation thresholds, etc.)	10-1: Thresholds are not updated after equipment condition degrades (e.g., sensor drift), so controllers operate with unreliable assumptions and reduced separation margins. [Ha1, Ha5]	10-2: Safe-height threshold is set too low for the environment, increasing collision risk during approach/landing near IGV or nearby objects. [Ha1]	10-3: Deviations thresholds are changed during an active approach sequence, causing premature acceptance/rejection and wrong-order prompts. [Ha5]	N/A
CA11: Gate handoff eligibility ("allow landing/hoist" prompts based on CPS/OCR/pose readiness)	111: The system does not generate the "allow landing/hoist" prompt when readiness is met, prompting operators to improvise or override outside validated readiness. [Ha5]	11-2: The system issues "allow landing" even though CPS alignment or OCR match is not satisfied, enabling unsafe handoff and potential loss of container control. [Ha3, Ha5]	11-3: "Allow landing/hoist" is sent before IGV has stabilized at the stopping point, leading to unsafe proximity during the final motion segment. [Ha1, Ha2]	N/A
CA12: FMS assigns IGV mission + provides global route/constraints (new UCAs, non-duplicative)	12-1: FMS omits mission constraints for DG-marked containers (e.g., speed/route restrictions), increasing transport risk and escalation to DG leakage/explosion under worst-case events. [Ha6]	12-2: FMS sends a global route based on stale map/zone availability (outdated occupancy), routing IGV through a temporarily unsafe corridor. [Ha2]	12-3: FMS dispatches a route update after comms quality has degraded and IGV has already stopped, creating partial delivery and inconsistent mission context. [Ha5]	N/A
CA13: FMS re-dispatch / cancel / reassign missions (fleet balancing under abnormality)	13-1: FMS does not cancel a mission when the target handoff point becomes unavailable, causing IGV queuing in conflict-prone areas. [Ha2]	13-2: FMS reassigns a mission to the wrong IGV (ID mismatch), producing container pickup/delivery confusion and instruction inaccuracy. [Ha5]	13-3: Mission cancellation arrives after IGV has entered the crane working envelope, creating sudden braking/avoidance and proximity violations. [Ha1, Ha2]	N/A
(3) Operational layer				
CA14: Remote QC operator confirms precise alignment & landing (press/hold at safe height)	14-1: Operator confirmation is not provided when required, leading to timeouts and manual workarounds that bypass validated readiness gates. [Ha5]	14-2: Operator confirms landing without verifying "ID match + alignment readiness," enabling incorrect handoff and potential container instability. [Ha3, Ha5]	14-3: Confirmation is given too early (before IGV is stable at stopping point), allowing descent while separation is not assured. [Ha1, Ha2]	N/A

(continued)

Table A. (*continued*)

CAs	UCAs			
	Not providing the CA leads to a hazard	Providing the CA leads to a hazard	Providing a potentially safe CA at inappropriate time	Improper CA duration
CA15: Remote QC operator performs manual fine adjustment when deviation exceeds threshold	15-1: Operator does not intervene despite deviation alarm, leading to misalignment and loss of controlled placement. [Ha3]	15-2: Manual adjustment overshoots into adjacent envelope (nearby objects/people/vehicles), reducing safe distance during automated operation. [Ha1]	15-3: Adjustment occurs after the system already initiated final deceleration/landing, creating conflicting motion intent and control inaccuracies. [Ha5]	N/A
CA16: Remote QC operator issues "resume work" after obstacle-triggered E-stop (post verification)	16-1: Operator does not issue resume after obstacle clearance, increasing ad-hoc onsite interventions and instruction inconsistency across the chain. [Ha5]	16-2: Operator resumes without verifying obstacle removal, allowing immediate motion into an occupied zone. [Ha1]	16-3: Resume is issued before the system re-establishes safe height / stable pose, causing unsafe restart conditions. [Ha1, Ha3]	N/A
CA17: Remote RTG operator confirms yard environment safe; permits hoist/landing	17-1: RTG operator does not provide the required safety permit, leading to bypass/override behaviors that erode control instruction correctness. [Ha5]	17-2: Permit is granted while a person/vehicle remains within the working area, reducing safe distance during automated operation. [Ha1]	17-3: Permit is issued before laser+vision confirms IGV pose readiness, allowing hoist/landing under misalignment. [Ha3, Ha5]	N/A
CA18: IGV onboard autonomy executes local perception/prediction/decision & local path planning (mixed traffic)	18-1: IGV does not execute local stop/avoid behavior when obstacle/person appears, violating safe separation in mixed-traffic. [Ha2, Ha1]	18-2: IGV performs an unnecessary abrupt evasive maneuver in a constrained corridor, destabilizing the carried container during transportation. [Ha3]	18-3: Local avoidance triggers too late (late prediction), resulting in near-collision proximity before braking/steering takes effect. [Ha2]	N/A
CA19: Dispatcher/remote control room approves recovery after comms interruption stop	19-1: Recovery approval is not issued after comms restoration, prompting manual workarounds and inconsistent mission state reconciliation. [Ha5]	19-2: Recovery approval is given while comms is still unstable (flapping), leading to partial command delivery and inaccurate multilevel instructions. [Ha5]	19-3: Approval is issued before IGV and FMS re-synchronize mission/position, causing wrong-order movement on restart. [Ha5, Ha2]	N/A
CA20: On-site responders clear obstacle / isolate area / reset equipment and report readiness	20-1: Obstacle is not cleared or isolation is not established, but readiness is reported, causing unsafe restart into occupied space. [Ha1, Ha2]	20-2: Responders clear obstacles while automation is not fully paused/locked-out, increasing exposure near active equipment. [Ha1]	20-3: Readiness is reported before the area is fully verified and equipment reset is complete, enabling premature resume. [Ha1, Ha5]	N/A
(4) Physical layer				
CA21: QC emergency stop relay activation (motion cut-off)	21-1: E-stop is not activated upon confirmed obstacle detection, allowing continued motion and unsafe separation. [Ha1]	21-2: E-stop is activated during a critical container control phase (e.g., near-contact landing) without need, increasing risk of losing stable control of the load. [Ha3]	21-3: E-stop triggers after the controller should have entered controlled deceleration, resulting in insufficient stopping distance. [Ha1]	21-4: E-stop is cleared too soon (before verification/authorization), allowing re-motion while hazards persist. [Ha1]
CA22: QC spreader lock/unlock (twistlock state command)	22-1: Unlock command is not issued when required (after stable placement), leading to forced manual interventions and inconsistent execution states. [Ha5]	22-2: Unlock is issued while the container is not fully supported/stable, causing loss of container control. [Ha3]	22-3: Lock is commanded after lift begins (wrong order), creating unstable pickup and control inaccuracies. [Ha3, Ha5]	N/A
CA23: IGV emergency braking / safety stop activation	23-1: Emergency braking is not triggered when a person/vehicle intrudes into the path, violating safe separation during transportation. [Ha2]	23-2: Emergency braking triggers repeatedly due to false positives, increasing abrupt stops and raising the likelihood of container instability during transit. [Ha3]	23-3: Emergency braking triggers too late in mixed traffic, resulting in near-contact proximity before stop. [Ha2]	23-4: Safety stop is held too long after clearance (no timely re-authorization), creating congestion that drives unsafe ad-hoc maneuvers. [Ha2]

References

1. de la Peña Zarzuelo, I., Freire Soeane, M.J., López Bermúdez, B.: Industry 4.0 in the port and maritime industry: a literature review. J. Ind. Inf. Integr. **20**, 100173 (2020)
2. Xiao, G., Pan, L., Lai, F.: Application, opportunities, and challenges of digital technologies in the decarbonizing shipping industry: a bibliometric analysis. Front. Mar. Sci. **12** (2025)

3. de la Peña Zarzuelo, I.: Cybersecurity in ports and maritime industry: reasons for raising awareness on this issue. Transp. Policy. **100**, 1–4 (2021)
4. Mayo, K.E.: Graph analytics in cyber intelligence: enhancing security for smart seaports (2024)
5. Xu, M., Peng, P., Yang, Y., Wang, S., Chen, J.: Identification of critical sets of ports in cascading failures of global liner shipping network. Transp. Res. Part E Logist. Transp. Rev. **200**, 461–478 (2025)
6. Metin, B., Dirik, H., Memis, S., Aydin, M.: Business logic vulnerabilities in the digital era: a detection framework using artificial intelligence. Information. **16**(7) (2025)
7. McIntosh, R.D., Becker, A.: Applying MCDA to weight indicators of seaport vulnerability to climate and extreme weather impacts for U.S. North Atlantic ports. Environ. Syst. Decis. **40**(3), 356–370 (2020)
8. Abdelhafez, M.A., Ellingwood, B., Mahmoud, H.: Vulnerability of seaports to hurricanes and sea level rise in a changing climate: a case study for Mobile, AL. Coast. Eng. **167**, 103884 (2021)
9. Wood, N.J., Good, J.W.: Vulnerability of port and harbor communities to earthquake and tsunami hazards: the use of GIS in community hazard planning. Coast. Manag. **32**(3), 243–269 (2004)
10. Jirouskova, N., Majumdar, A., Angeloudis, P.: A top-down integrated approach to define key vulnerability indices for disaster risk management at container terminals. In: Gazetas, G., Anastasopoulos, I. (eds.) 2nd International Conference on Natural Hazards & Infrastructure. National Technical University of Athens, Chania, Greece (2019)
11. Lasse, R.T.N., Kotzab, H., Münch, C., Wagenitz, A.: Investigating digital collaboration: human factors in artificial intelligence-driven collaboration platforms for resilient port operations. Oper. Supply Chain Manag. **18**(2), 185–200 (2025)
12. Cao, X., Lam, J.S.L.: A fast reaction-based port vulnerability assessment: case of Tianjin Port explosion. Transp. Res. A Policy Pract. **128**, 11–33 (2019)
13. Heng, L.: Study on the Vulnerability of Dalian Port Container Terminal in the Context of COVID-19. Dalian Maritime University, Dalian (2022)
14. Messner, S., Murphy, C., Langeland, O.: Climate change and sea level rise impacts at ports and a consistent methodology to evaluate vulnerability and risk. WIT Trans. Ecol. Environ. **169**, 141–153 (2013)
15. Chhetri, P., Corcoran, J., Hall, G.: Seaport resilience to climate change: mapping vulnerability to sea-level rise. J. Spat. Sci. **60**, 65–78 (2015)
16. Vairo, T., Pontiggia, M., Fabiano, B., Curro, F.: Understanding the vulnerability of complex systems: an integrated approach. Chem. Eng. Trans. **99**, 277–282 (2023)
17. Leveson, N.G., Thomas, J.P.: System-Theoretic Process Analysis (STPA) Handbook 2018. Available from: http://www.flighttestsafety.org/images/STPA_Handbook.pdf (2018). Accessed 25 Aug 2025
18. Su, D.T., Hsieh, C.H., Tai, H.H.: Container hub-port vulnerability: Hong Kong, Kaohsiung and Xiamen. J. Mar. Eng. Technol. **15**(1), 19–30 (2016)
19. Jiang, M., Lu, J., Fu, X., Zhang, M.: Port vulnerability assessment from a supply chain perspective. Ocean Coast. Manag. **213**, 105851 (2021)
20. Hsieh, C.-H., Tai, H.-H., Lee, Y.-N.: Port vulnerability assessment from the perspective of critical infrastructure interdependency. Marit. Policy Manag. **41**(6), 589–606 (2014)
21. Qian, Y., Wang, H.: Vulnerability assessment for port logistics system based on DEMATEL-ISM-BWM. Systems. **11**(12), 567 (2023)
22. Patriarca, R., Di Gravio, G., Costantino, F.: Resilience engineering: current status of the research and future challenges. Saf. Sci. **102**, 79–100 (2018)

23. Hans, O., Radlak, K., Adamy, J.: Human-factor focused application of STPA to remotely driven vehicles. In: 2024 IEEE World Forum on Public Safety Technology (WFPST), pp. 120–125. IEEE (2024)
24. Williams, A.D.: Beyond a series of security nets: applying STAMP & STPA to port security. J. Transp. Secur. **8**(3), 139–157 (2015)
25. Zhang, C.W., Yang, Y., Jin, Z., Wang, Y.: A hybrid SgDT framework for risk analysis of container-handling operations at automated container terminals. Ocean Coast. Manag. **257** (2024)
26. Melendez, E., Goerlandt, F.: A STAMP-informed framework for classifying interorganizational risk management challenges in ports. Saf. Sci. **192**, 107000 (2025)
27. Xu, F.Y., Zhang, H., Wang, F., Yu, D.: Large language model applications in disaster management: an interdisciplinary review. Int. J. Disaster Risk Reduct. **127** (2025)
28. Ding, W.P., Zhao, L., Li, S., Zhang, Y.: Large language models for cyber resilience: a comprehensive review, challenges, and future perspectives. Appl. Soft Comput. **170** (2025)
29. Lu, T., Zhang, M., Chen, J., Liu, Y.: Human-AI interactive framework for smart evacuation safety analysis in large infrastructures. Reliab. Eng. Syst. Saf. **266** (2026)
30. Qi, Y., Wang, L., Zhao, X., Zhang, H.: Safety analysis in the era of large language models: a case study of STPA using ChatGPT. Mach. Learn. Appl. **19**(000) (2025)
31. Kaya, G.K., Aydin, M., Demir, S., Yilmaz, U.: Large language models powered system safety assessment: applying STPA and FRAM. Saf. Sci. **191** (2025)
32. Alenjareghi, M.J., Fotouhi, M., Mousavi, A., Li, W.: Proactive safety reasoning in human-robot collaboration in disassembly through LLM-augmented STPA and FMEA. Robot. Comput. Integr. Manuf. **98** (2026)
33. Li, M.: Intelligent transformation of traditional container terminals at Xiamen port. J. Port Sci. Technol. **10**(3) (2020)
34. Wang, L.: Application of sensor fusion technology in autonomous driving horizontal transport equipment at container terminals. Xiamen Technology. **31**(3), 48–51 (2025)

Expert Versus LLM Decision Support: Insect-Based Bioconversion Case Study

Rallou Thomopoulos[1(✉)], Joachim Galy[1,2], Stéphane Couture[2], and Erwan Engel[3]

[1] IATE, Université de Montpellier, INRAE, Institut Agro, Montpellier, France
rallou.thomopoulos@inrae.fr
[2] MIAT, INRAE, Toulouse, France
[3] QuaPA, INRAE, Theix, France

Abstract. This paper presents a workflow that transforms a customized literature corpus into decision support regarding the implementation of innovations, here insect-based biowaste valorization, for animal feed (scenario S1) or for biofuel production (scenario S2). From the literature, the workflow builds graphs representing system variables and influence relationships, constructs scenarios that combine the values of key graph variables, and structures pro/con arguments that evaluate the scenarios by criteria (safety, social acceptance, ...) and stakeholder groups (biowaste producers, insect breeders, ...), producing collective attitude scores. We compare expert results with LLM results. While expert analysis built S1 and S2 from three key variables (output-product, input-substrate, insect used), LLM retrieval suggests additional ones (microbiome modulation, frass value) to refine scenarios. Manual analysis identified 337 arguments for assessing scenarios, versus 478 (+42%) with LLM. Both ranked S1 highest and identified chemical safety as a critical bottleneck. Perspective is hybrid expert–LLM decision-support.

Keywords: Knowledge retrieval · Large Language Model (LLM) · Argumentation · Scenario Analysis · Systemic Approach

1 Introduction

Sustainability transitions require decisions under uncertainty, where scientific knowledge is incomplete, multidisciplinary, and rapidly evolving [28]. In various fields, such as agri-food systems and the bioeconomy, technologies are being developed with the concern of better preservation of resources and the environment.

Among these, insect-based biowaste valorization consists of raising insects as an ecological means of recycling and valorizing biowaste, for example in the form of proteins for animal feed, fertilizers, biofuels, etc. In urban and peri-urban settings, where waste management has become more critical than ever, insect-based biowaste valorization offers an attractive solution but also raises specific

A. P. Cabral Seixas Costa et al. (Eds.): ICDSST 2026, LNBIP 582, pp. 287–301, 2026.
https://doi.org/10.1007/978-3-032-25146-6_19

questions, particularly regarding the heterogeneity of biowaste, control of safety risks, regulatory and social constraints. Currently in Europe, studies are focused on evaluating the potential of this solution that is still under development, while simultaneously anticipating the benefits and risks of recycling through insect-based valorization of different types of substrate and particularly biowaste [17]. While initial work focused on authorized substrates based on plant products, the current slowdown in the sector is prompting a new generation of work that focuses on currently unauthorized biowaste ([32], in waste management), which could allow regulations to evolve and the sector to gain a second wind.

To evaluate the potential of such emerging technologies, a first issue to overcome is the identification and collection of relevant information to delimit and analyze possible decisions [9,40]. A second issue consists of the intelligent processing of data and knowledge, as well as the interpretability of the results obtained, which plays a major role in supporting decision-making and guiding informed sustainable development [30,43].

The paper addresses the following research questions (RQs):

- RQ1: How to envision plausible scenarios for sustainable innovation? (in the present case, scenarios of usage of biowaste valorization, e.g. in animal feed, biofuels, etc.)
- RQ2: What is the overall attractiveness of each scenario?
- RQ3: How to efficiently extract relevant information from the identified sources for these purposes?

This paper proposes a methodological workflow that transforms large collections of scientific publications into structured qualitative decision support. The central idea is to use an interaction graph as a bridge between unstructured literature and prospective scenario analysis: variables and their relations are extracted from a customized literature corpus, aggregated into a graph representation of the insect-based bioconversion system, and analyzed to identify key variables whose possible values combined together define scenarios. Each scenario is then evaluated through structured, bipolar arguments (pros and cons) organized by criteria and stakeholder groups, and summarized using a collective attitude score using argumentation-based decision models.

The workflow is illustrated with a qualitative case study on insect-based bioeconomy, focusing on urban and peri-urban insect-based conversion of biowaste using *Hermetia illucens*. In this case study, the workflow is first manually processed through expert analysis. The need for an automated process came from the observation that the insect-based biowaste valorization domain was rapidly evolving, with respect to both initiatives in the field and research studies. Thus, the literature corpus has to be updated regularly to take into account new publications. On the other hand, the manual retrieval process is time-costly and cannot easily be repeated regularly. Consequently, the automation of the corpus update using scientific monitoring, and its exploitation using LLM knowledge retrieval, was set up.

The contributions of the paper are threefold:

- A **generic workflow** transforming a literature corpus into decision support using interaction graphs, key variable identification, scenario generation, and argumentation-based evaluation, applicable across emerging technology domains.
- An **insect-based bioconversion case study**, contrasting manual expert retrieval versus LLM/RAG analysis (+42% coverage from 143 abstracts), revealing complementary insights into safety bottlenecks, economic trade-offs, and novel variables (microbiome, frass).
- A **hybrid expert-LLMs perspective** demonstrating how LLMs scale knowledge volume while experts ensure coherence with stakeholders' concerns, enabling operational multi-criteria decision support for sustainable transitions.

The paper is structured as follows: Sect. 2 presents the methodological workflow, detailing in part 2.2 the creation of the literature corpus, the construction of scenarios using interaction graphs and key variable identification, and in part 2.3 qualitative benefit-risk analysis using argumentation frameworks. Section 3 reports the results of both manual expert retrieval and LLM-based retrieval, including scenario rankings, argument distributions, and the LLM-derived interaction graph. Section 4 discusses the comparative strengths of both approaches and their implications for decision support, followed by limitations and perspectives for hybrid expert-LLM pipelines.

2 Methodological Workflow

The methodological workflow consists of three steps: (i) Corpus construction and knowledge retrieval methods; (ii) Scenario building, including the representation of an interaction graph with system variables and influence relationships, and the construction of prospective scenarios from key variable values; (iii) Qualitative benefit-risk analysis of the scenarios based on argumentation.

2.1 Construction of a Validated Specific Scientific Corpus and Knowledge Retrieval Method

For manual retrieval, expert information sources consist of 14 webinars organized within the Insect4City consortium [17], during which 18 researchers and 4 industrialists intervened, addressing various aspects: insect biology, transformation processes, biowaste recycling, animal and human nutrition, chemical and microbiological safety, consumer acceptability, environmental, economic, and social sustainability, multicriteria evaluation. These webinars and the information shared within the network made it possible to constitute an initial, validated specific scientific corpus.

For LLM retrieval, the information sources result from domain-specific scientific monitoring launched every two months. It is focused on research and development within Europe. It centers on the use of two edible insect species (*T. molitor* and *H. illucens*) in recycling organic waste and agricultural by-products, for industrial applications such as animal feed, fertilizers, ... The monitored sources include scientific papers, project results, regulatory documents, and updates from the private sector. Each launch provides on average 200 results that are manually verified by a researcher in the field. In the present study, only the abstracts of the latest launch were analyzed, corresponding to 143 abstracts from 160 documents in total.

Characteristics of the corpus and consequences for the retrieval method are as follows:

- customized corpus, thus mobilizing the so-called RAG (Retrieval Augmented Generation) framework [3,19,27];
- possibly large. While the 143 abstracts remain manageable online, larger volumes need local tools, e.g. if the complete papers are explored;
- public sources, leaving the option to use online LLM tools;
- unknown document structure, favoring engineered prompts over RAG reprogramming [4,19,23].

Comparative experiments using several online LLM tools as well as local tools and institutional servers are ongoing. The results presented in Sect. 3.2 were obtained using Perplexity AI. Defining correct and validated prompts to extract the expected outputs (e.g., interaction graphs, see Sect. 2.2) is part of the methodology. To evaluate the result of the prompt regarding a manually obtained "gold standard", we use text samples as test corpus and compare the prompt results with manually obtained ones.

2.2 Scenario Building

The objective of scenario-building is exploring diverse futures for implementing novel insect-based technologies. The approach employed is known as the "scenario method"—or Godet method [20]. From the scientific corpus presented in Sect. 2.1, it proceeds through these sequential steps: (i) identify variables and influence relationships from the scientific corpus; (ii) construct interaction graphs, where nodes represent variables and links represent influence relationships; (iii) compute for each variable its influence (outgoing links) and dependence (incoming links); (iv) select key variables (with high influence and dependence); (v) generate scenarios by combining key variable values.

The method takes its source in the domain of system structural analysis [22]. The notions of "variable" and "influence relationship" are meant accordingly. A variable refers to any element that has a value (quantifiable or not) that can change. An influence relationship is a directed link expressing that one variable has a direct effect on another.

Scenarios are used classically in foresight methodologies [36]. They are characterized as states of critical variables at particular moments, derived from transitions between current conditions and potential future outcomes [33]. Various scenario categories exist, including predictive, explorative, and normative types [8,29]. Among established foresight techniques, the "scenario method" proves particularly appropriate for agri-food chain applications. As a multi-actor approach, it relies on exchanges among diverse stakeholders involved [21]. Nevertheless, adaptations were proposed in [11,12] and led to the use of individual interviews or specialized publications as input, as in the present study. This methodology has demonstrated effectiveness across multiple sectors and scales over time [16,26], showing relevance in agri-food contexts [11–13,39]. Its core principle involves constructing plausible future trajectories for the system under examination. The representation of the different viewpoints involved ensures the relevance of the scenarios to system actors [2,5,6], who often have divergent views and priorities across criteria (economic, social, environmental, sensory, technical, safety, etc.) [18,34]. This is why capturing the dynamics of agri-food chains requires modeling from multiple viewpoints [10,31].

2.3 Qualitative Benefit-Risk Analysis

The benefit-risk assessment utilized *MyChoice* software [41], previously applied in various agri-food contexts [14,25,39,44]. For each scenario obtained from the scenario-building step (Sect. 2.2), arguments commenting favorably or unfavorably the scenario, based on its variable values (variables of Sect. 2.2), from the viewpoint of a given stakeholder and a given consideration (denoted "criterion", e.g. safety, environment, ...), are identified from the scientific corpus and structured into an argument database.

From this argument database, graphical visualizations allow for exploring pro/con argument repartition among stakehoders and criteria for each scenario. *MyChoice* also computes collective attitude scores (0=favorable, 1=unfavorable), based on the proportions of positive arguments and the frequency of the arguments. These scores allow for ranking the scenarios either globally or per stakeholder/criterion.

MyChoice implements artificial intelligence argumentation theory [15], representing arguments and their contradictions ("attacks") to derive coherence-satisfying argument subsets. This theory establishes argumentation as a reasoning paradigm centered on the construction and evaluation of interacting arguments. Several works integrate argumentation within multicriteria and multi-actor frameworks to explain collective decisions [7,9,24,35,37,38,41,42]. Decision-oriented frameworks [1] define arguments as statements that support or oppose alternatives according to the estimated capacity of these alternatives to fulfill or not specific objectives. In the approach of *MyChoice*, objectives (e.g., "ensuring a consistent level of safety throughout the seasons") are classified in higher-level analytical dimensions denoted criteria (e.g., "microbial safety") [41].

Compared to conventional methods, *MyChoice* incorporates distinctive features that are well suited to practical application-oriented use and provides software implementation. It features detailed argument encoding (scenario, bipolarity, stakeholder, criterion, aim, justification, etc.). It also introduces and computes collective attitude.

3 Results and Discussion

3.1 Manual Retrieval

Scenarios. From the collected information, three key variables (output-product, input-substrate, insect used) were expertly identified by the Insect4City consortium (see Sect. 2.1): following a series of interdisciplinary webinars, several workshops for project development led to simplified diagrams and to the choice of the key variables to be considered. From their value combinations, three scenarios were retained for analysis:

- Scenario 1 (S1) considers insect-based biowaste valorization in animal feed. Fertilizer (in the form of frass) is a coproduct in all scenarios that implement insect-based technologies. It is defined by the key variable combination (output-product = animal feed + fertilizer, input-substrate = agricultural by-products, insect used = *H.illucens*);
- Scenario 2 (S2) considers insect-based biowaste valorization in biofuels. It is defined by the key variable combination (output-product = biofuels + fertilizer, input-substrate = organic biowaste, insect used = *H.illucens*);
- Scenario 3 (S3) is the "business-as-usual" (or "reference") scenario where no innovation is implemented ((∅, ∅, ∅) combination).

337 arguments were obtained: 178 for S1, 111 for S2, 48 for S3.

Benefit-Risk Analysis.

Overall Attractiveness and Scenario Ranking. Applying the method of Sect. 2.3, a collective attitude measure is calculated for each scenario. The results are presented in Table 1, indicating – based on collected data – an attractiveness for both insect-based biowaste valorization scenarios, more pronounced for Scenario 1, and rejection of Scenario 3 "Business-as-usual".

Table 1. Collective attitude obtained for each scenario.

	Scenario 1	Scenario 2	Scenario 3
Collective attitude	0,67	0,62	0,41

Quantity of Arguments Identified by Criterion or by Actor. The quantitative distribution of arguments across different criteria reveals varying concerns between scenarios 1 and 2. For scenario 1, the majority of arguments concern, in decreasing order of importance, chemical safety (23% of arguments), economic profitability (17%), microbiological safety (15%), environmental impact (12%), nutritional interest (11%), social acceptability (10%), with other criteria representing less than 10% of arguments. For scenario 2, economic profitability raises the most concerns (21% of arguments), while nutritional aspects are eliminated due to non-food use (biofuels). The stakeholders involved are more diverse in scenario 1 (insect breeders, but also citizens, other breeders, biowaste suppliers, ...).

Attractiveness of Scenarios by Criterion or by Actor. In Fig. 1, positive arguments (green circles) and negative arguments (red circles) are distinguished, and criterion-by-criterion attitude computation enables classification of criteria into different categories according to the favorable, unfavorable or mixed opinions that emerge. In scenario 1, criteria related to regulation, nutritional interest, and environmental impact receive very positive evaluations. The same analysis conducted at a finer level (sub-criteria or aims) provides explanatory elements on argument content. For example, the chemical safety criterion, which falls into the mixed class, groups three types of concerns: avoiding chemical risks in frass used as fertilizer (very favorable opinion), avoiding chemical contamination impacting insect health (mixed opinion), and avoiding chemical contamination in finished products (mixed opinion). The same analysis can also be performed by actor.

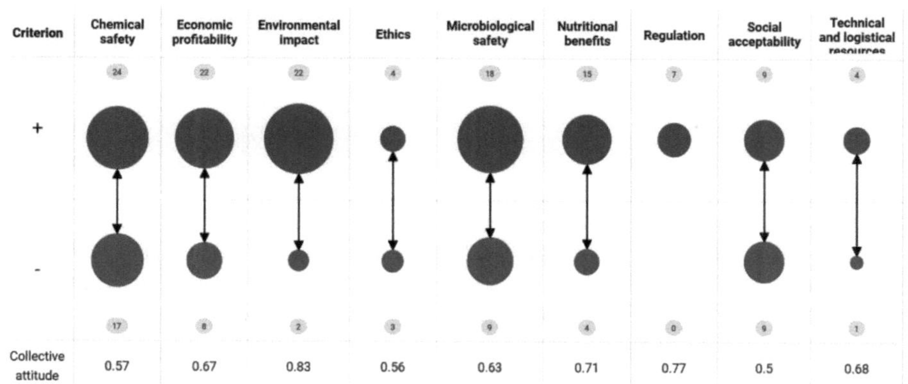

Fig. 1. Analysis of scenario 1 attractiveness by criterion. The number of corresponding pro arguments is indicated above each green circle and the number of con arguments below each red circle.

Complementary Analyses: Scenario Comparison by Criterion or Actor, Cross-Analysis Criteria/Actors. Complementary visuals (Fig. 2) allow us to comparatively identify the strengths and weaknesses of different scenarios. Figure 2(a) shows that the chemical safety and social acceptability criteria are mixed in all

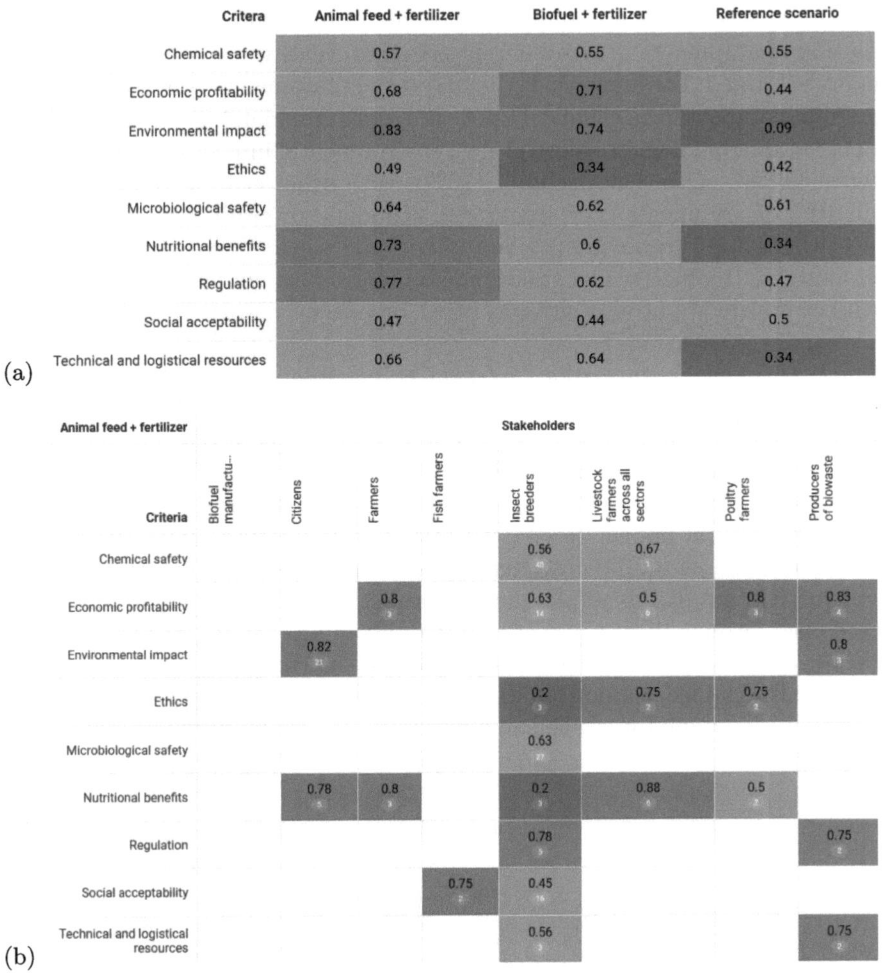

Fig. 2. Detailed analysis: (a) Scenario comparison by criterion and (b) Cross-analysis. The dark green, light green, orange, and red colors indicate respectively, very positive, positive, mitigate, and negative collective attitudes. Cross-analysis additionally indicated the number of corresponding arguments. (Color figure online)

scenarios. It also highlights differences between the two insect-based valorization scenarios on certain criteria: nutritional, since scenario 1 is for food use, while scenario 2 targets biofuel production; but also regarding regulation, more structured in scenario 1, economic profitability more promising in scenario 2, or ethics regarding animal welfare. In Fig. 2(b), computing attitudes jointly by criterion and actor shows, for scenario 1, that most concerned actors anticipate only benefits from insect-based valorization development, except insect breeders who will face implementation difficulties for this new sector. For example, with regard to insect nutrition (deemed very unfavorable), they will need to adapt to using new biowaste food substrates to properly feed the insects.

3.2 LLM Retrieval

LLM Prompting. Prompt 1 was defined iteratively and used to generate the interaction graph. It is dedicated to the extraction of variables (then grouped into higher-level variables) and the influence relationship between variables.

```
Prompt 1:
From [list of abstracts], extract entomoconversion variables (e.g.
urban waste heterogeneity, safety risk, regulatory barriers)
as nodes.
Group these variables into higher-level variables.
List qualitative influences between variables as directed edges
(A→B if A drives B).
```

Scenario Refinement. 28 variables were identified and then grouped into 8 higher-level ones. Figure 3 shows the graph resulting from the higher-level variables and the influence relationships identified.

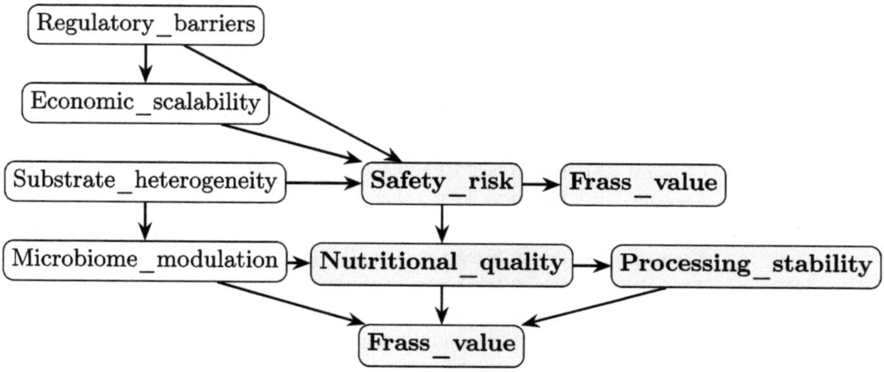

Fig. 3. Interaction graph built from LLM retrieval

Yellow nodes correspond to outputs/targets (distinguished from input/challenge factors in white). Nevertheless, in view of the entire insect-based valorization system, these yellow target variables can be considered as input variables for the next steps and actors in the chain: distributors and users of fertilizers, biofuels or feed products. As such, these variables can be considered both influent and dependent and thus key variables of the system. It seems that the selected literature corpus concentrates on the insect-based valorization process rather than on its downstream steps.

Compared with the scenarios previously obtained manually (see Sect. 3.1), "frass value" can be noticed as a difference in the variables identified to distinguish between possible scenarios. In LLM results, initial scenarios 1 and 2 could be developed into sub-scenarios depending on their frass values. This is confirmed in the following steps. Another difference is the "microbiome modulation" variable which is novelly identified.

Elements of Benefit-Risk Analysis

Arguments. 478 arguments were derived from the 143-abstracts corpus. Each argument is attached with a confidence score [45] based on: (i) its frequency, accounting for 30% of the score; (ii) its explicit (versus implicit) formulation, accounting for 40% of the score; (iii) its consensus among the literature entries,

Table 2. Top 30 LLM arguments ranked by confidence (decreasing). S1=Animal feed (+fertilizer), S2=Biofuels (+fertilizer).

ID	Scenario	Criterion	Type	Conf.	Source Excerpt
1	S1 (frass)	Chemical safety	Con	0.91	Cd bioaccumulation from digestate→frass
2	S2	Regulation	Con	0.89	Regulatory delays investment
3	S1 (frass)	Regulation	Con	0.89	Regulatory barriers→Economic scalability
4	S2	Chemical safety	Con	0.89	Fenpropimorph accumulation in larvae
5	S1	Chemical safety	Con	0.85	Substrate heterogeneity→pesticide risk
6	S1	Economic	Pro	0.85	Regional supply density (11,300t/yr)
7	S1 (frass)	Microbial safety	Pro	0.84	Microbiome recruits SCFA producers
8	S1	Technological	Con	0.82	BSFL oil oxidizes >180°C (peroxides)
9	S1	Nutrition	Pro	0.82	Alcalase hydrolysis 61% protein recovery
10	S1	Chemical safety	Pro	0.78	PUFA:SFA <1.3 (favorable ratio)
11	S1	Economic	Pro	0.78	FCR 1.95-2.23 vs conventional feeds
12	S1 (frass)	Microbial safety	Con	0.76	Pathogen risk E.coli/Staph in frass
13	S2	Environment	Pro	0.76	Sugarcane trash→high bioconversion
14	S1	Technological	Pro	0.74	Polyhalite supplementation improves FCR
15	S2	Economic	Pro	0.74	Substrate reduction efficiency high
16	S1	Chemical safety	Con	0.72	Seasonality increases contamination risk
17	S2	Social	Con	0.72	Consumer disgust remains strong barrie"
18	S1	Economic	Pro	0.71	N/P/K nutrient-rich frass fertilizer
19	S2	Nutrition	Con	0.71	Non-food use eliminates nutrition criterion
20	S1	Environment	Pro	0.70	Land use -64% vs soybean meal baseline
21	S1	Technological	Pro	0.70	Two-stage alkaline extraction 75% recovery
22	S1 (frass)	Chemical safety	Con	0.70	ARG risk if substrate contaminated
23	S2	Economic	Con	0.68	Biofuel certification delays scaling
24	S1	Regulation	Con	0.68	Novel food approval process lengthy
25	S1	Environment	Con	0.67	GWP +3.6E+07 kg CO_2eq vs biogas
26	S1	Social	Pro	0.67	Communication overcomes acceptance barriers
27	S1	Nutrition	Con	0.65	Feed additives optimization needed
28	S2	Technological	Con	0.65	Lipid extraction efficiency variable
29	S1 (frass)	Microbial safety	Pro	0.91	N/P/K nutrient-rich co-product value
30	S1	Economic	Con	0.65	High seasonality logistical challenges

for 20% of the score; and (iv) the score is reinforced if the argument expresses a necessary condition to fulfill the criterion, for 10%.

Table 2 shows the top-30 arguments with the highest confidence scores, ranked by decreasing confidence.

Scenario Ranking. The attitude scores obtained for each scenario from the 478 LLM-derived arguments lead to the same global ranking of arguments, as summarized in Table 3

Table 3. Scenario attitudes: Manual vs. LLM retrieval

Scenario	Manual	LLM
S1	0.67	0.71
S2	0.62	0.68
S3	0.41	0.39

3.3 Comparison Between both Method Results

In the LLM-identified arguments, safety concerns dominate, with Cd/Pb bioaccumulation emerging as the highest concern, reflecting apprehensions about chemical contaminant transfer from waste substrates to insect biomass. Economic considerations are balanced, weighing industrial scaling potential against regulatory barriers that hinder commercialization. Although there are fewer environmental arguments, the benefits of reducing waste from bioconversion processes are strongly highlighted. Overall, the Pro:Con ratio indicates a slightly

Table 4. Argument coverage comparison: LLM (478 args) vs Manual (337 args).

Criterion	LLM Args	% LLM	Top LLM Gain vs Manual
Chemical safety	92 (19%)	+18%	Heavy metals (Cd/Pb bioaccumulation)
Microbial safety	42 (9%)	+12%	ARG[1]/frass pathogens
Economic	105 (22%)	+20%	Regional scaling (11kt/yr)
Environment	86 (18%)	+35%	Waste reduction
Nutrition	76 (16%)	Stable	PUFA:SFA[2] ratios
Technological	28 (6%)	+8%	Processing stability
Regulation	25 (5%)	+15%	Novel food delays
Social	24 (5%)	+10%	Consumer disgust
Total	**478**	**+42%**	

[1] Antibiotic Resistance Genes

[2] Poly-Unsaturated Fatty Acids : Saturated Fatty Acids

optimistic attitude toward the viability of insect-based biowaste valorization despite challenges.

Table 4 summarizes the coverage of the different criteria in the arguments for LLM versus manual retrieval.

4 Conclusion

This study demonstrates the efficacy of the proposed workflow in transforming heterogeneous scientific literature into scenarios and decision analysis about these scenarios for sustainable transitions (research question RQ1). Applied to urban/peri-urban insect-based biowaste valorization, manual expert retrieval yielded 337 structured arguments across various criteria (chemical safety, economic profitability, microbiological safety, etc.), producing scenario rankings (research question RQ2) that highlight environmental, regulatory, and nutritional strengths against persistent safety trade-offs.

LLM-based retrieval from 143 recent abstracts substantially expanded coverage (+42% arguments to 478), with chemical safety dominating top-expressed concerns, balanced economic arguments, strong environmental waste-reduction benefits, and a slightly optimistic global attitude. The LLM-derived interaction graph also identified leverage points absent from manual analysis, mainly frass value subscenarios, and modulation of insect larvae microbiome.

Comparative analysis confirms complementary strengths of both methods (research question RQ3): manual expert retrieval ensures that scenarios make sense in practice, that results remain interpretable, and stakeholder positions are taken into account; customized LLM/RAG retrieval scales knowledge volume while preserving domain specificity through iterative prompt engineering. Both approaches converge on scenario 1 slight superiority, in light of current literature, and reveal chemical safety as the pivotal bottleneck, validating previous projects' benefit-risk analysis framework.

The workflow is generic across emerging technologies requiring multi-criteria and multi-actor foresight under uncertainty. Limitations include field immaturity, since the impacts are still emerging, and potential pro-innovation bias in sources; mitigation involves continuous corpus updates and diverse stakeholder inputs. Future enhancements target hybrid expert-LLM pipelines to combine the complementary strengths of both approaches.

Acknowledgments. This research was supported by INRAE Metaprogramme BETTER, and by the French National Research Agency (contract No. ANR-25-CE21-5431; "REFLY: Recycling of unauthorized biowaste by the black soldier fly: benefits and risks").

References

1. Amgoud, L., Prade, H.: Using arguments for making and explaining decisions. Artif. Intell. **173**(3–4), 413–436 (2009)
2. Arce-Gomez, A., Donovan, J.D., Bedggood, R.E.: Social impact assessments: developing a consolidated conceptual framework. Environ. Impact Assess. Rev. **50**, 85–94 (2015). https://doi.org/10.1016/j.eiar.2014.08.006
3. Arslan, M., Ghanem, H., Munawar, S., Cruz, C.: A survey on rag with llms. Proc. Comput. Sci. **246**, 3781–3790 (2024). https://doi.org/10.1016/j.procs.2024.09.178
4. Asai, A., Wu, Z., Wang, Y., Sil, A., Hashimoto, K., Lewis, P.: Self-rag: learning to retrieve, generate, and critique through self-reflection. arXiv preprint arXiv:2310.11511 (2023). https://arxiv.org/abs/2310.11511
5. Becker, D.R., Harris, C.C., McLaughlin, W.J., Nielsen, E.A.: A participatory approach to social impact assessment: the interactive community forum. Environ. Impact Assess. Rev. **23**(3), 367–382 (2003). https://doi.org/10.1016/S0195-9255(02)00098-7
6. Becker, D.R., Harris, C.C., Nielsen, E.A., McLaughlin, W.J.: A comparison of a technical and a participatory application of social impact assessment. Impact Assess. Project Appraisal **22**(3), 177–189 (2004). https://doi.org/10.3152/147154604781765932
7. Bisquert, P., Croitoru, M., Karanikolas, N.: A qualitative decision-making approach overlapping argumentation and social choice. In: Rothe, J. (ed.) Algorithmic Decision Theory, Lecture Notes in Computer Science, vol. 10576, pp. 344–349. Springer, Cham (2017). https://doi.org/10.1007/978-3-319-67504-6_25
8. Börjeson, L., Höjer, M., Dreborg, K.H., Ekvall, T., Finnveden, G.: Scenario types and techniques: towards a user's guide. Futures **38**(7), 723–739 (2006). https://doi.org/10.1016/j.futures.2005.12.002
9. Bourguet, J.R., Thomopoulos, R., Mugnier, M.L., Abécassis, J.: An artificial intelligence-based approach to deal with argumentation applied to food quality in a public health policy. Expert Syst. Appl. **40**(11), 4539–4546 (2013). https://doi.org/10.1016/j.eswa.2013.01.059
10. Brugha, R., Varvasovszky, Z.: Stakeholder analysis: a review. Health Policy Plan. **15**(3), 239–246 (2000). https://doi.org/10.1093/heapol/15.3.239
11. Chaib, R.L., Macombe, C., Thomopoulos, R.: Adaptation of a participatory system-modeling method to the constraints of remote working. In: Conference on Complex Systems-France. Dijon, France (2021). preprint available at https://easychair.org/publications/preprint/l8VF
12. Chaib, R.L., Macombe, C., Thomopoulos, R.: Adapting a participatory modelling method to forecast food system scenarios: a case study on the pork value-chain. Econom. Agro-alimentare/Food Econ. **24**(3), 1–37 (2022). https://doi.org/10.3280/ecag2022oa14488
13. Chaib, R.L., Macombe, C., Thomopoulos, R.: Designing the future of agri-food chains: comparison of prospective analysis built 40 years ago and today. In: FoodSim 2022, Ghent, Belgium (2022). https://hal.inrae.fr/hal-03770738
14. Chaib, R.L., Macombe, C., Thomopoulos, R.: Structuring ontologies from natural language for collaborative scenario modeling in agri-food systems. Front. Artif. Intell. **5**, 1056989 (2022). https://doi.org/10.3389/frai.2022.1056989
15. Dung, P.M.: On the acceptability of arguments and its fundamental role in nonmonotonic reasoning, logic programming and n-person games. Artif. Intell. J. **77**, 321–357 (1995)

16. Duperrin, J.C., Godet, M., Puiseux, L.: Les scénarios du développement de l'énergie nucléaire à l'horizon 2000: application de la méthode SMIC 74. CEA (1975). https://hal-lara.archives-ouvertes.fr/hal-02185147
17. Engel, E., et al.: The project INSECT4CITY: assessment of benefits and risks of insect-based bioconversion for recycling bio-waste from urban and peri-urban areas. Proc. Nutr. Soc. **81**(OCE2), E73 (2022). https://doi.org/10.1017/S0029665122000969
18. Funtowicz, S.O., Martinez-Alier, J., Munda, G., Ravetz, J.R.: Information tools for environmental policy under conditions of complexity. Environ. Issues Ser. No. 9 (1999). https://www.andreasaltelli.eu/file/repository/envissue09.pdf
19. Gao, Y., et al.: Retrieval-augmented generation for large language models: a survey. arXiv preprint arXiv:2312.10997 (2024), available at https://arxiv.org/abs/2312.10997
20. Godet, M.: Strategic foresight: use and misuse of scenario building. LIPSOR Working Paper (10), 143 (2008). https://prospectivayfuturo.files.wordpress.com/2017/09/godet-strategic-foresight-n10-2009.pdf
21. Godet, M., Durance, P.: La prospective stratégique pour les entreprises et les territoires. Dunod, Paris, France (2001)
22. Harary, F., Norman, R.Z., Cartwright, D.: Structure Models: An Introduction to the Theory of Directed Graphs. John Wiley & Sons, New York (1965)
23. Jiang, Z., Liu, A., Zhang, D., Xu, X., Dai, Y.: Customization and personalization of large language models for engineering design. CIRP Ann. **74**(1), 191–195 (2025). https://doi.org/10.1016/j.cirp.2025.03.001
24. Karanikolas, N., Bisquert, P., Buche, P., Kaklamanis, C., Thomopoulos, R.: A decision support tool for agricultural applications based on computational social choice and argumentation. Int. J. Agric. Environ. Inf. Syst. **9**(3), 54–73 (2018). https://doi.org/10.4018/IJAEIS.2018070104
25. Kurtz, A., Thomopoulos, R.: Safety vs. sustainability concerns of infant food users: French results and european perspectives. Sustainability **13**(18) (2021). https://doi.org/10.3390/su131810074
26. Lesourne, J., Godet, M., Barré, R., Chapuy, P., Fèvre, J., et al.: L'industrie de l'aluminium à la fin du siècle: réflexions prospectives. Centre national de l'entrepreneuriat (1986). https://hal-lara.archives-ouvertes.fr/hal-02185189v1
27. Lewis, P., et al.: Retrieval-augmented generation for knowledge-intensive nlp tasks. Adv. Neural Inf. Process. Syst. **33**, 9459–9474 (2020)
28. Malekpour, S., Walker, W.E., de Haan, F.J., Frantzeskaki, N., Marchau, V.A.: Bridging decision making under deep uncertainty (dmdu) and transition management (tm) to improve strategic planning for sustainable development. Environ. Sci. Policy **107**, 158–167 (2020). https://doi.org/10.1016/j.envsci.2020.03.002
29. Marini, C., Blanc, I.: Towards prospective life cycle assessment: How to identify key parameters inducing most uncertainties in the future? application to photovoltaic systems installed in spain. In: Murgante, B., et al. (eds.) ICCSA 2014, LNCS, vol. 8581, pp. 691–706. Springer, Cham (2014). https://doi.org/10.1007/978-3-319-09150-1_51
30. Morin, M., et al.: Explaining the results of an optimization-based decision support system – a machine learning approach. ITM Web Conf. **14**, 00002 (2017). https://doi.org/10.1051/itmconf/20171400002
31. Munda, G.: Social multi-criteria evaluation: methodological foundations and operational consequences. Eur. J. Oper. Res. **158**(3), 662–677 (2004). https://doi.org/10.1016/S0377-2217(03)00369-2

32. Papin, M., Pattier, M., Davranche, M., Borel, P., Engel, E., Planche, C.: Chemical contamination of black soldier fly larvae raised on eu-authorized or unauthorized substrate. Waste Manage. **210**, 115219 (2026). https://doi.org/10.1016/j.wasman.2025.115219
33. Pesonen, H.L., et al.: Framework for scenario development in lca. Int. J. Life Cycle Assess. **5**(1), 21–30 (2000). https://doi.org/10.1007/BF02978555
34. Rosen, R.: Complexity as a system property. Int. J. Gen Syst **3**(4), 227–232 (1977). https://doi.org/10.1080/03081077708934768
35. Salliou, N., Thomopoulos, R.: Modelling multicriteria argument networks about reduced meat consumption. In: FOODSIM'2018, pp. 46–51 (2018)
36. Schoen, A., Könnölä, T., Warnke, P., Barré, R., Kuhlmann, S.: Tailoring foresight to field specificities. Futures **43**(3), 232–242 (2011). https://doi.org/10.1016/j.futures.2010.11.002, special Issue: Future-oriented Technology Analysis
37. Sohn, J., et al.: Argumentation corrected context weighting-life cycle assessment: a practical method of including stakeholder perspectives in multi-criteria decision support for lca. Sustainability **12**(6), 2170 (2020). https://doi.org/10.3390/su12062170
38. Thomopoulos, R.: A practical application approach to argumentation for multicriteria analysis and decision support. EURO J. Decis. Process. **6**(3–4), 237–255 (2018). https://doi.org/10.1007/s40070-018-0087-2
39. Thomopoulos, R., Chaib, R.L., Dervilly, G.: Benefit-risk analysis of prospective scenarios implementing new-generation food safety management technologies: qualitative, stakeholder-driven study of the infant food chain. Food Res. Int. **214**, 116687 (2025). https://doi.org/10.1016/j.foodres.2025.116687
40. Thomopoulos, R., Charnomordic, B., Cuq, B., Abécassis, J.: Artificial intelligence-based decision support system to manage quality of durum wheat products. Qual. Assur. Safety Crops Foods **1**(3), 179–190 (2009). https://doi.org/10.1111/j.1757-837X.2009.00029.x
41. Thomopoulos, R., Cufi, J., Le Breton, M.: A generic software to support collective decision in food chains and in multi-stakeholder situations. In: FoodSim 2020, Ghent, Belgium (2020). https://hal.science/hal-02484363
42. Thomopoulos, R., Paturel, D.: Multidimensional analysis through argumentation? contributions from a short food supply chain experience. In: Benferhat, S., et al. (eds.) Advances in Artificial Intelligence: From Theory to Practice, LNAI, vol. 10351, pp. 268–274. Springer, Cham (2017). https://doi.org/10.1007/978-3-319-60045-1_29
43. Vinuesa, R., Azizpour, H., Leite, I., Balaam, M., Dignum, V., Domisch, S., Felländer, A., Langhans, S.D., Tegmark, M., Nerini, F.F.: The role of artificial intelligence in achieving the austainable development goals. Nat. Commun. **11**, 233 (2020). https://doi.org/10.1038/s41467-019-14108-y
44. Vivas, C., Planche, C., Macombe, C., Borel, P., Engel, E., Thomopoulos, R.: Une analyse systémique et interdisciplinaire de scénarios de développement durable basés sur l'entomoconversion. Revue Francophone du développement durable **20** (2022). https://hal.inrae.fr/hal-03999567
45. Wang, M., Liu, B., Zhang, S., et al.: Causalrag: integrating causal graphs into retrieval-augmented generation. arXiv preprint arXiv:2503.19878 (2025)

Understanding the Business Intelligence Landscape: A Scientometric Analysis Using Large Language Models

Peter B. Keenan[1] and Ciara Heavin[2(✉)]

[1] Management Information Systems, UCD Business School, University College Dublin, Dublin, Ireland
peter.keenan@ucd.ie
[2] Business Information Systems, Cork University Business School, University College Cork, Cork, Ireland
C.Heavin@ucc.ie

Abstract. This study maps the business intelligence (BI) landscape through a scientometric analysis augmented by large language models (LLMs). One limitation of scientometric analysis is that searches may include many articles that are only peripherally related to the field of interest. In this paper, we test the ability of widely used LLM models to remove less relevant papers from a test dataset. Using a dataset downloaded from Scopus, we conducted a classification exercise to identify if the paper abstracts described a BI paper. We first used an experienced expert as a baseline and then used several LLMs to conduct a similar classification. Our study shows that LLMs can contribute to refining a scientometric dataset, but that the quality of the prompt used plays a vital role in this process. From this initial investigation, we propose a forward-looking research agenda for further synthesis of AI and scientiometrics, which will advance scholarship in the BI field and in other academic disciplines.

Keywords: large language models (LLMs) · Generative AI (GenAI) · Scientometrics · Business Intelligence (BI)

1 Introduction

Business intelligence (BI) is a key component within the wider decision support systems (DSS) landscape, which also encompasses personal decision support systems, group support systems, executive information systems, online analytical processing tools, and data warehousing solutions. It is an umbrella term that combines architectures, tools, databases, analytical tools, applications, and methodologies [1]. BI's major objective is to enable interactive access to data, to enable manipulation of data, and to allow business managers and analysts conduct appropriate analyses [2]. By analysing historical and current data, situations, and performances, decision makers receive valuable insights that enable them to make more informed and better decisions [3]. The BI field has expanded rapidly, driven by advances in data analytics, artificial intelligence (AI),

A. P. Cabral Seixas Costa et al. (Eds.): ICDSST 2026, LNBIP 582, pp. 302–313, 2026.
https://doi.org/10.1007/978-3-032-25146-6_20

and digital transformation [4]. As a result, the BI research landscape has become increasingly broad, interdisciplinary, and fragmented [1]. Understanding the BI landscape is increasingly important as organisations navigate sustainable and resilient transitions, where transition-oriented decision support plays a critical role in shaping evidence-based strategies in dynamic environments.

The boundaries of the BI domain are also blurred because it is heavily influenced by industry, vendor terminology, and rapidly changing technological trends rather than a stable, singular academic theory [5]. Large-scale bibliometric analysis can identify influential authors, institutions, journals, and research clusters across decades of BI research. LLMs provide new opportunities to enhance scientometric research. LLMs can process and summarise large corpora of text, identify latent themes, and support more nuanced interpretation of bibliometric clusters. Combining LLMs with scientometric analysis enables a more complete and semantically rich view of the BI field. The growing integration of LLMs into scientometric research demonstrates their capacity to enhance literature analysis, topic identification, and the design of analytical tools for knowledge-intensive domains such as BI. Such capabilities strengthen BI's role in supporting sustainable transitions, as DSS increasingly rely on advanced analytics and AI-enabled insights to guide organisational change and resilience efforts.

Subsequently, the objective of this research is to test the ability of LLMs to contribute to large-scale scientometric analysis to better understand the BI field. The remainder of this article is organised as follows. First, we outline key concepts in scientiometrics. Next, we examine the role of LLMs and provide an overview of the BI field. We then describe the design of our experiment, present the resulting findings, and conclude with a discussion of the implications of our work.

2 Analysis Approach

2.1 Scientometric Analysis

Scientometrics is the quantitative study of scientific communication through the medium of publications [6]. This investigation has benefited from the growth and widespread adoption of bibliographic databases. While modern computer capacity allows scientometric measures across all domains [7], scientometric research usually involves only subdomains of interest. The selection of publications in this subdomain is a crucial step in the research. This selection can use subject classifications for journals, keywords associated with each article, and free-text search within titles and abstracts, or combinations of these.

Both Scopus and WOS contain '*author keywords*' which are supplied by the author to the publisher. Scopus then adds '*indexed keywords*' which are derived from thesauri that Elsevier owns or licenses; these should use standard terms which are less diverse than the author keywords. Title, keyword, and abstract-based searches can be used to identify a useful subset of papers for the scientometric examination of an academic field. However, inconsistent use of language often results in search datasets that include papers that are not central to the researcher's interests. Traditional scientometric research can restrict datasets more precisely using additional criteria such as the presence of specific text or cited references. Nevertheless, scientometric analysis would benefit from the ability to identify a more accurate initial dataset, and the application of AI to analyse abstracts for supplementary information would facilitate this process.

2.2 Large Language Models as a Scientometric Tool

LLMs are advanced machine learning (ML) models employed for natural language processing tasks [8]. They are trained using vast amounts of text data and use billions of parameters to understand, generate, and analyse human language. LLMs can be used for tasks such as summarisation, translation, text generation, and answering questions. LLMs can now be used directly for academic search [9] and academic databases are now adding AI capabilities, including Scopus AI and the Web of Science Research Assistant.

The attraction of LLMs is that they have already been trained on a large corpus of text and do not need specific training on a text dataset [8]. The extensive range of text used to train LLMs enables them to understand context and nuance, enabling them to follow instructions accurately and to explain their reasoning [10]. LLMs can be used to classify large datasets [8]. where the volume of work may exceed human resources, and where humans are likely to lose concentration and show declining performance. Törnberg [10] tested GPT-4, a well-known LLM, on a Twitter dataset and showed that it achieved higher accuracy than both supervised models and human coders.

As a literature search is based on the text of abstracts or articles, the LLM approach is potentially valuable in scientometrics. Joos, Keim and Fisher [11] assessed the ability of several LLMs to assist in an academic literature search, making the point that a keyword search can often throw up an unmanageably large number of papers, which then must be further filtered for relevance to provide an appropriate subset for scientometric analysis.

As literature search is a significant component of academic work, these new techniques can facilitate research [12]. In medicine, a systematic review (SR) is a type of research that uses a structured, transparent, and reproducible method to summarise and critically appraise existing studies addressing a clearly formulated question [13]. Recent systematic literature reviews have used LLMs to analyse article abstracts [14]. Guo et al. [15] tested the ability of the GPT-4 LLM to replicate the outcomes of several previously published systematic literature reviews. They found that it performed well and suggested that LLMs would be valuable as an initial filtering tool, allowing human screeners to focus on a smaller, more relevant subset of papers. Delgado-Chaves et al. [16] tested the ability of 18 LLM models to select all relevant articles for three systematic reviews, based on the article title and abstracts. Their work identified a reliable performance by different models but also showed variation in their strengths and weaknesses. In the DSS field, Keenan [17] examined the ability of GPT to identify data-driven or model-driven systems from the abstract of the article.

3 The Business Intelligence Field

BI was popularised by Howard Dresner of the Gartner Group in 1989 to describe a set of concepts and methods to improve business decision-making by using fact-based support systems [18]. BI consists of business users accessing data from the data warehouse to perform enterprise reporting, OLAP, querying, and predictive analytics [19].

Between 2010 and 2020, the rise of Web 2.0 technologies and social media profoundly reshaped the IT landscape, expanding BI beyond traditional internal datasets to include externally generated, user-driven, and collaborative web content [4]. This era saw BI systems incorporating real-time data from blogs, social networks, online reviews, and other participatory platforms, reflecting Web 2.0's emphasis on user empowerment, community collaboration, and mashup-driven information integration [4]. Existing research emphasises that modern BI must draw insights not only from internal organisational systems but also from the wider web ecosystem, including competitor information, user-generated content, collective intelligence, and social network interactions. This shift reflects the evolution of BI from a traditional, internally focused, reporting function toward a more adaptive and externally oriented decision-support environment [20]. Social media rapidly became a strategic knowledge source, prompting organisations to integrate user-generated content, online interactions, and social signals into BI workflows to strengthen CRM, market sensing, and performance management practices [21]. As this integration matured, firms recognised that social media and BI were becoming inseparable for sustaining competitive advantage, driving the emergence of social media intelligence as a critical extension of the BI ecosystem [20].

More recently, BI has been augmented by innovation in analytics, conversational interfaces, and agentic AI systems that not only interpret data but also recommend and even initiate actions. BI platforms increasingly integrate natural-language querying, automated insight generation, and AI, reflecting an industry-wide shift toward more accessible and context-rich decision support. Although advanced agentic BI capabilities are emerging, user trust in AI-generated insights remains limited [22].

The DSS and BI literature is related [23], both discussing the completed system and issues such as decision-making enhancement, human-computer interfaces, system integration, and organisational adoption challenges. Journals such as *Expert Systems with Applications* and *Decision Support Systems* are important both for DSS and BI. However, BI papers also appear in journals that have little or no relevance for the general DSS field: including *Journal of Intelligence Studies in Business*, *Information Systems Management*, and the *International Journal of Data Warehousing and Mining*. This includes cases dealing with a single system and review papers discussing multiple systems. However, keywords like DSS and BI also appear in papers that do not discuss complete systems, but rather analytics or computer science techniques that could contribute to a BI. The object of this research is to test the ability of an LLM to classify a paper as a BI paper based on its title and abstract.

4 Analysis

4.1 Data

For this exercise, we downloaded papers whose title, abstract, or keywords contained the term "Business Intelligence."

```
TITLE-ABS-KEY ("Business Intelligence") AND PUBYEAR > 2005 AND
PUBYEAR < 2020
```

We selected the years from 2006 to 2019, we did not select more recent articles, as we wished to assess the extent to which papers were cited. In our test dataset, we included papers with 30 or more citations. We then removed a small number of papers with missing data or very short abstracts and proceeded to use the 375 remaining records. These were drawn from 195 publications, illustrating the diversity of BI; the only journals in the dataset with more than 10 articles were Decision Support Systems, Expert Systems with Applications, and the International Journal of Information Management.

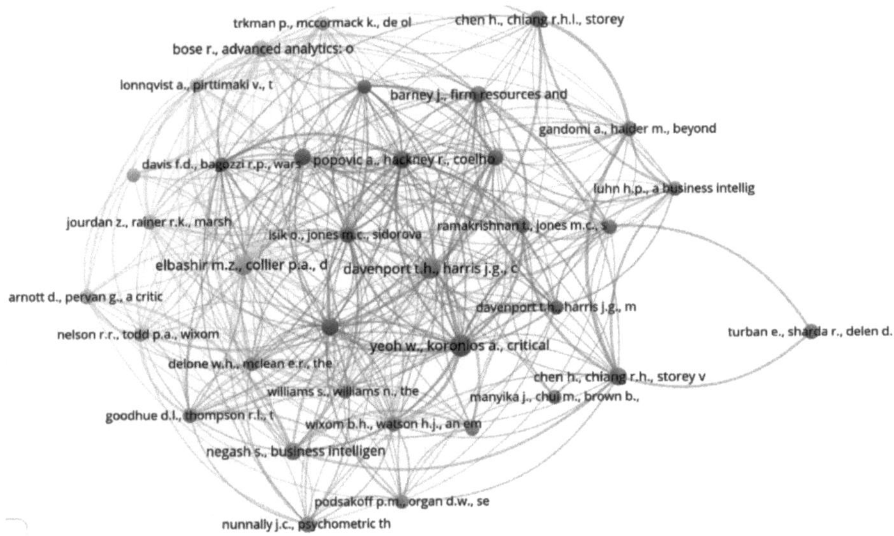

Fig. 1. Major papers cited by our dataset.

These 375 papers cite earlier seminal papers in the field (Fig. 1) and have been cited by more recent papers (Fig. 2). Note that the key citing journals shown in Fig. 2 have a broadly similar pattern to the distribution of journals in the dataset. The journals Decision Support Systems, Expert Systems with Applications, and the International Journal of Information Management play a central role, with many other journals with a small number of citations.

In this analysis, we used a traditional keyword search to identify a set of relevant papers. However, keyword search has some limitations in building a dataset for Scientometric analysis. Keyword search may produce papers of different types, and each type of paper may have a distinct role in the analysis. For instance, the BI keyword may be used for papers describing complete systems and used for papers discussing technologies that could contribute to a BI, if included in a system.

4.2 Method

This research compares different LLM models using OpenRouter.ai, a unified platform that provides developers with a single access point to hundreds of different LLMs and AI

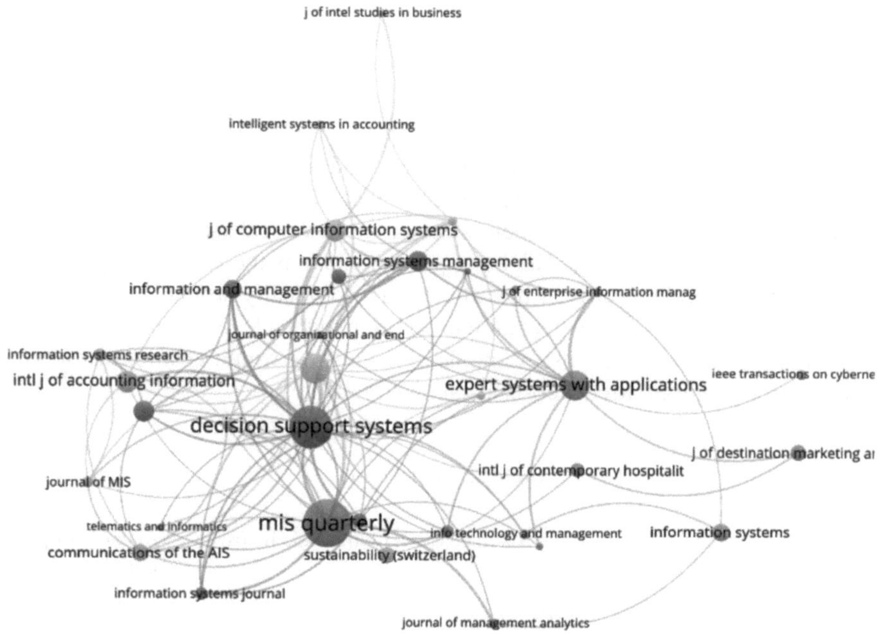

Fig. 2. Journals citing our dataset.

providers. We used the Openrouter Application Programming Interface (API) (https://openrouter.ai/api) accessed from Python code. The various LLMs on this platform have different costs. LLMs process text in fundamental units known as **tokens**, which may represent words, sub-words, or individual characters. Model pricing on OpenRouter is based on the number of **prompt tokens** (input) and **completion tokens** (output). Prompt tokens are the text supplied by the user, including the problem description and any system-level instructions (for example, instructions specifying how the model should format its output). Completion tokens are the tokens generated by the model in response to that input. They include the model's direct answer, any explanation (if included), and any additional text produced as output.

In OpenRouter, there are different prices for prompt and completion tokens, typically expressed as US$*x* per 1 million tokens. Prices change frequently as new models are introduced, and in early 2026, they ranged from free to US$600 per 1 M completion tokens. As an overhead contribution, Openrouter also charges a 5.5% payment handling charge in addition to the per token payments.

5 Results

5.1 Classification

The performance of an LLM is determined by the model used and by the prompt used. In this exercise, we used the interactive versions of two market leading LLMs, ChatGPT and Gemini, to identify the main characteristics of BI described in the literature. We then asked them to devise an appropriate prompt.

```
what definition of Business Intelligence is used in the key aca-
demic publications in the field?
can you propose a prompt for an LLM to identify business intelli-
gence academic articles from their title, keywords and abstract?
```

Informed by various interactions with LLMs and their output, we also manually devised a prompt which better reflected the rules that might be used.

These are reasonably priced models, suitable for large-scale scientometric analysis beyond the scale of our dataset. The input prompt contains the text of the prompt and the title, abstract and keywords of the article, this is typically around 1100 tokens. In our analysis, we asked for an explanation in addition to the classification, so the output prompt was typically 900 tokens. A simple classification would cost less as it would have less verbose output. Model costs are shown in Table 1.

Table 1. Model costs

Model	Input token cost per 1 m	Output token cost per 1 m	Typical Cost per record
google/gemini-2.5-flash-lite	$0.10	$0.40	$0.0002
google/gemini-2.5-flash	$0.30	$2.50	$0.0020
openai/gpt-5-nano	$0.05	$0.40	$0.0005
openai/gpt-5-mini	$0.25	$2.00	$0.0023

In our analysis, we used four prompts and four popular models; the results are shown in Table 2. We compared the models' classification with a human-coded classification, which classified 75.2% (282) of the records as BI. When running the LLMs, a small number of papers did not process correctly using certain combinations of prompt and model; we did not investigate these errors.

In general, using a LLM with simpler prompts classified a higher proportion of articles as BI, including many that the human classifier had thought were not BI. The more elaborate prompt that we designed accepted fewer abstracts as being BI than the human coder. All approaches rejected some abstracts that mentioned BI, but were not significantly about BI, for instance [24]. It is clearly useful that even the simplest approach can identify these outliers to remove them from the dataset.

Together, these findings highlight the importance of prompt design in calibrating LLM classification behaviour and ensuring that BI-related datasets remain focused and analytically coherent.

5.2 Discussion

LLMs offer a compelling cost advantage, particularly when used for large-scale text classification that would otherwise require substantial human labour. Our experiments show that running multiple prompt variations or re-classifying datasets incurs only minimal

Table 2. Results from LLM analysis

		gemini-2.5-flash-lite		gemini-2.5-flash		gpt-5-nano		gpt-5-mini	
Prompt	Classification	BI	Not BI	BI	Not BI	BI	Not BI	Bi	Not BI
1	BI	273	7	267	13	263	19	273	9
	Not BI	83	10	73	20	72	21	78	15
2	BI	278	4	279	3	251	31	280	2
	Not BI	85	8	88	5	56	36	88	5
3	BI	258	24	262	18	247	25	226	55
	not BI	67	24	69	24	56	30	54	39
4	BI	97	181	142	140	73	208	61	221
	Not BI	3	89	13	80	3	90	3	90

additional expense, making iterative experimentation financially feasible. This low operational cost positions LLMs as an attractive complement to human coding, especially in research contexts where efficiency and scalability are essential.

In our analysis, the simpler prompts produced many false positives, while the complex prompt had many false negatives. The acceptability of these errors is dependent on the objectives of the analysis. The simpler prompts do eliminate a small number of unsuitable papers but otherwise have similar results to the keyword search on Scopus. This may not be especially useful in scientometric analysis, which is based on aggregate analysis and can tolerate outliers to some extent. The designed prompt gives a smaller dataset, but this reduced dataset is likely to be closer to the core definition of BI and consequently more useful for analysis.

As illustrated in Table 3, the performance of different prompt designs revealed a clear trade-off between simplicity and precision. By contrast, the more complex and carefully structured prompt produced a smaller but more focused dataset. This version of the prompt was more selective and aligned more closely with the conceptual definition of BI. This behaviour underscores the inherent trade-offs in prompt engineering, where tightening conceptual boundaries can unintentionally narrow the resulting corpus too far.

Ultimately, the acceptability of these error patterns depends on the purpose of the analysis. Scientometric studies often tolerate some degree of noise since aggregate patterns tend to remain stable despite the inclusion of outliers. In such cases, a simple prompt may be sufficient, although not especially advantageous. In contrast, when the objective is to refine or validate the conceptual boundaries of a field, the smaller, more precise dataset generated by the complex prompt is likely to be more valuable. This curated set is closer to the core literature on BI and therefore offers a stronger foundation for analyses that require conceptual clarity rather than broad coverage.

Table 3. Prompt Type v Output Matrix.

Prompt Type	Typical Errors	Output Size	Output Characteristics	Usefulness for Scientometric Analysis
Simple Prompt	**High false positives** (includes many irrelevant papers)	Large	Broad, noisy, like keyword search; weak filtering	**Moderate–Low** Too many irrelevant results; adds little beyond Scopus search
Complex Prompt	**High false negatives** (misses relevant papers)	Smaller	More selective, closer to the conceptual core of BI	**High** More focused dataset better for conceptual/field delimitation

6 Conclusion

Gaining a clear understanding of the BI landscape is becoming increasingly crucial as organisations work through sustainable and resilient transitions, where transition-oriented decision support is essential for guiding evidence-based decision-making in rapidly evolving contexts. This study explores whether LLMs can move beyond surface-level keyword matching to nuanced interpretation, which is essential for their adoption in academic workflows like literature reviews, meta-analyses, and research planning. Our investigation indicated that LLMs can be of value, and at reasonable cost, but that the quality of the prompt used plays a key role in the analysis.

Further work is required to assess the full value of LLMs in scientometric analysis. This could include the value of keywords, as Scopus keywords differ from those on the Web of Science. Beyond assigning each abstract to a single class, LLMs could also support multi-label classification. For example, in this dataset an LLM might classify articles simultaneously into the Analytics, IS, or Computer Science domains rather than not BI. Researchers could also investigate how LLMs distinguish between different article types, such as review papers versus case studies.

Any further work would benefit from more sophisticated prompt design. The BI field is characterised by inconsistent use of language, and some rigour is needed in defining it for a LLM. The BI/DSS academic community might even engage in a project to build a customised LLM, using training data selected by the community and using hosting approaches which allow customised weights within the models, this would have considerable value for teaching and academic purposes.

In conclusion, LLMs can rapidly classify hundreds or thousands of articles, thereby accelerating the early stages of literature review or scientometric analysis by reducing manual processing. This paper has illustrated the value of popular LLM models for the classification of abstracts in the BI field. There is a great deal of potential for further work in this area.

Appendix 1: Prompts Used

Prompt 1 Basic Prompt

```
### ROLE
You are an academic expert in Business Intelligence and Decision
Support Systems.

For this task, identify if an article is Business Intelligence (BI)
from its title, keywords and abstract.

Output
Classification: BI or Not BI
Justification: why is it classified this way
```

Prompt 2 Proposed by Gemini

```
**Task: ** Act as a binary classifier for academic literature.
Evaluate the following Title, Keywords, and Abstract for relevance
to Business Intelligence research.
**Definition of BI:** The transformation of raw data into mean-
ingful and useful information for business analysis purposes.
Includes data warehousing, descriptive analytics, and decision
support systems.
```

Prompt 3 Proposed by ChatGPT

```
### ROLE
You are an academic expert in Business Intelligence and Decision
Support Systems.
For this task, classify an article as Business Intelligence (BI)
only if its central contribution concerns systems, processes, or
methodologies that:
Collect, integrate, store, or manage data for analytical purposes,
and Transform data into information, knowledge, or insights, and
Explicitly aim to support organisational or managerial decision-
making or performance management.
Articles that focus solely on data collection, data quality, basic
data analytics, or generic information systems without a decision-
support or performance orientation should not be classified as BI.
```

Prompt 4 Proposed by the Authors

```
### ROLE
You are an expert classifier for academic papers in Business Intel-
ligence (BI).
```

```
### TASK
Analyse the provided paper (title, abstract, keywords) and assign
exactly one category BI or NOT BI

Classify a paper as "BI" if and only if the paper substantially
describes an integrated BI system or architecture for end users
(managers, analysts, decision makers).
A paper qualifies as BI if it includes:
- A description of a complete or near complete BI system, includ-
ing multiple components such as raw data sources, integrated data
pipelines, centralised storage (e.g., data warehouse), and a user
interface for interacting with data or reports. One of these com-
ponents alone, with an end user interface, is not a complete BI
system,
- Evidence of real-world decision support (e.g., use cases, work-
flows, evaluation).
- BI adoption as the primary topic.
Constraint: BI must be the primary system under study, not an inci-
dental element.
A paper qualifies as NOT BI if it:
- Performs isolated data analysis, statistical modelling, or
machine learning without a full BI system.
- Focuses on theoretical algorithms or technical optimisations
without a user-facing interface.
- Only discusses particular individual technologies such as data
warehouses, data mining, semantic web, or social media, without a
completed BI system with a user interface being the main theme of
the paper.
- Describes general information systems in an organisation with
little specific focus on BI.
when uncertain, assign Not BI.
```

References

1. Liang, T.-P., Liu, Y.-H.: Research landscape of business intelligence and big data analytics: a bibliometrics study. Exp. Syst. Appl. **111**, 2–10 (2018)
2. Chaudhuri, S., Dayal, U., Narasayya, V.: An overview of business intelligence technology. Commun. ACM. **54**, 88–98 (2011)
3. Sharda, R., Delen, D., Turban, E., Aronson, J., Liang, T.: Business intelligence and analytics. Syst. Decis. Supp. **398**, 2014 (2014)
4. Chen, H., Chiang, R.H., Storey, V.C.: Business intelligence and analytics: from big data to big impact. MIS Q, 1165–1188 (2012)
5. Meredith, R., Remington, S., O'Donnell, P., Sharma, N.: Organisational transformation through business intelligence: theory, the vendor perspective, and a research agenda. J. Decis. Syst. **21**, 187–201 (2012)
6. Mingers, J., Leydesdorff, L.: A review of theory and practice in scientometrics. Eur. J. Oper. Res. **246**, 1–19 (2015)

7. Leydesdorff, L., Rafols, I., Chen, C.: Interactive overlays of journals and the measurement of interdisciplinarity on the basis of aggregated journal–journal citations. J. Am. Soc. Inform. Sci. Technol. **64**, 2573–2586 (2013)
8. Kumar, P.: Large language models (LLMs): survey, technical frameworks, and future challenges. Artif. Intell. Rev. **57**, 260 (2024)
9. Zhou, J. et al.: AcademicBrowse: benchmarking academic browse ability of LLMs. arXiv preprint https://arxiv.org/abs/2506.13784 (2025)
10. Törnberg, P.: Large language models outperform expert coders and supervised classifiers at annotating political social media messages. Soc. Sci. Comput. Rev. **43**, 1181–1195 (2025)
11. Joos, L., Keim, D.A., Fischer, M.T.: Cutting through the clutter: the potential of LLMs for efficient filtration in systematic literature reviews. arXiv preprint https://arxiv.org/abs/2407.10652 (2024)
12. Sami, A.M. et al.: System for systematic literature review using multiple AI agents: concept and an empirical evaluation. arXiv preprint https://arxiv.org/abs/2403.08399 (2024)
13. Moosapour, H., Saeidifard, F., Aalaa, M., Soltani, A., Larijani, B.: The rationale behind systematic reviews in clinical medicine: a conceptual framework. J. Diab. Metab. Disord. **20**, 919–929 (2021)
14. Syriani, E., David, I., Kumar, G.: Screening articles for systematic reviews with ChatGPT. J. Comput. Lang. **80**, 101287 (2024)
15. Guo, E., Gupta, M., Deng, J., Park, Y.-J., Paget, M., Naugler, C.: Automated paper screening for clinical reviews using large language models: data analysis study. J. Med. Internet Res. **26**, e48996 (2024)
16. Delgado-Chaves, F.M. et al.: Transforming literature screening: the emerging role of large language models in systematic reviews. Proc. Nat. Acad. Sci. **122**, e2411962122 (2025)
17. Keenan, P.: Using GPT to classify DSS. In: DSA ISC25: Intelligent Systems & Decision Making: Human Insights in the Era of AI. Springer (2025)
18. Power, D.J.: Decision Support Systems: Concepts and Resources for Managers. Quorum Books, Westport, CN (2002)
19. Watson, H.J., Wixom, B.H.: The current state of business intelligence. Computer. **40**, 96–99 (2007)
20. Yang, J., Xiu, P., Sun, L., Ying, L., Muthu, B.: Social media data analytics for business decision making system to competitive analysis. Inform. Process. Manag. **59**, 102751 (2022)
21. Dutot, V., Mosconi, E.: Social media and business intelligence: defining and understanding social media intelligence, vol. 25, pp. 191–192. Taylor & Francis (2016)
22. Hughes, L. et al.: AI agents and agentic systems: a multi-expert analysis. J. Comput. Inform. Syst., 1–29 (2025)
23. Shim, J.P., Warkentin, M., Courtney, J.F., Power, D.J., Sharda, R., Carlsson, C.: Past, present, and future of decision support technology. Decision Supp. Syst. **33**, 111–126 (2002)
24. Hunt, S.D., Madhavaram, S.: Teaching marketing strategy: using resource-advantage theory as an integrative theoretical foundation. J. Market. Educ. **28**, 93–105 (2006)

Decision Tools, Applications and Business Cases

A Decision Support Optimization Model to Reduce Congestion in Fitness Centres with an Account for Users' Preferences

Franklin Djeumou Fomeni[1,2,3](✉) and El Walid Yermeche[1]

[1] Université du Québec à Montreal (UQAM), 320 Rue Sainte Catherine Est, Montreal, Canada
djeumou_fomeni.franklin@uqam.ca

[2] Group for Research in Decision Analysis (GERAD), 3000, ch. de la Côte-Sainte-Catherine, Montréal, QC H3T 2A7, Canada

[3] Interuniversity Centre in Network, Transport and Logistics (CIRRELT), 2920, Chemin de la Tour, Université de Montréal, Montréal, QC H3T 1J4, Canada

Abstract. It is well known that physical activity plays a key role in maintaining good health. Over the past few decades, there has been a significant increase in both the number of fitness centres worldwide and the number of individuals who use them. In addition to contributing to public health, fitness centres also play an important role in the economy, generating revenue by attracting new members and retaining existing ones. During peak hours, however, the number of customers present in a facility often exceeds the available equipment capacity. As a result, some customers are forced to wait for equipment to become available or to deviate from their planned workout routines, which can lead to dissatisfaction and member attrition. In this paper, we propose a decision support optimization model designed to reduce peak-hour congestion in fitness centres while improving customer experience. The model is intended to support an online booking system in which customers submit provisional workout plans along with their preferences. The system then outputs a final, congestion-free workout plan for each customer that closely matches these preferences. The proposed model aims to maximize equipment utilization while ensuring that each customer can complete their workout with minimal idle time. The model is tested on realistic data sets, and the computational results show that, in addition to supporting the proposed booking system, it can reduce customer idle time by up to 19% compared to a first-come, first-served operational policy.

Keywords: Integer programming · Network Optimization · Decision support models

1 Introduction

It is well known that physical activity plays a key role in maintaining good health. The World Health Organization defines physical activity as any bodily movement

A. P. Cabral Seixas Costa et al. (Eds.): ICDSST 2026, LNBIP 582, pp. 317–332, 2026.
https://doi.org/10.1007/978-3-032-25146-6_21

produced by skeletal muscles that requires energy expenditure [3]. It also recognizes physical inactivity as the fourth leading risk factor for global mortality, accounting for approximately 6% of deaths worldwide. Moreover, physical inactivity is estimated to be responsible for approximately 21âĂŞ25% of breast and colon cancers, 27% of diabetes cases, and nearly 30% of the burden of ischaemic heart disease [3]. On the other hand, regular and adequate levels of physical activity provide numerous benefits. For adults, physical activity contributes to reducing the risk of hypertension, coronary heart disease, stroke, diabetes, breast and colon cancer, depression, and falls. It also helps improve bone and functional health and constitutes a key determinant of energy expenditure, which is fundamental to energy balance and weight control.

People around the world are becoming increasingly health-conscious and are engaging more frequently in physical activities. For example, in Canada, more than 16.2 million individuals aged 12 and over reported participating in at least 150 min of moderate- to vigorous-intensity aerobic physical activity per week [4]. This trend has been accompanied by a rapid growth in both the number of fitness centres worldwide and the number of their users. Fitness centres are health, recreational, and social facilities dedicated to exercise, sports, and other physical activities. They offer a wide range of services, including organized group classes such as spinning, yoga, and martial arts; organized and informal team sports; and individual fitness opportunities such as cardiovascular training, weight training, and swimming.

In addition to contributing to public health, fitness centres play a significant role in the economies of most developed countries. They generate revenue by recruiting new members and encouraging existing members to renew their memberships [5]. As in any business, individuals are more likely to join or renew a membership when they perceive sufficient value for their money, particularly in terms of service quality. To this end, fitness centre owners already employ various incentives to retain existing customers and attract new ones [5]. These efforts are often complemented by incentives offered by employers to encourage employees to remain physically active [6,7].

From the customers' perspective, Afthinos *et al.* [8] identify several variables that influence satisfaction with fitness centres. Among the most important is the *availability* of adequate space during exercise, implying that overcrowded facilities are likely to generate dissatisfaction. Indeed, peak hours at fitness centres are often characterized by customers waiting idly for equipment to become available or being forced to use equipment that was not part of their planned workout, thus deviating from their training routines. Although some fitness centres offer discounted off-peak memberships to mitigate this issue, overcrowding remains a persistent problem due to the growing popularity of fitness centres and the limited availability of many customers, who can often exercise only after work.

During the COVID-19 pandemic [9], the fitness industry was significantly affected, as many fitness centres were forced to close or operate under severe capacity restrictions. Nevertheless, this period prompted fitness associations in many countries to collaborate on the development of joint health and safety guidelines, some of which continue to benefit the industry beyond the pandemic. One such measure aimed at addressing overcrowding was the introduction of online booking systems (e.g., in Canada [10] and New Zealand [11]). When efficiently implemented, these systems can provide a long-term solution to the long-standing issue of peak-hour congestion. While online booking systems are already used in many fitness centres, their application is often limited to group classes with small capacities. Extending these systems to cover the entire training facility would require customers to reserve available time slots in advance, with fully booked slots becoming unavailable to additional users.

Although such booking systems allow fitness centres to control capacity and promote customer safety, we argue that mathematical optimization can be used to further enhance their efficiency. In practice, facility capacity is often calculated without considering customers' individual workout plans or preferences. For instance, consider a one-hour time interval during which 40 customers arrive at a facility equipped with only 10 treadmills, all intending to use a treadmill for their workout. In this case, 30 customers would be unable to follow their planned routines and might either switch to alternative exercises or leave without exercising. If workout plans were shared within the booking system, however, it would be possible to reassign 30 of these customers to different time slots and allocate the available treadmill capacity to customers with different exercise plans. Consequently, incorporating customers' workout plans into the booking system is essential for improving efficiency through mathematical optimization. This problem bears similarities to the assignment of trajectories in air traffic flow management (ATFM). In ATFM, airlines independently plan optimal trajectories for their flights, but because these trajectories share network resources with limited capacities—such as airports and airspace sectors—they must be coordinated and optimized to produce feasible final trajectories [12–17].

In this paper, we propose a network optimization model designed to support online booking systems for fitness centres. More specifically, the model is meant to support a transition from a "First Come First Served" booking system to an "optimization-driven" booking system. The model incorporates customers' individual workout plans and seeks to (i) minimize total delay, defined as the difference between a customer's intended workout start time and the actual time at which they are admitted to the facility; (ii) minimize deviations from customers' workout plans; and (iii) maximize facility utilization, measured by the number of customers who can be accommodated.

Our main contribution is the development of a network-based mathematical optimization model that enhances the efficiency of fitness centre booking systems. In the proposed model, a fitness centre is represented as a fully connected network in which clusters of similar equipment form the nodes. Each customer's workout plan is modeled as a path through this network, with each node corresponding to a specific exercise type. The model is solved using a commercial solver, and computational results based on realistic data sets demonstrate that it can reduce total delay by up to 17.5% and customers' idle time by up to 19%.

The remainder of this paper is organized as follows. Section 2 provides a detailed description of the problem and the underlying assumptions. The proposed optimization model is presented in Sect. 3. Computational results are discussed in Sect. 4, and concluding remarks are given in Sect. 5.

2 Problem Description

In general, the fitness industry consists of companies that own one or more fitness centres or operate franchise networks. The business model of each fitness centre typically revolves around a set of activities, such as physiotherapy, group classes, personal training, and free workouts. Among these, free workout activities usually occupy the largest portion of the available space. The free workout area of a fitness centre is generally organized into clusters of similar equipment, such as treadmills, elliptical machines, stationary bikes, rowing machines, free weights, and other types of exercise equipment. Each cluster provides a specific set of benefits to customers.

Customers subscribe to different types of memberships, which grant access to various services offered by the facility, such as saunas, hydro-massage beds, or showers, or provide flexibility to use multiple fitness centres operated by the same company. In this paper, we consider a highly capacity-constrained setting involving a fitness company that owns several fitness centres located within a given geographical region and within reasonable travel distance for its customers. A representative example is a network of fitness centres belonging to the same company and located within a large city, where customers may choose which facility to attend on a given day.

On the demand side, we consider a set of customers who regularly use the facilities to perform workouts. Depending on the type of membership held, a customer may be allowed to use the facility on any day of the week, at any time, or even at different branches of the same fitness company. The choice of a specific fitness centre on a given day may depend on several factors, including personal preferences, proximity, perceived attractiveness of the facility, availability of training partners, and the expected level of congestion.

We assume that fitness centres have implemented online booking systems to better control facility capacity. This assumption is related to the fact that our model is designed to support the digital transition from a "First Come First Served" booking system to an "Optimization-driven" booking system. Such systems are intended to determine which customers are allowed to enter a facility during each time interval. Unlike traditional booking systems that rely solely on aggregate facility capacity—defined, for example, by the total number of equipment items that can be used while respecting health and safety guidelines—we consider a booking system that also accounts for customers' workout plans. Specifically, customers are assumed to be willing to share their workout plans in advance, and the system admits customers whose plans are not in conflict, thereby reducing congestion. We argue that the effectiveness of such a system can be significantly enhanced through mathematical optimization. Accordingly, we propose a model that disaggregates facility capacity to the level of equipment clusters, allowing utilization to be maximized based on workout plans rather than on aggregate facility capacity.

We focus on a tactical-level planning problem, covering one or more days in advance, in which customers submit their workout plans for a specific day through the online booking system. For each customer, a workout plan consists of (i) an intended arrival time, (ii) the sequence of equipment to be used, (iii) the planned duration of use for each piece of equipment, and (iv) the preferred fitness centre location, provided that the customer's membership allows for such flexibility. Customers may also specify alternative equipment options to be used if their preferred equipment is unavailable, as well as a set of alternative arrival times. Supported by the optimization model, the booking system assigns a final workout plan to each customer while ensuring that the capacity of each equipment cluster is not exceeded at any time. The assigned workout plan may be identical to the originally submitted plan or may include deviations. We consider two types of deviations: time deviations, which affect arrival times or durations of equipment use, and exercise deviations, which occur when a customer is assigned to alternative equipment instead of the initially planned one. In addition, the final workout plan may be assigned to a fitness centre other than the customer's preferred location, provided that such reassignment is allowed by the membership and acceptable to the customer.

Time is a critical component of the problem. The booking system operates on a day-by-day basis and considers the business hours of each fitness centre. Accordingly, we define a planning horizon that covers the operating hours of the facilities, which may span up to 24 h. To capture variations in equipment usage and customer activity, the planning horizon is discretized into short time periods, typically representing intervals of 5 or 10 min. This level of granularity allows customers to specify workout start times that closely match their preferences and provides managers with improved visibility into equipment utilization at the cluster level. Disaggregating capacity in this manner also offers operational advantages, such as the ability to schedule equipment cleaning by cluster through temporary capacity adjustments, without requiring full facility shutdowns.

3 The Capacity Constrained Network Flow Model

We model this problem as a capacity constrained network flow problem, where the flow is determined by the customers using the fitness centre equipment in a sequential manner. The model is aimed at a fitness company that has one or more fitness centres located within a given geographical region. We will denote the set of these fitness centres as $\mathcal{J}$ with each centre indexed by j. In our mathematical formulation, we represent this setting as a *disconnected graph* where each subgraph element corresponds to the layout of a single fitness centre represented as a network. In fact, in our model consideration, each fitness centre of the company is represented as a fully connected network, where the nodes corresponds to the equipment in the centre and the arcs represent the possibility of moving from one equipment to another. We also model the workout plan of each customer as a path which starts at the entry point and ends at the exit point of the facility. The sequence of the nodes on the path then represents the sequence of execution of the customer's workout plan. The problem setting can thus be represented as a collection of disjoint directed graphs $G = \bigcup_{j \in \mathcal{J}} G_j$. Each sub-graph $G_j = (\mathcal{N}_j; \mathcal{E}_j)$, wherein $\mathcal{N}_j$ is the set of nodes (equipment) and $\mathcal{E}_j$ is the set of arcs connecting the nodes, will be the graph representing the network for the fitness centre j. This representation as a disconnected graph is due to the fact that a workout plan can only start and end in the same fitness centre, while any centre can be selected for a workout plan. The time horizon considered will be discretized into time periods. For example, a planning horizon of 24 h can be discretized into time periods of 5 min lengths, yielding the set of discretized time periods to contain 288 periods. The details about the model and its parameters are given below.

3.1 Data Notation

The data for our mathematical model is composed of the following:

- $\mathcal{J}$: is the set of the fitness centres owned by the company.
- $\mathcal{I}$: is the set of all the customers of the fitness company.
- $\mathcal{T}$: is the set of (discretized) time periods.
- i, j, k, n and t: are respectively the indexes of customers, fitness centre, cluster of equipment, node (equipment or entry/exit point) and time period.
- $\mathcal{L}_i$: is the set of fitness centres feasible for the training of customer $i \in \mathcal{I}$.
- $\mathcal{N}_j$: is the set of all the equipment in the fitness centre $j \in \mathcal{J}$.
- $\mathcal{K}_j$: is the set of the clusters of equipment in the fitness centre $j \in \mathcal{J}$.
- $\mathcal{S}_{k,j}$: is the set of the equipment belonging to cluster $k \in \mathcal{K}_j$ at the fitness centre $j \in \mathcal{J}$.
- a_j and d_j: are respectively the check-in and the check-out points of the fitness centre $j \in \mathcal{J}$.
- $\underline{t}_j^i$ and $\overline{t}_j^i$: are respectively the planned arrival time and departure time of customer $i \in \mathcal{I}$ at the fitness centre $j \in \mathcal{L}_i$.
- $\mathcal{N}_j^i$: is the set of feasible equipment in the workout plan of customer $i \in \mathcal{I}$ at the fitness centre $j \in \mathcal{L}_i$. This includes the equipment in the preferred workout plan as well as alternative equipment.
- $\Delta_{i,j}^+(n)$ and $\Delta_{i,j}^-(n)$: are the set of the nodes that are respectively subsequent and preceding to node $n \in \mathcal{N}_j^i$ in the workout plan of customer $i \in \mathcal{I}$ at the fitness centre $j \in \mathcal{L}_i$.
- $T_{n,j}^i = \left[\underline{T}_{n,j}^i, \overline{T}_{n,j}^i\right]$: is the feasible time interval for customer $i \in \mathcal{I}$ to use equipment $n \in \mathcal{N}_j^i$ at the fitness centre $j \in \mathcal{L}_i$. For example, if the customer's plan is to start using equipment n **from period 10 for 20 min in a 5-minutes discretization setting, then $T_{n,j}^i = \{10; 11; 12; 13\}$.**
- $T_j^i = \left[\underline{T}_{a_j,j}^i, \overline{T}_{d_j,j}^i\right]$: is the feasible time interval for customer $i \in \mathcal{I}$ to workout at the fitness centre $j \in \mathcal{L}_i$.

- $\alpha^i_{n,n',j}$: is the number of time period planned by customer $i \in \mathcal{I}$ to use equipment $n \in \mathcal{N}^i_j$ before moving onto equipment $n' \in \Delta^+_{ij}(n)$ at the fitness centre $j \in \mathcal{L}_i$.
- $A^t_{k,j}$: is the capacity of cluster $k \in \mathcal{K}_j$ at the fitness centre $j \in \mathcal{J}$ during time period $t \in \mathcal{T}$.
- $C^i_{n,j}$: is the penalty cost incurred when customer $i \in \mathcal{I}$ utilizes equipment $n \in \mathcal{N}^i_j$ at the fitness centre $j \in \mathcal{L}_i$. This cost is 0 if the customer had planned to used the equipment $n \in \mathcal{N}^i_j$.
- γ^i: is the maximum total amount of time acceptable by customer $i \in \mathcal{I}$ to be idle while waiting for equipment to be free.
- P^i_j: is the penalty of cost (dissatisfaction) if the workout plan of customer $i \in \mathcal{I}$ cannot be fulfilled at the fitness centre $j \in \mathcal{L}_i$.

3.2 The Decision Variables

We define the decision variables for our model following an idea that is well known in the network flow literature [12–14,17]. Namely, these variables will define, for each customer, the equipment that is being used at a specific fitness centre for each time period.

$$x^i_{n,j}(t) = \begin{cases} 1 & \text{if customer } i \in \mathcal{I} \text{ is planned to use equipment } n \in \mathcal{N}^i_j \\ & \text{at the fitness centre } j \in \mathcal{L}_i \text{ by time period } t \in \mathcal{T}^i_{n,j} \\ 0 & \text{otherwise ,} \end{cases}$$

It is assumed that a customer will use each equipment type for at least one time period. This is representative of a duration of about 5 to 10 min, which is very compatible with the reality.

3.3 The Objective Functions

In this section, we present two objective functions that can be used either individually or simultaneously as key performance indicators to guide the optimization model. These objectives are the total delay and the total deviation from customers' workout plans.

3.4 Delay

The aim of this objective function is to minimize the total delay in the start time of customers workouts. In other word, this is the difference between the time that they intended to start their workout and the time at which they are actually allowed into the fitness centre. For example, if a customer would ideally like to start her training at 6pm, the model will aim at assigning her a starting time that is as close as possible to 6pm.

$$\sum_{i \in \mathcal{I}} \sum_{j \in \mathcal{L}_i} \sum_{t \in T^i_{a_j,j}} (t - \underline{t}^i_j) \left(x^i_{a_j,j}(t) - x^i_{a_j,j}(t-1) \right). \tag{1}$$

3.5 Deviation from Workout Plan

One of the main motivation of this paper is that the customers share their workout plans in advance to allow the fitness company to make a better planning of their fitness facilities. In return, these customers would expect minimal changes to their workout plan. This means that they would want to be allowed to used the equipment that they intend to use and spend enough time on these equipment, within a certain allowed limit. Thus any change of equipment to be used will result in the dissatisfaction for the customers. Usually deviation from workout plan will occur when the number of customers wanting to use equipment in a cluster is larger than the number of equipment available in that cluster.

$$\sum_{i\in\mathcal{I}}\sum_{j\in\mathcal{L}_i}\sum_{n\in\mathcal{N}_j^i} C_{n,j}^i x_{n,j}^i\left(\overline{T}_{n,j}^i\right). \tag{2}$$

Note that the cost parameter $C_{n,j}^i$ is 0 if customer $i\in\mathcal{I}$ had planned to used the equipment $n\in\mathcal{N}_j^i$ at the fitness centre $j\in\mathcal{L}_i$. The cost is also modulated to account for the preferences of the customer in terms of utilizing alternative equipment in the event that their preferred equipment are not available.

3.6 The Constraints

In this section we present the constraints of our model, whose role are to ensure that each customer is assigned at most a single workout plan at one fitness centre and that the number of people present in each cluster of equipment do not exceed the capacity of the cluster.

1. *Time connectivity constraints*:
   The first set of constraints is concerned with the time connectivity of the workout plans. These constraints follow naturally from the definition of the variables. Indeed, if a customer is using an equipment n *by* the time period t^*, then $x_{n,j}^i(t)$ must have a value of 1 for all later time periods ($t\geq t^*$). They are stated as

$$x_{n,j}^i(t-1)-x_{n,j}^i(t)\leq 0, \qquad \forall i\in\mathcal{I}, j\in\mathcal{L}_i, n\in\mathcal{N}_j^i, t\in T_{n,j}^i. \tag{3}$$

2. *Clusters capacity constraints:*
   The second group of constraints are the clusters capacity constraints. These constraints ensure that the number of people using the equipment in each cluster do not exceed the cluster's capacity at any given time period. These constraints allow the fitness centre to enforce some health and safety measures by limiting the number of customers present in the centre. Note that by making the capacity of the clusters time dependent provides an opportunity to efficiently schedule the cleaning of the equipment by cluster, without having to shut down the entire facility.

$$\sum_{i\in\mathcal{I}}\sum_{n\in\mathcal{S}_{k,j}\cap\mathcal{N}_j^i}\max\left\{0, x_{n,j}^i(t) - \sum_{n'\in\Delta_{i,j}^+(n)} x_{n',j}^i(t)\right\} \le A_{k,j}^t, \qquad \forall j\in\mathcal{J}, k\in\mathcal{K}_j, t\in\mathcal{T}. \tag{4}$$

The left-hand side of the inequality (4) represents the number of customers that are using an equipment in the cluster k at each time period. In fact, when a customer enters a cluster, $x_{n,j}^i(t)$ becomes 1 with $\sum_{n'\in\Delta_{i,j}^+(n)} x_{n',j}^i(t) = 0$ until the customer moves on to use a subsequent equipment, in which case, this customer can no longer be counted as present in the cluster (now $\sum_{n'\in\Delta_{i,j}^+(n)} x_{n',j}^i(t) = 1$). On the other hand, one should note that the term $x_{n,j}^i(t) - \sum_{n'\in\Delta_{i,j}^+(n)} x_{n',j}^i(t)$ will take the value -1 if the customer starts using an equipment that is adjacent to n without using equipment n. It will also be equal to -1 for any time period t that is beyond the feasible time interval of equipment n, i.e., $t > \overline{T}_{n,j}^i$ and $t \le \overline{T}_{n',j}^i$. Thus, the maximum function is therefore introduced to handle these special cases in the constraints as in [13]. A linearized form of the constraint is given in Appendix 2.

3. *Network flow constraints:*
The third group of constraints are the network flow constraints, which ensure that each assigned workout plan is unique and that the flow of customers within the facilities are as smooth as possible.

$$\sum_{j\in\mathcal{L}_i} x_{d_j,j}^i(\overline{T}_{d_j,j}^i) = 1, \qquad \forall i\in\mathcal{I}, \tag{5}$$

$$x_{n,j}^i(t) \le \sum_{n'\in\Delta_{i,j}^-(n)} x_{n',j}^i(t-\alpha_{n',j}^i), \qquad \forall i\in\mathcal{I}, j\in\mathcal{L}_i, n\in\mathcal{N}_j^i\setminus\{a_j\}, t\in T_{n,j}^i \tag{6}$$

$$x_{n,j}^i(\overline{T}_{n,j}^i) \le \sum_{n'\in\Delta_{i,j}^+(n)} x_{n',j}^i(\overline{T}_{n',j}^i), \qquad \forall i\in\mathcal{I}, j\in\mathcal{L}_i, n\in\mathcal{N}_j^i\setminus\{d_j\}, \tag{7}$$

$$\sum_{n'\in\Delta_{i,j}^+(n)} x_{n',j}^i(\overline{T}_{n',j}^i) \le 1, \qquad \forall i\in\mathcal{I}, j\in\mathcal{L}_i, n\in\mathcal{N}_j^i\setminus\{d_j\} \tag{8}$$

The constraints (5) ensure that the workout plan of each customer should start and end in one single fitness centre. Constraints (6) state that a customer cannot start exercising with an equipment n at time t if she had not spend at least $\alpha_{n',j}^i$ time periods on the preceding equipment according the current workout plan. The constraints (7) state that each customer must have moved on to one of the subsequent equipment by the latest time period at which she is allowed to use equipment n. Finally, constraints (8) ensure that the customers can only use one equipment after another.

4. *Maximum workout time*:

$$x^i_{a_j,j}(t) - x^i_{d_j,j}(t + \bar{t}^i_j - \underline{t}^i_j + \gamma^i) \leq 0, \qquad \forall i \in \mathcal{I}, j \in \mathcal{L}_i, t \in T^i_{a_j,j}, t + \bar{t}^i_j - \underline{t}^i_j + \gamma^i \in T^i_{d_j,j}. \quad (9)$$

The fourth group of constraints ensure that the total amount of time spent in a fitness centre by each customer do not exceed the maximum time allowed by either the customer or the fitness centre. Indeed, a fitness may centre may set a time limit, of 1 h for example, for each customer, while some customer may not want to spend more than a time limit which is less than that of the centre (45 min for example).

4 Computational Experiments

The above mathematical model is a binary integer programming model in which, the order of magnitude of the decision variables is approximately $\mathcal{O}(|\mathcal{I}| \times |\mathcal{N}| \times |\mathcal{J}| \times |\mathcal{T}|)$. Solving such models for a complete scale with a large number of variables may necessitates the development of sophisticated solution approaches or the use of heuristics or metaheuristic methods. However, in order to test the ability of our mathematical model in generating optimized and feasible schedules, we consider a small and tractable example, but realistic. Indeed, our proposed optimization is very useful in highly capacity constrained situations. Thus, we have considered an example of peak-hour in a gym represented by a three-hours time interval that has been discretized in time periods of 15 min, for a total 12 time periods. We consider a set of 60 customers planning their workouts in a single gym facility. The workout plans of each customer is randomly assigned to contain the planned use of about 3 to 6 equipments. The layout of the fitness centre facility under consideration is shown in Fig. 3, which is a replication of a real gym based in Montreal (Canada). The type of equipment and their numbers are shown directly in this figure. It is assumed the workout plan of each of the 60 customer should start and end within the 3-hour planning horizon.

The optimization model was coded using the R Package version 4.3.2 [1] environment and solved with the open source GLPK solver version 5.0 [2] on an Intel Core *i*7-13620 @ 2.40 GHz machine with 64 GB of memory and running on Windows 11. In the implementation we mainly focus on analyzing three main results, namely, the total delay the total deviation from the customers' workout plans and the "*first in first out*" (FIFO) policy which represents a case where optimization is not used by the management. The latter policy allows us to better assess the contribution of the optimization model in supporting the online booking system.

The resulting scheduling for the 60 customers are shown in Figs. 1, 2a and 2b. In these figures, each horizontal line represent the scheduled workout plan of each customer. The different colours show the clusters in which they are working out at each time period, the "•" represents the estimated start time of

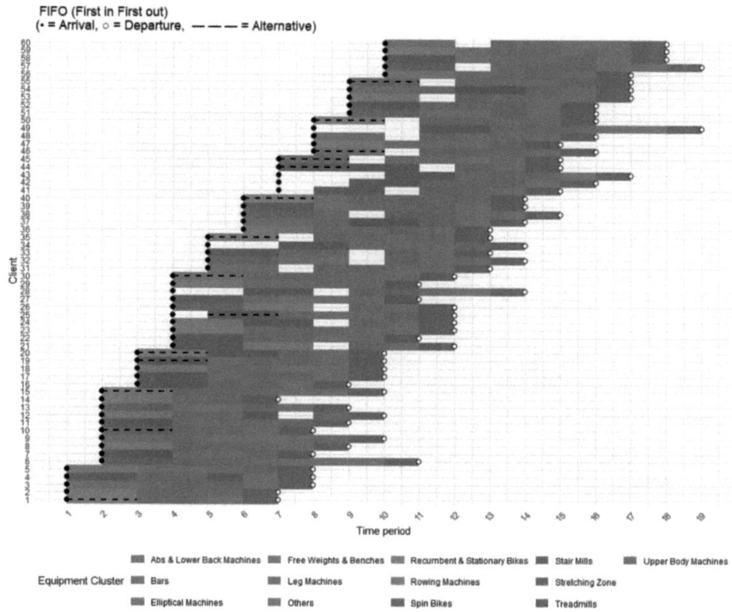

Fig. 1. The scheduling under the FIFO planning

the workout, while the "○" represents the exit. The blank space between two workout represents idle time for the customer while waiting for a machine to become available so that they can use. Finally, the dashed line represent when a customer is using an alternative equipment from their planned schedule. In these graphs, Fig. 1 show the scheduling under the FIFO booking policy, i.e., the scheduling if optimization is not used to support the booking system. In this graph, it can be seen that all the customers are arriving on time, but because of the capacity restriction, a significant number of customers have to idle in between two workouts (blank spaces), which either extend their time in the facility or negatively affect their satisfaction with regarding the gym.

The graph in Fig. 2a shows the scheduling when optimization is supporting the booking schedule with a focus on minimizing the total delay. It can be seen in this graph that customers' workout sessions are more compact, with less idleness in between. It can also be observed that the total duration of the makespan is reduced compared to FIFO. Furthermore, it appears that for a majority of customers, the model suggests a slightly early start of the session than the initially planned arrival time: these customers are therefore informed of a new arrival time, earlier but still within the defined tolerance. Conversely, only a small number of customers see their schedule slightly delayed, which helps to smooth the congestion across clusters while avoiding unacceptable idleness. Finally, The graph in Fig. 2b shows the scheduling when optimization is supporting the booking schedule with a focus on minimizing the total deviation from customers' workout plans. This graphs presents a structure very similar to

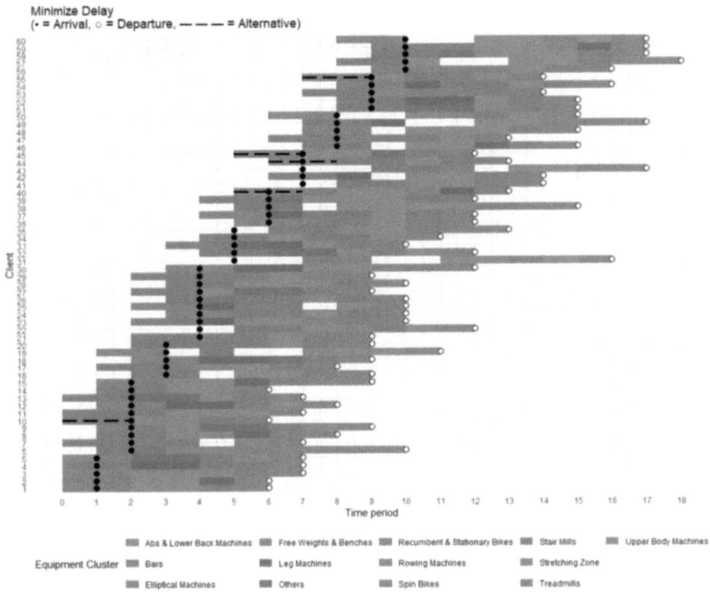

(a) The scheduling when minimizing the total Delay

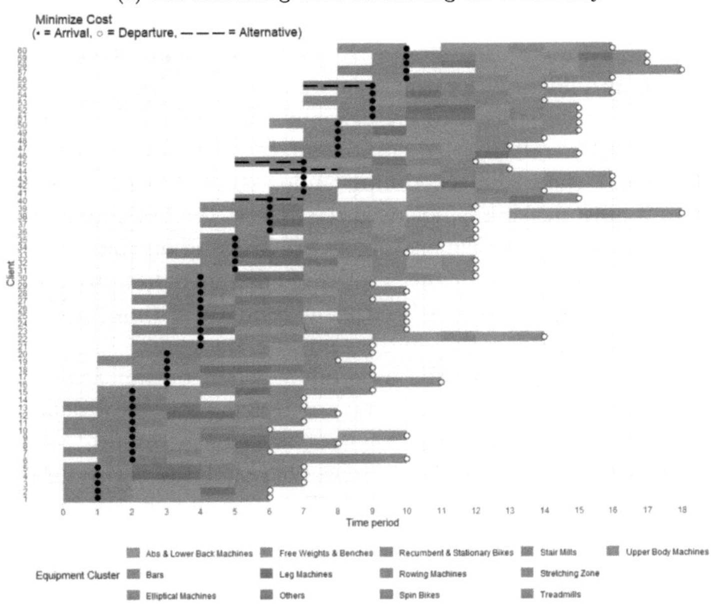

(b) The scheduling when minimizing the total deviation

Fig. 2. (a) The scheduling when minimizing the total Delay (b) The scheduling when minimizing the total deviation

that obtained when minimizing the delay. As before, the scheduling is generally compact and adheres to the same principles of anticipating or slightly shifting sessions within the clients' tolerance limits. The main difference lies in the use of alternatives: a slightly lower number of dashed segments are observed, reflecting a more sparing use of alternative equipment, consistent with the objective of minimizing the total deviation from customers' workout plans.

Overall, one of the most obvious conclusion is the clear superiority of the two optimization models compared to the FIFO approach used as a benchmark. This superiority is reflected in most performance indicators. In terms of customer experience, the optimized models reduce the average idle from approximately 39 to up to 12 min per customer.

5 Conclusion

In this paper, we addressed the problem of overcrowding in fitness centres during peak hours by proposing a mathematical optimization framework to support the transition from a "First Come First Served" booking system to an "optimization-driven" booking system. Motivated by the increasing demand for fitness facilities and the negative impact of congestion on customer satisfaction and workout quality, we modeled the interaction between customers' individual training plans and the limited capacity of fitness centre equipment. By representing a fitness centre as a network of equipment clusters and interpreting workout plans as paths within this network, the proposed model enables a more informed and efficient allocation of customers to time slots and resources.

Computational experiments conducted on realistic data sets demonstrate the effectiveness of the proposed approach. The results show that incorporating customers' training plans into the booking and allocation process can significantly reduce total delays and idle times, while simultaneously improving overall facility utilization. In particular, the model achieved reductions of up to 17.5% in total delay and up to 19% in customers' idle time, highlighting the potential of optimization-based decision support tools to improve both customer experience and operational efficiency in fitness centres.

Several directions for future research naturally emerge from this work. These include extending the model to handle stochastic arrivals and no-shows, incorporating fairness considerations among customers, and studying dynamic or real-time re-optimization as booking requests evolve. Additionally, integrating behavioral aspects, such as customers' willingness to accept alternative exercises or time slots, could further enhance the practical applicability of the model. Overall, this work illustrates how mathematical optimization can play a meaningful role in addressing real-world challenges in the fitness industry and beyond.

Acknowledgments. This work was partly supported by the Canadian Natural Sciences and Engineering Research Council under grant 2021-03307 This support is greatly appreciated.

Disclosure of Interests. The authors have no competing interests to declare that are relevant to the content of this article.

Appendix 1: Layout the the Test Facility

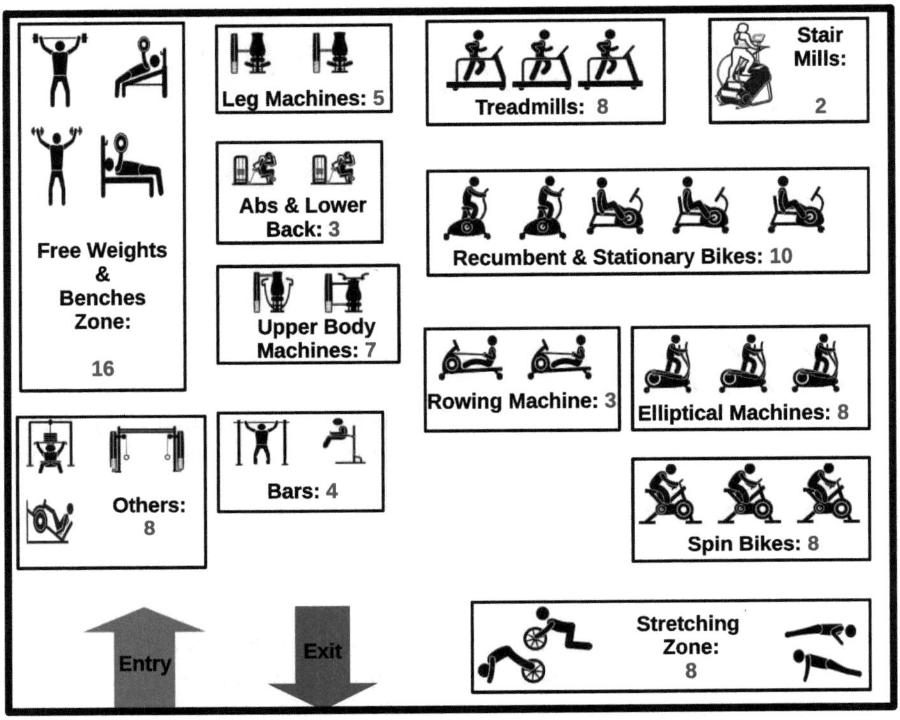

Fig. 3. A realistic fitness centre layout: the numbers are the capacity of the clusters

Appendix 2: Linearization of the Clusters Capacity Constraint (4)

It should be noted that the left hand-side of the clusters capacity constraints (4) is nonlinear. It can be linearized using the standard procedure of introducing an auxiliary variable $z_{i,n}^{k,j,t}$ for all $i \in \mathcal{I}, n \in \mathcal{S}_{k,j} \cap \mathcal{N}_j^i, j \in \mathcal{J}, k \in \mathcal{K}_j, t \in \mathcal{T}$, which will replace the max term in the constraint to become the following linear constraint

$$\sum_{i \in \mathcal{I}} \sum_{n \in \mathcal{S}_{k,j} \cap \mathcal{N}_j^i} z_{i,n}^{k,j,t} \leq A_{k,j}^t, \qquad \forall j \in \mathcal{J}, k \in \mathcal{K}_j, t \in \mathcal{T}.$$

The following linking constraints have to be added to make the linearization complete:

$$z_{i,n}^{k,j,t} \geq x_{n,j}^{i}(t) - \sum_{n' \in \Delta_{i,j}^{+}(n)} x_{n',j}^{i}(t) \quad \forall i \in \mathcal{I},\ n \in \mathcal{S}_{k,j} \cap \mathcal{N}_{j}^{i}, j \in \mathcal{J}, k \in \mathcal{K}_{j}, t \in \mathcal{T}$$

and

$$z_{i,n}^{k,j,t} \geq 0 \quad \forall i \in \mathcal{I},\ n \in \mathcal{S}_{k,j} \cap \mathcal{N}_{j}^{i}, j \in \mathcal{J}, k \in \mathcal{K}_{j}, t \in \mathcal{T}.$$

References

1. R Core Team. R: A language and environment for statistical computing. R Foundation for Statistical Computing (2025). https://www.r-project.org
2. GNU Project. GLPK (GNU Linear Programming Kit) (2012). https://www.gnu.org/software/glpk/
3. The World Health Organization. Physical Activity (2024).https://www.who.int/dietphysicalactivity/pa/en/
4. Statistics Canada. Health Reports: Comparison of self-reported and accelerometer-measured physical activity in Canadian adults (2018). https://www150.statcan.gc.ca/n1/daily-quotidien/181219/dq181219d-eng.htm
5. Masse, A.: Analyse des déterminants de la fréquentation dans les centres de conditionnement physique au Québec. *Mémoire de Maîtrise en Économie*, Université du Québec à Montréal (2012). https://archipel.uqam.ca/5323/1/M12746.pdf
6. Abraham, J.M., Crepin, D., Rothman, A.: Initiation and maintenance of fitness centre utilization in an incentive-based employer wellness program. J. Occup. Environ. Med. **57**(9), 952–957 (2015)
7. Hooker, S.A., Wooldridge, J.S., Ross, K.M., Masters, K.S.: Do monetary incentives increase fitness centre utilization? it depends. Am. J. Health Promot. **32**(3), 606–612 (2018). https://doi.org/10.1177/0890117116689321
8. Afthinos, Y., Theodorakis, N.D., Nassis, P.: Customers' expectations of service in Greek fitness centres: gender, age, type of sport centre, and motivation differences. Manag. Serv. Qual. Int. J. **15**(3), 245–258 (2005)
9. The World Health Organization. Coronavirus Disease (COVID-19) Pandemic (2019). https://www.who.int/emergencies/diseases/novel-coronavirus-2019
10. The fitness industry council of Canada. Framework for excercise facilities operating in a Covid-19 environment in Canada (2020). https://s3-us-west-2.amazonaws.com/ficdn-files.ca/files/covid_resources/FIC_Framework_Facilities_+CANADA.pdf
11. Exercise Association of New Zealand. Framework for exercise facilities in New Zealand operating within a COVID-19 environment (2020). https://exercisenz.org.nz/wp-content/uploads/2020/05/2020-May-6-CV19-Framework-Exercise-Industry.pdf
12. Bertsimas, D., Stock Patterson, S.: The air traffic management problem with en-route capacities. Oper. Res. **46**, 406–422 (1998)
13. Bertsimas, D., Lulli, G., Odoni, A.: An integer optimization approach to large-scale air traffic flow management. Oper. Res. **59**(1), 211–227 (2011)

14. Agustın, A., Alonso-Ayuso, A., Escudero, L.F., Pizarro, C.: On air traffic flow management with rerouting. Part I: deterministic case. Eur. J. Oper. Res. **219**, 156–166 (2012)
15. Djeumou Fomeni, G., Zografos, L.K.: An optimization model for assigning 4D-trajectories to flights under the TBO concept. Twelfth USA/Europe Air Traffic Management Research and Development Seminar (ATM2017). Seattle, Washington, USA June 26–30 (2017)
16. Dal Sasso, V., Djeumou Fomeni, F., Lulli, G., Zografos, K.: Planning efficient 4D trajectories in Air Traffic Flow Management. Eur. J. Oper. Res. **276**(2), 676–687 (2019)
17. Dal Sasso, V., Djeumou Fomeni, F., Lulli, G., Zografos, K.: Incorporating stakeholders' priorities and preferences in 4d trajectory optimization. Transp. Res. Part B Methodol. **117**, 594–609 (2018)

Optimisation and Energy Evaluation of Batch Pan Scheduling in a White Sugar Refinery

Ibrahim Dan Dije[1(✉)], Franklin Djeumou Fomeni[1], Montaz Ali[2], and Richard Loubser[3]

[1] CIRRELT, GERAD and Université du Québec à Montréal (UQAM), Department of Analytics, Operations and Information Technologies, Montréal, QC H3C 3P8, Canada
dan_dije.ibrahim@courrier.uqam.ca, djeumou_fomeni.franklin@uqam.ca

[2] School of Computer Science and Applied Mathematics, University of the Witwatersrand (Wits), Private Bag 3, Johannesburg 2050, South Africa
montaz.ali@wits.ac.za

[3] Sugar Milling Research Institute (SMRI), University of KwaZulu Natal, Durban, South Africa
rloubser@smri.org

Abstract. The Batch Pan Scheduling Problem represents a major operational challenge in white sugar refineries, where production planning must balance throughput objectives against high and fluctuating energy demands driven by crystallisation. This paper examines batch pan scheduling from an integrated production and energy-aware perspective. A discrete-time mixed-integer linear programming model and a mixed-integer multi-objective formulation are developed to represent the initiation of batch crystallisation cycles across sequential pan stages while accounting for total energy consumption. The models capture material flow dependencies, stage capacity constraints, and the temporal structure of industrial sugarhouse operations. The single-objective model maximises the number of completed batch cycles at the final pan stage within a given planning horizon, directly representing final sugar production. Building on this solution, the multi-objective model incorporates the achieved production level as a constraint with controlled relaxation to minimise total energy consumption. Numerical experiments investigate the influence of planning horizon length on scheduling flexibility and energy performance. Results indicate that extending the planning horizon enhances temporal coordination, enabling a more even distribution of operations and yielding substantial reductions in energy consumption without compromising production output. Furthermore, the multi-objective formulation achieves consistently stable energy performance, with total energy requirements comparable to or lower than those obtained from the single-objective model under the same horizon length. These findings demonstrate the critical role of planning horizon selection in energy-aware batch scheduling and show that energy savings and variability reduction can be realised through improved temporal coordination. The proposed framework provides a transparent and effective

A. P. Cabral Seixas Costa et al. (Eds.): ICDSST 2026, LNBIP 582, pp. 333–351, 2026.
https://doi.org/10.1007/978-3-032-25146-6_22

decision-support tool for energy-efficient production planning in batch process industries.

Keywords: Batch scheduling · Sugar refinery · Mixed-integer linear programming · Multi-objective optimisation

1 Introduction

White sugar production is a large-scale industrial activity characterized by complex material flows, stringent product quality specifications, and exceptionally high energy consumption. Among all processing units in a sugar refinery, the crystallisation section, commonly referred to as the sugarhouse, plays a decisive role in determining both the production efficiency and the energy performance. Sucrose crystallisation is carried out in batch operated vacuum pans, where evaporation and crystal growth occur simultaneously under reduced pressure. Operating under vacuum enables crystallisation at relatively low temperatures. This limits sucrose degradation, reduces the formation of colored compounds, and improves the quality and uniformity of sugar crystals. Each vacuum pan batch produces a semi solid mixture known as massecuite, typically composed of approximately 50% crystalline sugar and 50% mother liquor, which is typically less concentrated than the liquor fed into the pan, depending on the crystallisation technique employed. After crystallisation, the massecuite is discharged into strike receivers and subsequently processed in centrifugals, where sugar crystals are separated from the mother liquor. Additional water and steam are frequently required during centrifugation for washing and dissolution of fine crystals, thereby further coupling production decisions with steam or energy demand [12].

Crystallisation in white sugar refineries is inherently a multi stage process. To satisfy the stringent purity requirements of white sugar, refineries operate cascaded crystallisation stages in which the liquor discharged from one stage serves as the feedstock for the subsequent stage. As a result, downstream operations are conditioned on the completion of multiple upstream batch cycles, giving rise to strong inter stage dependencies and increasingly scheduling constraints.

We define the problem that coordinates these batch operations as the Batch Pan Scheduling Problem (BPSP). The problem seeks to determine the initiation times of batch cycles over a finite planning horizon such that material availability (mother liquor), equipment capacity, and operational constraints are satisfied while optimizing key performance indicators. These indicators typically include production of the final sugar, energy efficiency and energy demand (steam) stability. In this study, throughput is measured as the number of completed batch cycles at the final pan stage within the planning horizon, corresponding to the production of final sugar.

From an operational perspective, the BPSP is particularly challenging due to the dominant role of steam. Steam is consumed intensively during evaporation and centrifugation, and poorly coordinated batch starts can generate large fluctuations in energy demand (steam). As a result, batch process scheduling increasingly emphasizes production planning that explicitly accounts for energy considerations, whereby production decisions are coordinated to smooth energy demand (steam) profiles over time.

In addition to energy considerations, scheduling decisions must ensure sufficient liquor availability prior to pan initiation, adequate strike receiver capacity before massecuite discharge, and proper coordination of centrifugals to achieve the required sugar color specifications. The target sugar color is attained upon completion of the batch cycle in the final crystallisation stage. These requirements show the need for optimisation scheduling frameworks capable of simultaneously capturing material availability constraints and the logical constraints that govern the feasibility of initiating each stage.

Batch process scheduling has been studied within the operations research and chemical engineering communities. Classical scheduling theory focuses on sequencing and timing decisions for jobs processed on one or more machines, with objectives such as makespan minimization, tardiness reduction, and resource utilization [10, 11]. While these models provide valuable theoretical insights, they often assume independent jobs and neglect material-flow and energy interactions, making them insufficient for complex process industries such as sugar manufacturing.

To overcome these limitations, specialized modeling frameworks have been developed for batch and semi continuous process industries. The State Task Network (STN) and Resource Task Network (RTN) paradigms explicitly represent tasks, resources, and material states, enabling the modeling of batch operations, intermediate storage, and shared utilities [6, 7]. These frameworks have become the foundation for many short term scheduling models in the chemical and food processing industries. A critical modeling choice in batch scheduling concerns the representation of time. Discrete time formulations divide the planning horizon into uniform time intervals and associate binary variables with task initiation decisions. Although such models may result in large mixed integer linear programs (MILPs), they are particularly well suited for systems dominated by logical rules and event driven behavior. Continuous time formulations, by contrast, reduce the number of time points but introduce additional complexity related to task sequencing and overlap [1]. The sugar industry has motivated a number of specialized scheduling and control models due to its unique combination of batch crystallisation, strong inter stage dependencies, and high energy intensity. A seminal contribution is the work of [12], who developed a discrete event and Mixed Logical Dynamical (MLD) framework for modeling and optimisation of sugar plants. Their approach transformed logical pan program rules into mixed integer constraints and embedded them within a Model Predictive Control (MPC) framework, demonstrating significant improvements in energy demand (steam) stability.

Subsequent studies explored the integration of scheduling and control in sugar refineries. [9] proposed a set based formulation for scheduling mixed batch continuous sugar processes with variable cycle times. [2, 3] demonstrated the effectiveness of MPC for managing energy consumption and product quality in industrial sugar factories. These contributions underline that energy-aware scheduling is not merely a cost-saving measure, but a fundamental requirement for sustainable and stable refinery operation [4].

Beyond operational efficiency and economic considerations, the sugarcane processing industry also faces pressure to reduce its environmental footprint in response to climate change mitigation objectives. Sugar manufacturing is a highly energy intensive industrial activity, relying predominantly on steam for evaporation, crystallisation, and downstream separation, which contributes to greenhouse gas emissions when fossil fuels are used [13]. As decarbonization policies and energy efficiency targets tighten, reducing energy consumption in sugar refineries has become a critical lever for mitigating climate impacts. In this context, optimisation based batch scheduling offers a practical and cost effective approach to energy reduction at the operational level. By coordinating batch initiation decisions across multiple crystallisation stages and smoothing steam demand profiles over time, energy-aware scheduling can lower energy consumption and associated emissions without requiring major capital investments for green energy technology. [4] addressed this challenge by proposing strategies for monitoring energy consumption in sugarcane processing factories, highlighting the role of continuous measurement in supporting energy efficient operations.

More recently, energy-aware scheduling has emerged as a major research direction across process industries. Reviews such as [5] highlight the growing importance of explicitly incorporating energy considerations into scheduling models, particularly in the context of decarbonization and industrial sustainability. Nevertheless, many existing approaches rely on complex hybrid models or detailed process simulations that may limit scalability and applicability at the planning level.

This work contributes to the literature by proposing a discrete-time mixed integer linear programming (MILP) formulation designed to reflect the structural characteristics of white sugar refinery operations. The proposed model captures essential inter-stage material dependencies and batch initiation logic. In addition, the study introduces a multi-objective modelling framework that explicitly incorporates energy-related objective, enabling the simultaneous consideration of production performance and energy efficiency within the scheduling decision process.

This paper is organized as follows: Sect. 1 introduces the problem context, Sect. 2 presents mathematical formulations, including single objective and multi-objective model, Sect. 3 discusses the numerical results, and Sect. 4 concludes the work.

2 Problem Presentation and Mathematical Model

We consider a white sugar refinery operating four crystallisation stages, each associated with a dedicated vacuum pan. A stage is defined as the completion of a full batch cycle resulting in the production of a specific sugar color.

The stages are organized sequentially with dependencies. Stage 1 corresponds to the completion of a batch cycle in pan 1, producing the first sugar grade. Stage 2 completes a batch cycle in pan 2 and requires fine liquor from two completed stage 1 cycles. Stage 3 similarly requires liquor from two stage 2 cycles, while stage 4 produces the final sugar color using liquor obtained from two stage 3 cycles. Each pan cycle consists of five main processing steps with fixed duration: filling, evaporation, discharge, centrifugation, and sugar extraction. Steam consumption is concentrated in the evaporation and centrifugation steps, making the timing of pan initiations a critical driver of energy demand. The strong coupling between material flows and steam usage motivates the explicit inclusion of energy considerations in the scheduling process.

The batch pan scheduling problem addressed in this work is to determine the initiation times of batch cycles for each stage over a finite planning horizon such that material availability constraints are satisfied, pan capacities are respected, production is maximized, and operational stability at the end. The following schemes present a typical multi stage crystallisation process in a refinery (Tables 1 and 2).

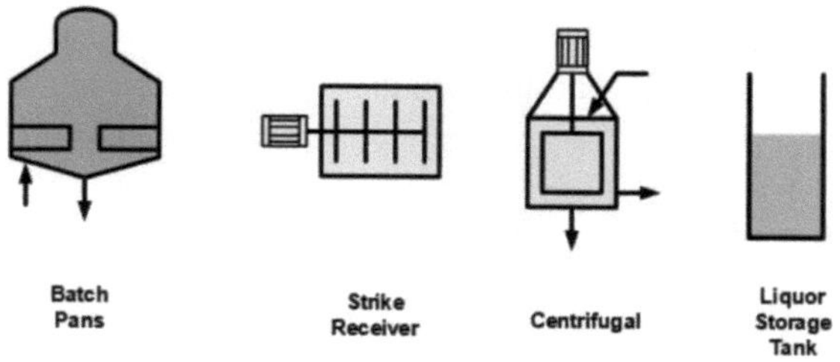

Fig. 1. Equipment required per stage.

We now present a discrete time mixed integer linear programming formulation of the Batch Pan Scheduling Problem.

2.1 Sets and Parameters

- T: planning horizon length.
- $K = \{1, 2, 3, 4\}$: set of crystallisation stages.
- C: capacity of each vacuum pan.
- D^k: energy demand (steam) associated with initiating a batch at stage k.

2.2 Decision Variables

- X_t^k: takes 1 if a batch at stage $k \in K$ is initiated at time $t = 0, \ldots, T$, and 0 otherwise.
- V_t^k: volume of liquor available at stage $k \in K$ at time $t = 0, \ldots, T$.

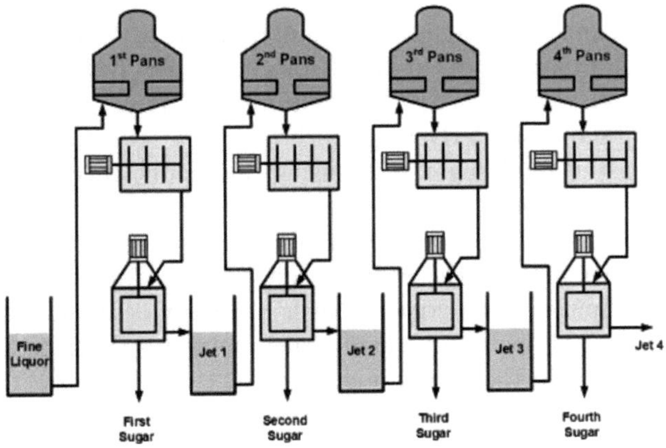

Fig. 2. Multiple stage crystallisation process.

2.3 Single-Objective Model

The original operational objective is to minimize the variability of energy demand (steam) over the planning horizon, which can be expressed as the variance of total steam consumption given by the following:

$$\mu = \frac{1}{|T|}\sum_{t=0}^{|T|}\left(\sum_{k=1}^{4} D^k X_t^k - \frac{1}{|T|}\sum_{t=0}^{|T|}\sum_{k=1}^{4} D^k X_t^k\right)^2 \tag{1}$$

However, this objective is nonlinear. To preserve linearity and maintain computational tractability, we adopt an objective related to maximizing sugar production, as effective scheduling of operations in process industries has the potential to generate substantial economic returns [9]. Since the final sugar product is obtained only when stage 4 is initiated, maximizing the number of stage 4 batch starts serves as an indicator for maximizing the sugar output rate:

$$\max Z = \sum_{t=0}^{T} X_t^4 \tag{2}$$

After each batch initiation, a distinct sugar grade is produced whose colour reflects the colour of the remaining mother liquor; as nearly 50% of sucrose is crystallised at each stage, the liquor colour changes and the sugar quantity decreases progressively from first to fourth sugar, after which all sugars are blended to produce a final sugar of average quality while satisfying liquor balance constraints.

Material Balance Constraints.

$$V_{t+1}^{k+1} = V_t^{k+1} - CX_t^{k+1} + \frac{C}{2}X_t^k, \quad k = 1, 2, 3,\ t = 0, \ldots, T-1. \tag{3}$$

Availability Constraints.

$$X_t^k \leq \frac{V_t^k}{C}, \quad k = 1, 2, 3, 4, \ t = 0, \ldots, T. \tag{4}$$

Variable Bounds.

$$0 \leq V_t^k \leq C, \quad k = 1, 2, 3, 4, \ t = 0, \ldots, T. \tag{5}$$

These constraints ensure material feasibility, enforce inter stage dependencies, and prevent infeasible batch initiations. For the previous model, we introduce additional constraints when a lower bound $\underline{D}$ and an upper bound $\overline{D}$ on energy consumption are specified. These constraints are formulated as follows.

Total energy constraint.

$$\underline{D} \leq \sum_{t=0}^{T} \sum_{k \in K} D^k X_t^k \leq \overline{D} \tag{6}$$

The single objective model is used to maximise the product. Using that model, the energy consumption will be evaluate by time horizon size. This evaluation helps to determine a flexible time horizon with less energy consumption.

2.4 Assumptions and Modeling Remarks

This section clarifies the underlying assumptions adopted in the proposed formulation and provides additional modeling remarks that justify the structural choices made in the mathematical model. Explicitly stating these assumptions improves model transparency and facilitates interpretation of the numerical results, as recommended in the batch scheduling and process optimisation literature [5,7].

Modeling Assumptions. The following assumptions are made throughout this study:

1. **Discrete time representation.** The planning horizon is discretized into uniform time periods. This choice is motivated by the event driven nature of batch pan operations, where decisions are naturally associated with batch start times. Discrete time formulations are well suited for capturing logical constraints and inter stage dependencies in batch processes [7].
2. **Deterministic processing times.** The durations of the five pan cycle steps are assumed to be fixed and known in advance. While processing times may vary in practice, deterministic approximations are commonly used at the planning level [9].
3. **Single batch per pan at a time.** Each vacuum pan can process at most one batch at any given time, consistent with standard industrial operation.

4. **Fixed pan capacity.** All pans are assumed to have identical and constant capacity C.
5. **Aggregated material representation.** Liquor volumes are tracked in aggregate form without explicit compositional detail. This level of abstraction is appropriate for scheduling decisions [12].
6. **Deterministic inter stage conversion.** Two upstream batches are required to initiate one downstream batch, represented through fixed conversion ratios.
7. **No equipment failures.** Equipment is assumed to be continuously available; maintenance and breakdowns are not considered.

Modeling Remarks. The discrete time MILP formulation reflects a deliberate trade off between model fidelity and computational tractability. While hybrid and continuous time models offer higher resolution, they often lead to large-scale and difficult to solve optimisation problems. The proposed formulation captures the dominant structural features of the batch pan scheduling problem while remaining suitable for medium term to long term planning.

The surrogate objective function that maximizes the number of stage 4 batch starts. Although it does not explicitly minimize steam variability, it aligns with industrial production goals and preserves linearity. Extensions to multi objective formulations incorporating explicit energy smoothing terms are straightforward. Finally, the model is intended for planning and scheduling rather than real time control. Detailed dynamic behavior and quality control are better addressed by lower level control strategies such as Model Predictive Control, which complement the planning level decisions generated by the proposed framework [2,12].

2.5 Multi-objective Model

In this section, we present a linear multi-objective model. Instead of minimizing the energy variation, we optimize the total energy, defined as

$$\min Z_2 = \sum_{t=0}^{T} \sum_{k=1}^{4} D^k X_t^k. \tag{7}$$

Simultaneously, we maximize the total number of stage 4 batch starts:

$$\max Z_1 = \sum_{t=0}^{T} X_t^4. \tag{8}$$

These two objectives are conflicting: improving one generally leads to the deterioration of the other. The new model considers the same constraints as the single-objective model, together with the following additional constraint, which keeps the total number of stage 4 batches initiation near optimal:

$$\sum_{t=0}^{T} X_{4,t} \geq (1-\epsilon)\, Z_1^{\star}. \tag{9}$$

Here, $Z_1^\star$ denotes the optimal value obtained from the single-objective model. This formulation allows an $\epsilon\%$ degradation of the primary objective. This method is known as the ϵ-constraint method for multi-objective linear programming [8]. More formally, let us define the feasible region of the model as

$$\mathcal{F} = \{(X, V) : (3), (4), (5)\}. \tag{10}$$

First, we optimize the total number of stage 4 batches while ignoring energy consumption:

$$Z_1^\star = \max_{(X,V)\in\mathcal{F}} \sum_{t=0}^{T} X_{4,t}. \tag{11}$$

After obtaining $Z_1^\star$, we redefine the admissible solution set as

$$\mathcal{F}_\epsilon = \{(X, V) \in \mathcal{F} : Z_1(X, V) \geq (1-\epsilon) Z_1^\star\}. \tag{12}$$

The feasible set $\mathcal{F}_\epsilon$ contains solutions that maintain near-maximal production while allowing at most an $\epsilon\%$ loss in output of the first objective. Finally, we minimize the total energy consumption:

$$Z_2^\star = \min_{(X,V)\in\mathcal{F}_\epsilon} \sum_{t=0}^{T} \sum_{k=1}^{4} D^k X_t^k. \tag{13}$$

This procedure first implicitly determines the Pareto frontier by maximizing production and then explores this frontier to identify the minimum-energy solution.

In the next section, we provide the numerical investigation to evaluate our model output.

3 Numerical Experiments

In this section, numerical experiments are conducted to evaluate the performance of the proposed scheduling models, namely the single-objective and multi-objective formulations, under different planning horizons. Detailed scheduling patterns are analyzed for two representative horizon lengths, $T = 40$ and $T = 42$, across the four crystallisation stages indexed by $k \in \{1, 2, 3, 4\}$. These two horizon lengths are selected to illustrate their impact on the scheduling structure and its visual representation. Each stage is associated with a fixed steam energy demand coefficient D^k, representing the energy consumption incurred when initiating a batch cycle at stage k. Specifically, the energy demand parameters are set to $D^1 = 8.0$, $D^2 = 6.5$, $D^3 = 9.0$, and $D^4 = 7.5$. The capacity of each vacuum pan is assumed to be constant and equal to $C = 100$.

In the first set of experiments, we present the optimal solution of the single-objective model without energy optimisation. We then analyze the multi-objective model, which determines the energy-optimal solution for a given total number of scheduled batches with the parameter $\epsilon = 0.01$. Using the solutions obtained from both models, we evaluate the energy variation according to formula (1) for different planning horizons.

In addition to the detailed scheduling analysis, the impact of the planning horizon length on energy performance is systematically examined by computing the total energy consumption for horizons ranging from $T = 10$ to $T = 170$ for both models. This horizon sensitivity analysis allows us to assess how extending the planning horizon affects the optimal scheduling policy and the resulting cumulative energy consumption, thereby providing insights into the trade-off between temporal flexibility and energy efficiency in multi-stage batch production systems.

Finally, the model is solved under explicit bounds on total energy consumption to investigate the effect of energy constraints on scheduling decisions. In particular, lower and upper bounds on total energy consumption, denoted by $\underline{D} = 0$ and $\overline{D} = 500$ or $\overline{D} = 1000$, are imposed for planning horizons ranging from $T = 10$ to $T = 170$. These constrained experiments illustrate how restricting the admissible energy range influences the timing and coordination of batch initiations across stages, further clarifying the interaction between energy limits and production scheduling decisions. All models are solved using the Gurobi Optimizer, version 12.0.1.

3.1 Scheduling and Total Energy by Time Horizon

Figures 3 and 4 present the optimal scheduling patterns obtained for two different planning horizons, namely $T = 40$ and $T = 42$. Each figure illustrates the initiation times of batch cycles across the four crystallisation stages and highlights how scheduling decisions evolve as the available planning horizon changes.

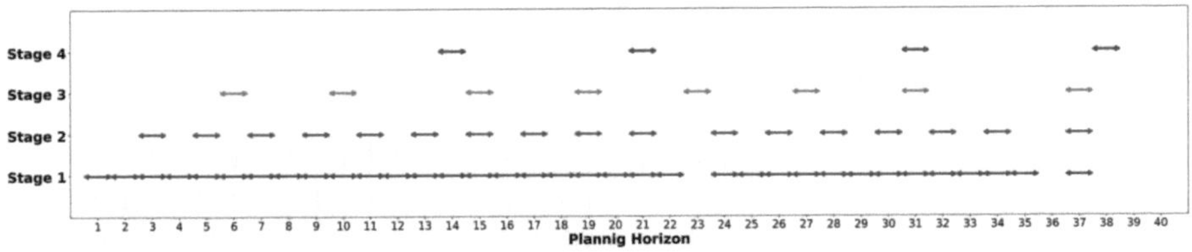

Fig. 3. Scheduling with time horizon T = 40.

For $T = 40$, batch initiations particularly in the downstream stages are relatively concentrated in time. This clustering reflects the limited temporal flexibility available to the optimizer, which forces downstream batches to be initiated as soon as material availability constraints are satisfied. Consequently, steam intensive operations overlap more frequently, resulting in a total energy consumption of 492.5.

When the planning horizon is extended to $T = 42$, the optimizer exploits the additional temporal flexibility to redistribute batch initiations more evenly over

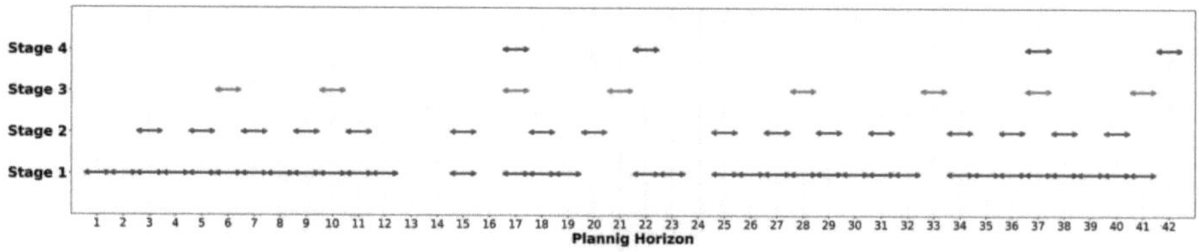

Fig. 4. Scheduling with time horizon T = 42.

time. In particular, the initiation of stage 3 and stage 4 batches is slightly delayed and better spaced, reducing the simultaneous execution of steam demanding operations. This improved temporal coordination leads to a lower total energy consumption of 478.0, while maintaining the same number of completed final stage batches.

These results highlight that total energy consumption is not determined solely by the number of completed final stage batches, but is strongly influenced by the temporal distribution of batch initiations across the planning horizon. Even modest extensions of the horizon can enable smoother scheduling patterns that reduce energy usage without compromising production output. The energy variation in this case is 37.17 for $T = 40$ and 31.87 for $T = 42$. Without energy optimisation, the flexible schedule provides lower total energy consumption and lower variation.

Figures 5 and 6 present the schedules obtained under the energy-optimised model while enforcing the same number of stage 4 starts. For the two planning horizons, the resulting total energy is 462.0, while the total variation equals 34.41 for $T = 40$ and 39.28 for $T = 42$. Compared with the schedule produced by the single-objective model, the inclusion of the energy criterion reshapes the timing of operations and yields lower total energy consumption. In addition, the variability of energy demand is slightly reduced for $T = 40$, whereas it increases for $T = 42$, suggesting that the smoothing effect of the optimisation depends on the horizon length. By comparing the total variation between the single-objective and multi-objective models, we observe that when $T = 42$, the smallest variation is achieved. Figures 7 and 8 illustrate the energy variation profile.

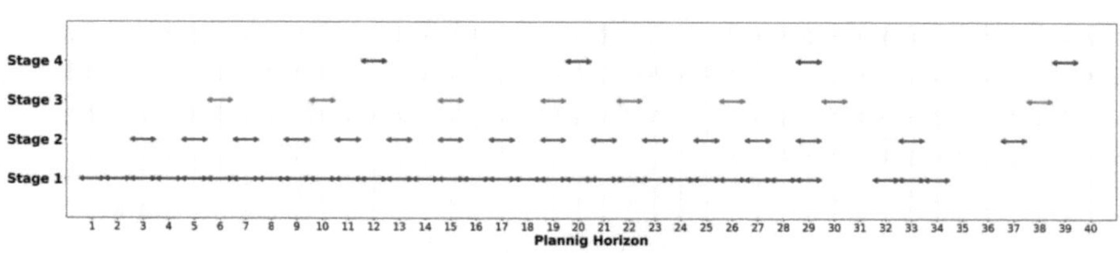

Fig. 5. Scheduling with time horizon $T = 40$ and energy optimisation.

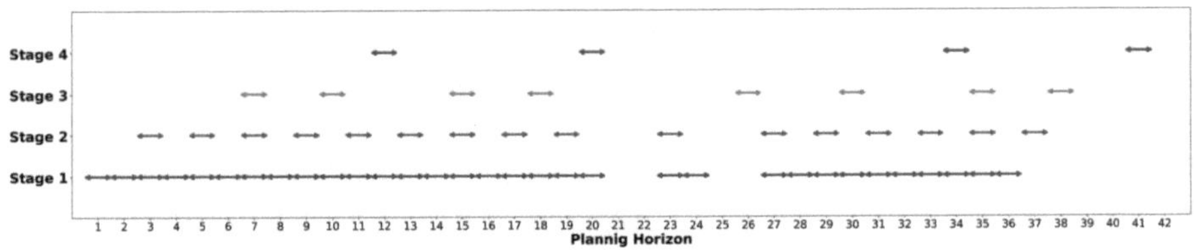

Fig. 6. Scheduling with time horizon $T = 42$ and energy optimisation.

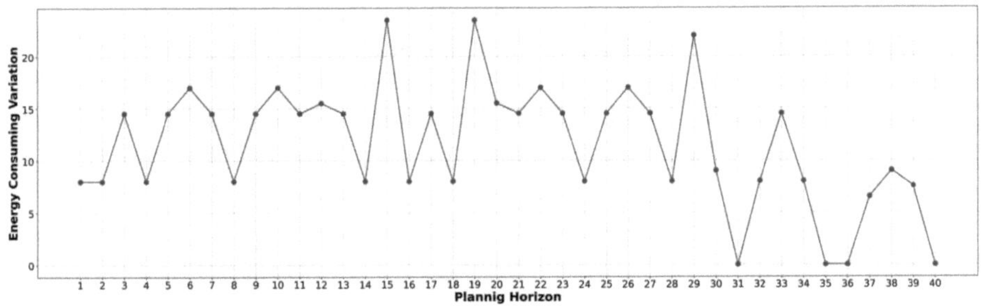

Fig. 7. Energy variation with time horizon $T = 40$ by using the multi-objective model.

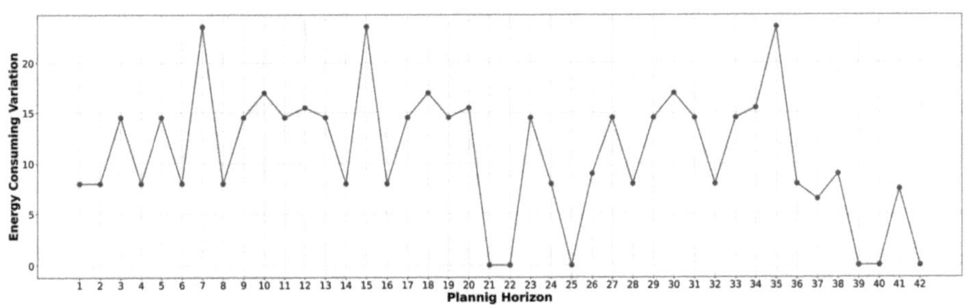

Fig. 8. Energy variation with time horizon $T = 42$ by using the multi-objective model.

These results confirm that energy performance in batch pan scheduling is primarily driven by temporal flexibility and multi-objective coordination, where time horizon extensions combined with energy optimisation yield significant reductions in both total energy consumption and demand variability while preserving production levels.

3.2 Total Energy Consumption by Time Horizon

This section examines the evolution of total energy consumption as a function of the planning horizon by considering single objective and multi-objective model. Figure 9 shows the total energy consumption together with the optimal number of stage 4 batch initiations for planning horizons ranging from $T = 10$ to $T = 170$ for our first model.

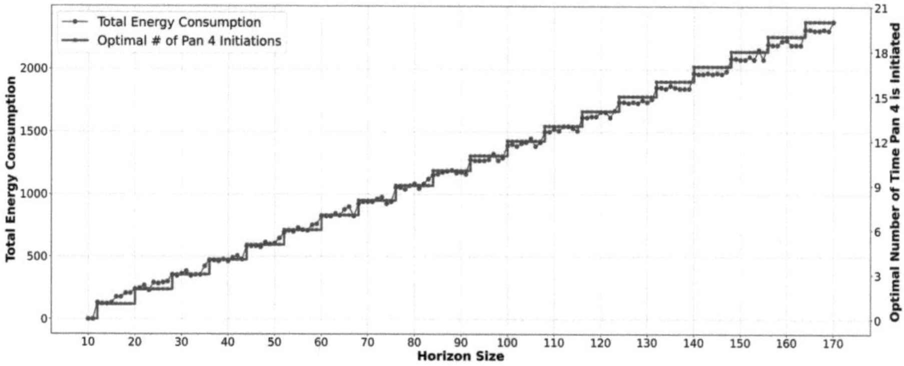

Fig. 9. Total energy consumption and the corresponding optimal number of Pan 4 initiations.

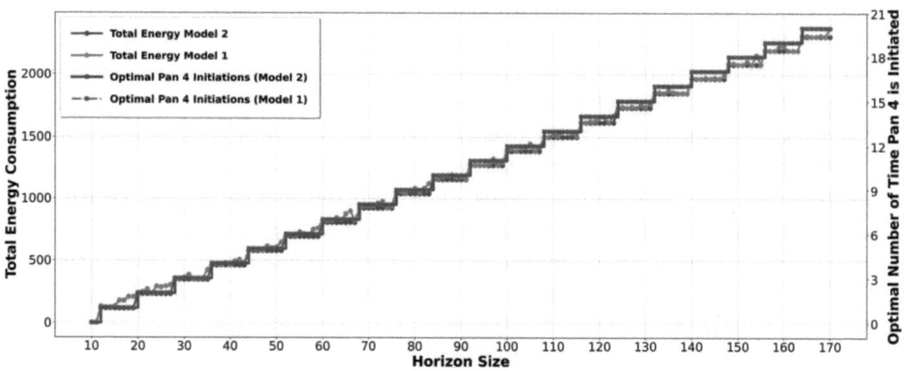

Fig. 10. Total energy consumption the corresponding optimal number of Pan 4 initiations: Single Objective vs Multi-Objective Model.

As expected, total energy consumption increases monotonically with the length of the planning horizon, reflecting the cumulative nature of steam usage over time. The number of stage 4 initiations exhibits a stepwise pattern, consistent with the discrete nature of batch production, where each increment corresponds to the completion of an additional batch cycle in the final crystallisation stage.

An important observation is that identical numbers of stage 4 initiations can be associated with different levels of total energy consumption across distinct

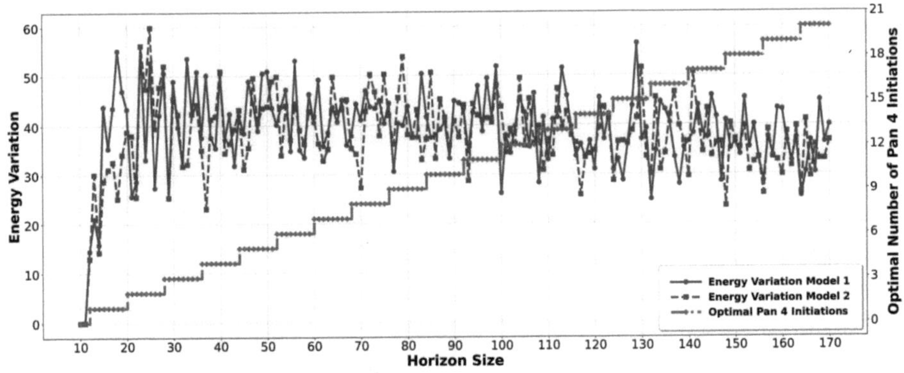

Fig. 11. Total energy variation and the corresponding optimal number of Pan 4 initiations: Single Objective vs Multi-Objective Model.

planning horizons. This confirms that energy usage depends not only on production volume but also on how batch initiations are temporally coordinated. Longer planning horizons provide greater scheduling flexibility, allowing batch starts to be distributed more evenly over time, which mitigates periods of concentrated steam demand and improves overall energy efficiency.

These findings reinforce the importance of considering time-horizon effects when evaluating energy performance in multi-stage batch scheduling with the model that do not include energy optimisation, as production metrics alone do not fully capture energy-related outcomes.

Figure 10 compares the total energy consumption and the optimal number of Pan 4 initiations obtained from the single-objective and multi-objective models as the planning horizon increases. The multi-objective model consistently achieves slightly lower and more stable total energy consumption than the single-objective model across all horizons, indicating improved energy efficiency while maintaining equivalent production performance. The stepwise evolution of the curves corresponds to discrete increases in the optimal number of Pan 4 initiations, shown on the secondary axis, which occur at identical horizon thresholds for both models. This alignment confirms that both formulations preserve the same production structure. Overall, the results demonstrate that the multi-objective model dominates by achieving comparable throughput decisions with systematically lower energy consumption, highlighting the benefit of the energy-aware optimisation strategy.

Regarding total variation, Fig. 11 illustrates the relationship between energy variation and the optimal number of Pan 4 initiations as the planning horizon increases for both models. For small horizons, energy variation rises sharply, reflecting limited scheduling flexibility and strong sensitivity to production horizon decisions. Beyond approximately 25–30 periods, both models enter a stable regime in which energy variation fluctuates around a nearly constant level despite the continuous increase in optimal initiations. The multi-objective model consistently exhibits slightly lower and smoother energy variation than the single-objective model, demonstrating improved energy balancing while maintaining identical production decisions, since both models generate the same initiation thresholds. Overall, the results show that increasing the number of optimal stage 4 initiations primarily drives total production expansion, whereas energy variation becomes largely independent of horizon length, with the multi-objective model achieving superior stability and energy efficiency along the optimal production trajectory.

3.3 Total Energy Consumption with Bounded Total Energy

We now investigate the effect of imposing explicit bounds on total energy consumption in both models. Figures 12 and 13 illustrate the total energy consumption and corresponding optimal number of stage 4 batch initiations as functions of the planning horizon, subject to an upper energy bound of 500 and 1000, respectively.

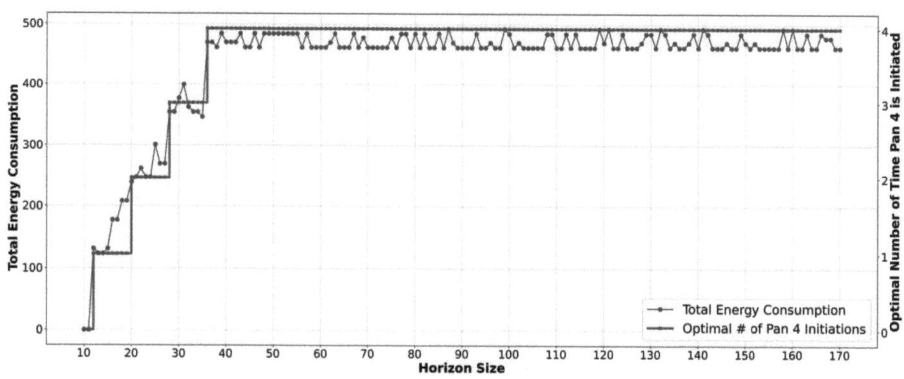

Fig. 12. Total energy consumption as a function of the planning horizon and the corresponding optimal number of pan 4 initiations with a bounded total energy 500.

As the planning horizon increases, the number of stage 4 initiations again exhibits a stepwise behavior. Beyond a certain horizon length, production reaches its maximum feasible level under the imposed energy constraint, and the number of final-stage batches remains constant. Despite this fixed production level, total energy consumption varies across different planning horizons. Larger planning horizons provide additional flexibility that allows steam-intensive operations to be scheduled more evenly over time, thereby reducing overlap in energy demand and improving overall energy efficiency, even under binding energy constraints.

Regarding the multi-objective model, Figs. 14 and 15 show that both models generate identical optimal Pan 4 initiation decisions as the planning horizon increases, indicating that production throughput is preserved under the imposed energy constraints. However, the multi-objective model consistently achieves

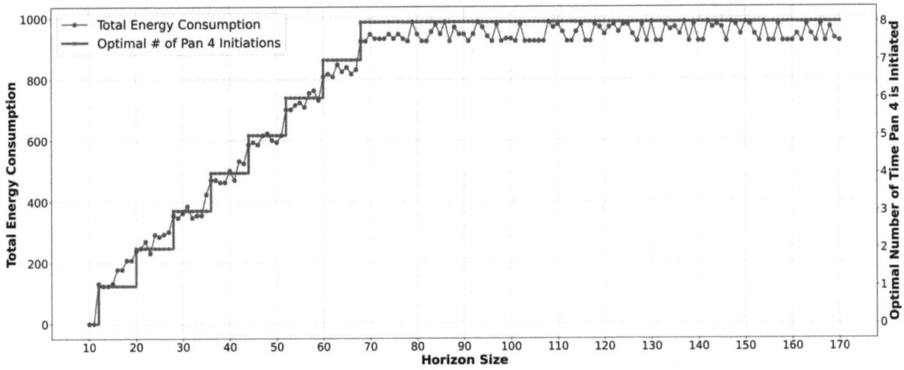

Fig. 13. Total energy consumption as a function of the planning horizon and the corresponding optimal number of pan 4 initiations with a bounded total energy 1000.

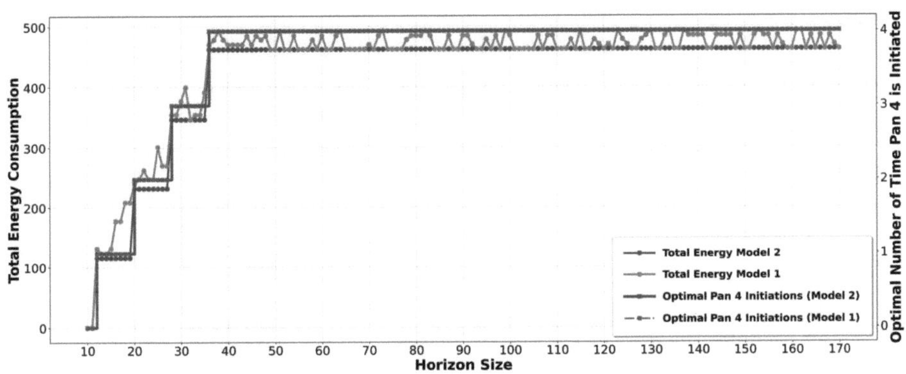

Fig. 14. Total energy consumption as a function of the planning horizon and the corresponding optimal number of Pan 4 initiations under a bounded total energy of 500: Single-Objective vs Multi-Objective Model.

lower and more stable total energy consumption than the single-objective model, demonstrating improved energy stability without sacrificing production performance.

Figures 16 and 17 illustrate the evolution of energy variation with respect to the planning horizon under different energy bounds. The multi-objective model consistently achieves lower and more stable energy variation than the single-objective model while preserving the same optimal number of Pan 4 initiations, thereby demonstrating improved energy smoothing.

Overall, the multi-objective model outperforms the single-objective formulation by maintaining identical production decisions while systematically reducing total energy consumption and energy variation across all planning horizons. These results confirm that incorporating energy optimisation into the optimi-

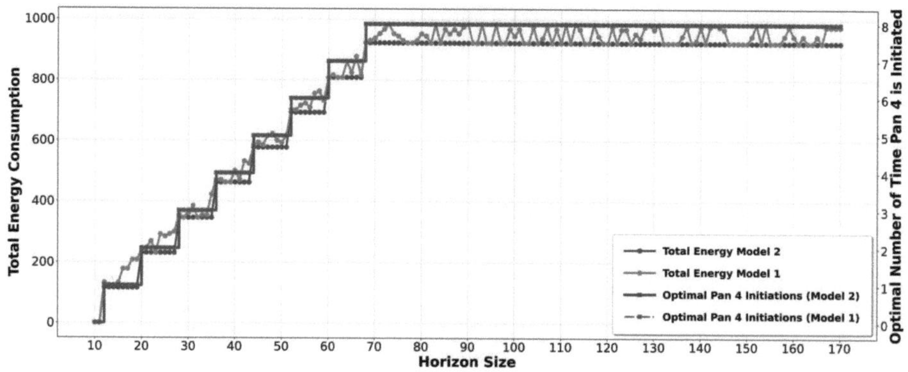

Fig. 15. Total energy consumption as a function of the planning horizon and the corresponding optimal number of Pan 4 initiations under a bounded total energy of 1000: Single-Objective vs Multi-Objective Model.

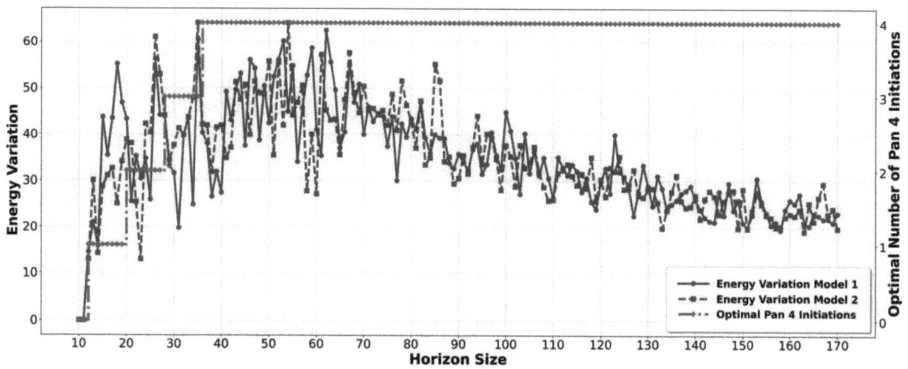

Fig. 16. Total energy variation as a function of the planning horizon and the corresponding optimal number of Pan 4 initiations under a bounded total energy of 500: Single-Objective vs Multi-Objective Model.

sation framework enhances operational stability and energy efficiency without compromising production output.

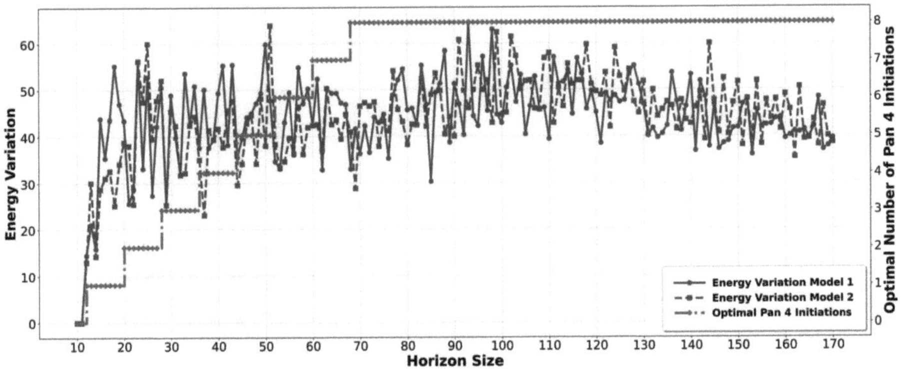

Fig. 17. Total energy variation as a function of the planning horizon and the corresponding optimal number of Pan 4 initiations under a bounded total energy of 1000: Single-Objective vs Multi-Objective Model.

4 Conclusion

This paper addressed the Batch Pan Scheduling Problem in white sugar refineries using an energy-aware planning perspective. A discrete time mixed integer linear programming formulation was developed to model batch initiation decisions across four sequential crystallisation stages, explicitly capturing inter stage material dependencies and capacity constraints. Numerical experiments showed that the planning horizon length has a significant impact on energy performance. Although production remained constant across different horizons, total energy consumption varied due to differences in the temporal distribution of batch initiations. Longer horizons allowed batch starts to be more evenly spaced, reducing the overlap of steam intensive operations and lowering cumulative energy consumption. These results demonstrate that energy efficiency in batch sugar production depends not only on the number of completed batches but also on the coordination of their initiation times when energy optimisation is not explicitly incorporated. By including total energy optimisation, in addition to maximising the number of stage 4 initiations, more stable results are obtained, providing greater flexibility in the selection of the planning horizon without increasing total energy consumption.

The proposed framework demonstrates the effectiveness of discrete-time scheduling models for analysing energy production trade-offs in multi-stage batch processes under both single-objective and multi-objective formulations. Future work will focus on extending the model to multi-objective formulations that explicitly incorporate energy smoothing objectives in addition to total energy minimisation. Furthermore, the inclusion of intermediate storage buffers will be considered to enhance the realism and flexibility of the scheduling problem, as well as to account for uncertainty in model parameters.

Acknowledgments. This research was partly made possible by the Centre of Excellence in Mathematical and Statistical Sciences (CoE MaSS) of South Africa through its sponsorship of the Mathematics in Industry Study Group. The second author would like to acknowledge the support of the Canadian Natural Sciences and Engineering Research Council (NSERC) under the grant 2021 03307. These supports are greatly appreciated.

Disclosure of Interests. The authors have no competing interests to declare that are relevant to the content of this article.

References

1. Badhotiya, G.K., Soni, G., Mittal, M.: Multi-site planning and scheduling: state-of-the-art review and future research directions. J. Global Oper. Strateg. Sourcing **13**(1), 17–37 (2020). https://doi.org/10.1108/JGOSS-02-2019-0014
2. Bordons, C., Camacho, E.F.: Applications of model predictive controllers in a sugar factory. In: Proceedings of ADCHEM 2000, pp. 329–334 (2000). https://doi.org/10.1016/S1474-6670(17)38561-0
3. De Prada, C., Alonso, C.J., Morilla, F., Bollain, M.H.: Supervision and advanced control in a beet sugar factory. In: Proceedings of ADCHEM 2000, pp. 341–346 (2000). https://doi.org/10.1016/S1474-6670(17)38563-4
4. Foxon, K.M., Loubser, R.C., Smith, G.T., Davis, S.B., Stolz, H.N.P.: Strategies for monitoring energy consumption in sugarcane processing factories. Coop. Sugar **48**(11), 33–44 (2017)
5. Harjunkoski, I., et al.: Scope for industrial applications of production scheduling models and solution methods. Comput. Chem. Eng. **62**, 161–193 (2014). https://doi.org/10.1016/j.compchemeng.2013.12.001
6. Maravelias, C.T.: General framework and modeling approach for chemical production scheduling. AIChE J. **58**(6), 1812–1828 (2012). https://doi.org/10.1002/aic.13801
7. Méndez, C.A., Cerdá, J., Grossmann, I.E., Harjunkoski, I., Fahl, M.: State-of-the-art review of optimization methods for short-term scheduling of batch processes. Comput. Chem. Eng. **30**(6–7), 913–946 (2006). https://doi.org/10.1016/j.compchemeng.2006.02.008
8. Mesquita-Cunha, M., et al.: New ϵ-constraint methods for multi-objective integer linear programming. Eur. J. Oper. Res. (2023). https://doi.org/10.1016/j.ejor.2022.07.044
9. Nott, P.H., Lee, P.L.: A sets formulation to schedule mixed batch/continuous process plants with variable cycle time. Comput. Chem. Eng. **23**(6), 875–890 (1999). https://doi.org/10.1016/S0098-1354(99)90009-1
10. Pinedo, M.: Scheduling: Theory, Algorithms, and Systems. Springer, 5 edn. (2016). https://doi.org/10.1007/978-3-319-26580-3
11. Potts, C.N., Kovalyov, M.Y.: Scheduling with batching: a review. Eur. J. Oper. Res. **120**(2), 228–249 (2000). https://doi.org/10.1016/S0377-2217(99)00153-8
12. van Wissen, M.E., Smeets, J.F.C., Muller, A., Verheijen, P.J.T.: Discrete event modeling and dynamic optimization of a sugar plant. In: Foundations of Computer-Aided Process Operations (FOCAPO 2003) (2003)
13. Zhao, D., Li, Y.R.: Climate change and sugarcane production: potential impact and mitigation strategies. Int. J. Agron. **2015**(1), 547386 (2015). https://doi.org/10.1155/2015/547386

Strategic-Tactical Integrated Planning in Blood Supply Chains

Virgínia Silva Magalhães(✉), Lásara Fabrícia Rodrigues(✉), and Luiz Ricardo Pinto(✉)

Universidade Federal de Minas Gerais, Belo Horizonte 31270-010, Brazil
virginiasmag@hotmail.com, lasara@dep.ufmg.br, luizricardopinto@gmail.com

Abstract. Blood Supply Chains (BSCs) are essential to ensuring transfusion safety and availability of blood products for patients in need. These chains face persistent challenges associated with the perishability of blood products, donation and demand uncertainty, and fragmented decision-making across strategic, tactical, and operational levels. Despite advances in applications of Operations Research in BSCs, the vertical integration of planning levels and the horizontal coordination among functional stages of the BSC remain a challenge to be addressed. This paper proposes an integrated decision-support framework that connects strategic and tactical planning by addressing facility location, demand forecasting, inventory management, production planning, and distribution optimization, thus fostering a more sustainable and resilient supply chain. The framework enables managers with a integrated tool to support decision-making across planning levels and BSC stages.

Keywords: Blood Supply Chain · Strategic-Tactical · Integrated Planning · Decision-Support System · Health Logistics

1 Introduction

Blood is an irreplaceable resource in healthcare, essential for surgeries and treatments, and there is currently no substitute for human blood despite technological advances [1]. Blood transfusions save lives and improve health outcomes, yet many patients lack timely access to safe blood, making transfusion activities critical for national health systems [2]. The World Health Organization (WHO) recommends that a safe and adequate blood supply be integrated into national health policies, as unsafe transfusion practices can lead to adverse reactions, infections, and chronic shortages [3].

Globally, blood donation rates remain low, with only 5% of eligible individuals donating blood. Approximately 118.5 million donations occur annually, 40% of which come from high-income countries that represent just 16% of the global population [3]. Donation rates average 3.15% in high-income countries, 2.3% in middle-income countries, and only 0.5% in low-income countries [3]. These disparities highlight the challenges of maintaining adequate and safe blood supplies, particularly in low-resource settings.

A. P. Cabral Seixas Costa et al. (Eds.): ICDSST 2026, LNBIP 582, pp. 352–365, 2026.
https://doi.org/10.1007/978-3-032-25146-6_23

Blood Supply Chains (BSCs) play a vital role in public health by ensuring the availability of blood products for critical situations. They support complex medical procedures [1], reduce waste [4], promote equity in healthcare access [5], and enhance resilience during emergencies [6]. Efficient BSC management ensures that the right products are available in the right quantities, minimizing shortages and wastage [4]. BSCs handle multiple components, such as red blood cells, platelets, plasma, and cryoprecipitate, each with distinct shelf lives and compatibility requirements [8].

Managing BSCs is complex due to perishability, uncertain supply and demand, and compatibility constraints. Integration across planning levels (strategic, tactical, and operational) is essential for efficiency and responsiveness [6,7]. Strategic decisions involve long-term network design and infrastructure, tactical decisions address inventory and logistics, and operational decisions manage day-to-day activities. Horizontal integration among collection centers, hospitals, and blood banks further improves resource utilization and resilience [9].

Despite significant advances in the application of quantitative methods and decision support technologies to Blood Supply Chains (BSCs), substantial gaps remain in achieving full vertical and horizontal integration. Most studies focus on isolated stages or specific objectives, leaving interdependencies underexplored [6].

This paper aims to propose an integrated decision-support framework that connects strategic and tactical planning. To do so, this research addresses: (i) facility location and allocation (strategic level), (ii) demand forecasting under uncertainty, (iii) tactical inventory management, (iv) tactical collection and production planning, and (v) tactical distribution optimization.

In doing so, the main research question that guide this research is: How to integrate strategic and tactical planning within the BSC stages?

The remainder of this paper reviews relevant literature with emphasis on integration gaps (Sect. 2), presents the conceptual framework (Sect. 3), details module specifications and formulations (Sect. 4), and discusses the integrated system and its practical application (Sect. 5).

2 Blood Supply Chain

Blood Supply Chains (BSCs) are complex multi-facility networks characterized by perishability, uncertain supply and demand, and compatibility constraints [5,6]. Efficient management requires integration across strategic, tactical, and operational planning levels, as well as coordination among collection, processing, storage, and distribution stages [7]. Despite advances in quantitative methods, significant gaps remain in achieving full vertical and horizontal integration and in addressing uncertainty and regional contexts [1,4,6]. Strategic planning studies address network design and facility location [18,20] but often isolate these decisions from tactical operations. Tactical planning research focuses on individual stages (collection [11], production [12], inventory [13], or distribution [14]) without capturing cross-stage interdependencies. Existing studies span developed countries (Norway, USA, France, Finland, Austria) and emerging economies (Turkey, India, Brazil, Colombia, and Iran).

Studies focusing on single stages include donor arrival prediction and bloodmobile routing for collection [11], platelet production optimization [12], and inventory policies for RBCs and platelets using simulation, stochastic programming, and dynamic programming [13]. Distribution studies address routing and allocation policies under perishability constraints [14].

Partial integration approaches combine two or more stages, such as collection and production [15], inventory and distribution [16], or production and inventory [17]. These models often use mixed integer linear programming (MILP), stochastic programming, or simulation to improve robustness and reduce shortages and wastage.

Complete integration models encompass the entire BSC, linking strategic network design with tactical and operational decisions. Examples include bi-objective MILP under uncertainty [18], simulation-optimization frameworks [19], and network competition models [20]. Recent studies address resilience and sustainability under disruptions and pandemics [21].

3 Conceptual Framework and System Architecture

The proposed decision-support system consists of an integrated blood supply chain network model structured into five modules. These modules address decisions at the strategic and tactical levels (vertical integration) and are also horizontally linked at each planning level (Fig. 1). At the strategic level of the blood supply chain, the network design problem is related to the design and structure of the chain [6] and directly affects all stages at the tactical level. The tactical level includes the interrelated problems of collection, production, inventory, and distribution planning.

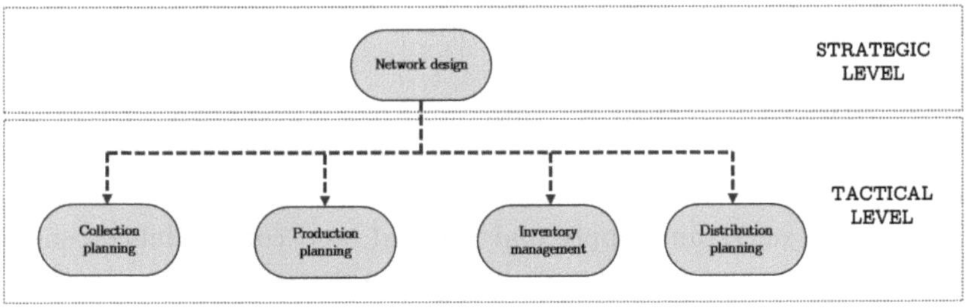

Fig. 1. BSC strategic and tactical planning level.

In Blood Supply Chains (BSCs), network design focuses on strategic decisions to determine the optimal location and capacity of facilities involved in the collection, testing, processing, storage, and distribution of blood. The facilities include fixed collection locations, mobile units, processing and storage centers, and distribution centers [4]. The strategic level addresses the long-term horizon, i.e., decisions that span years and shape the system in terms of infrastructure, management policies, and technologies.

Figure 2 shows the main inputs and outputs of the network design problem in BSCs. The inputs are related to the characteristics of municipalities, donation, demand, and production. The outputs of the network design problem are the location of facilities and the allocation of capacity to each facility. While the strategic level focuses on long-term decisions, the tactical level deals with medium-term decisions that are interconnected with strategic decisions. Here, each stage of the BSC is addressed in collection planning, production planning, inventory management, and distribution planning. At this level, the location and allocation of facilities, outputs of the strategic level, directly influence how resources are distributed and managed in the blood supply network [6].

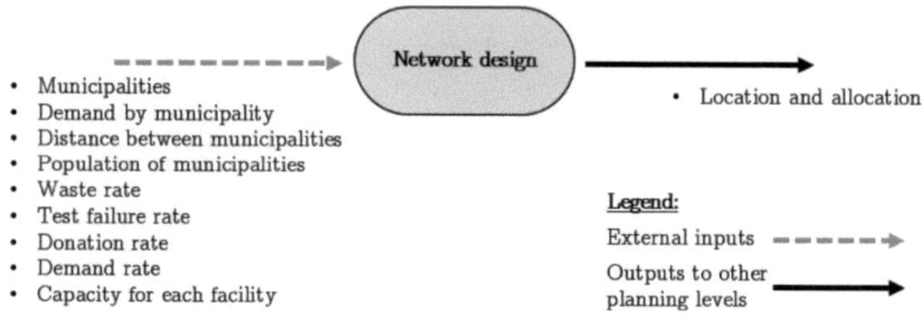

Fig. 2. Network design problem (input and output data).

Figure 3 shows the inputs and outputs for tactical problems: collection planning, production planning, inventory management, distribution planning. Blood collection planning is driven by uncertainty and randomness due to the unpredictable nature of blood donation, which requires great adaptability. Therefore, tactical decisions, such as donation campaigns and inventory management, are strongly influenced by blood availability, which determines resource allocation and medium-term action planning. The inputs of collection planning are divided into: (i) External inputs: related to external factors associated with donor population characteristics (age, sex, population of the donor's municipality, Human Development Index - HDI); (ii) Strategic inputs: obtained at the strategic level such as the location of the collection center and the allocation of municipalities served in that location; (iii) Collection needs: obtained from the production planning (same level) and can be determined based on the expected production needs to meet the demand for blood transfusions in hospitals. The outputs of the collection planning can be used at other planning levels or at the same planning level. The outputs are: (i) Collection forecast: monthly collection quantities to be used in operational planning; (ii) Order fulfillment: collected quantities to supply the production centers.

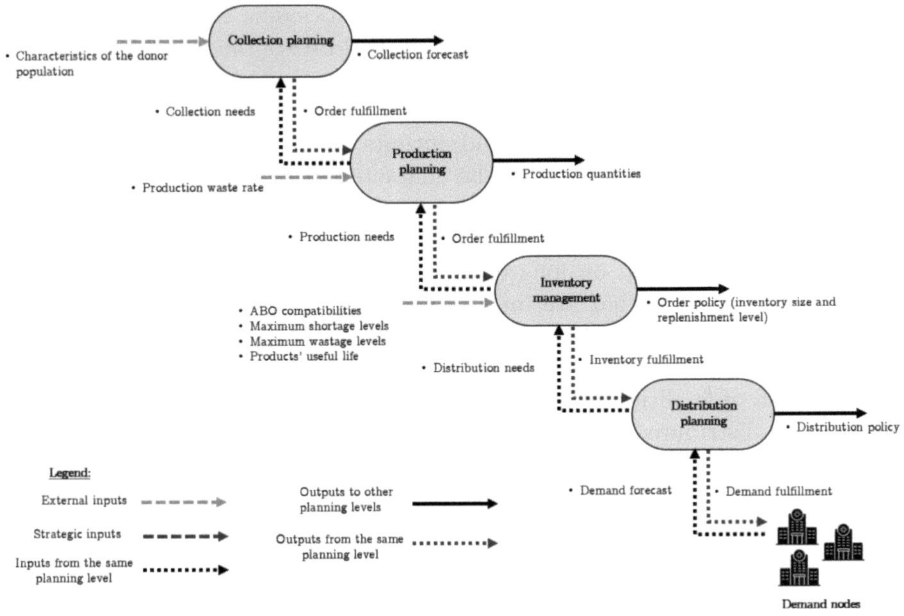

Fig. 3. Tactical problems (input and output data).

Production planning determines the quantities of blood components that blood centers must produce to meet demand over time. However, the amount of blood components produced depends on the number of donations collected (donor availability), and the production is planned based on collection planning and order fulfillment (blood component targets), in addition to the efficient resource allocation [6]. The main inputs of production planning are: (i) External inputs: production waste rates; (ii) Strategic inputs: location of production centers and allocation of collection centers to production centers; (iii) Order fulfillment: number of blood units available by collection centers (same level); (iv) Production needs: number of blood units that should be produced by production to supply the inventory. The outputs of the production planning are: (i) Production quantities: quantity of each product to be produced and the processing period (to be used as an input at the operational level); (ii) Order fulfillment: quantity produced to supply the inventories (same level); (iii) Collection needs: the quantity that must be collected in collection centers to meet production needs (same level).

Inventory management focuses on determining the ordering policy to be used for each blood product. This decision to ensure that the right quantities of blood components are available when needed, while minimizing overstocking that could lead to waste. Inputs of inventory management are: (i) Strategic inputs: location of storage centers, allocation of production centers to storage centers; (ii) External inputs: blood compatibility, shelf life of blood products and acceptable levels of wastage and shortages; (iii) Order fulfillment: number of units produced to supply the inventories; (iv) Distribution needs: number of units needed to supply the demand nodes. Outputs of the inventory management are: (i) Order policy: defines the appropriate inventory size and replenishment level policy to be used at the operational level; (ii) Inventory fulfillment: number of blood components to be sent to distribution (same level); (iii) Production needs: number of blood components that should be produced by production centers (same level).

Distribution planning involves critical decisions to ensure blood products are delivered to hospitals and health centers efficiently, meeting the forecasted demand of blood products at demand nodes. The main inputs of distribution planning include: (i) Strategic inputs: location of distribution centers and allocation of storage centers to distribution centers and allocation of distribution centers to demand nodes (hospitals and health centers); (ii) Inventory fulfillment: quantity of blood components supplied from inventory to fill the distribution centers (same level); (iii) Demand forecast: quantity of blood components forecast at demand nodes, based on demand historical data considering seasonal variations of blood transfusions, age, sex, HDI. The outputs of the distribution planning are: (i) Distribution policy: defines the ideal route from a distribution center to its demand nodes at the operational level; (ii) Distribution needs: quantities of blood units that should be inventoried to meet demand (same level); (iii) Demand fulfillment: quantity of blood components to be sent to demand nodes.

4 System Modules Specifications

The proposed framework is implemented through a decision-support system consisting of integrated modules designed to address various planning dimensions within the BSC, ensuring a comprehensive solution. Here, Demand Forecasting module plays a unique role. While not part of the core optimization framework itself (as discussed in Sect. 3), it is critical input provider. Its projections drive all subsequent strategic and tactical decisions, ensuring the entire planning process is grounded in reliable data.

4.1 Demand Forecast

The forecasting process follows a systematic workflow (Fig. 4), based on [23, 24]. It is addressing the problem definition, data collection from public sources, exploratory analysis, preprocessing (cleaning, outlier handling), model selection

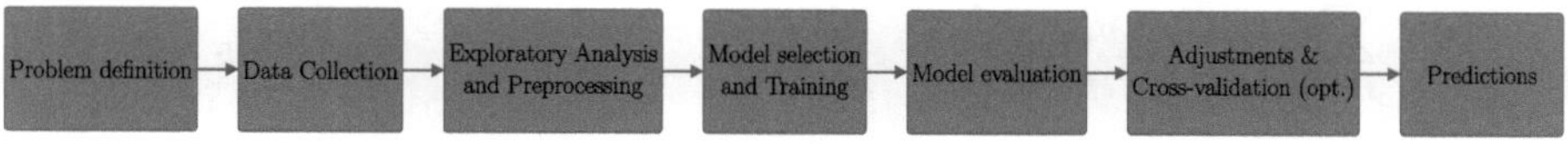

Fig. 4. Demand Forecast Steps. Source: Adapted from James et al. [23] and Geron [24].

(regression, time series, neural networks), train-test split, performance evaluation (RMSE, MAE, R^2), and cross-validation for refinement before deployment.

4.2 Network Design Problem

Network Design Problem module operates at the strategic level and determines the optimal configuration of the blood supply network. This includes decisions about the number, location, and capacity of collection, processing, storage, and distribution centers. The objective is to minimize total network costs, including installation and transportation, while ensuring adequate service levels. This module is formulated as a Mixed-Integer Programming (MIP) problem, which combines integer variables (e.g., flows) and binary variables (e.g., facility location decisions). By considering resource limitations and forecasted demand, the model delivers a cost-efficient and feasible network design that supports long-term planning [6]. The main sets, parameters, and decision variables used in this formulation are summarized in Table 1.

Table 1. Sets, parameters, and decision variables of the model.

Sets and indices
$I(i)$ - municipalities; $J(j)$ - collection; $W(w)$ testing; $K(k)$ - production; $L(l)$ - storage;
$B(b)$ - facility type (c - collection, t - testing, p - production, s - storage and distribution)
Parameters
CI_X - installation cost (CI_j, CI_k, CI_t, CI_l); pop_i - population; qo_i, qd_i - supply/demand;
$CT_{XX'}$ - transport cost ($CT_{ij}, CT_{jk}, CT_{jt}, CT_{kl}, CT_{li}$); Cap_b - capacity; $M = \sum_{i \in I} qo_i$;
$\text{wast}_{ad}, \text{wast}_t$ - discarded units
Decision variables
$a_{XX'}$ - binary flow ($a_{ij}, a_{jk}, a_{jt}, a_{kl}, a_{li}$); $f_{XX'}$ - flow ($f_{ij}, f_{jk}, f_{jt}, f_{kl}, f_{li}$);
y_{bX} - binary facility ($y_{cj}, y_{tw}, y_{pk}, y_{sl}$); Inv_l - extra stored units

The mathematical model is defined by equations (1)–(34).

$$z(\text{Total Cost}) = \min \left\{ \sum_{X,b} CI_X \, y_{bX} + \sum_{X,X'} CT_{XX'} \, a_{XX'} \right\}. \quad (1)$$

$$\sum_{j \in J} a_{ij} = 1, \quad \forall i \in I \quad (2)$$

$$a_{ij} \le y_{cj}, \quad \forall i \in I, \forall j \in J \quad (3)$$

$$a_{jk} \le y_{cj}, \quad \forall j \in J, \forall k \in K \quad (4)$$

$$a_{ij} \ge y_{cj}, \quad \forall i \in I, \forall j \in J \mid i = j \quad (5)$$

$$qo_{ij} \cdot a_{ij} = f_{ij}, \quad \forall i \in I, \forall j \in J \quad (6)$$

$$\sum_{i \in I} f_{ij} \leq M \cdot a_{ij}, \quad \forall i \in I, \forall j \in J \quad (7)$$

$$\sum_{i \in I} f_{ij} \leq \text{Cap}_c, \quad \forall j \in J \quad (8)$$

$$\sum_{i \in I} f_{ij} = \sum_{k \in K} f_{jk}, \quad \forall j \in J \quad (9)$$

$$\sum_{i \in I} f_{ij} = \sum_{w \in W} f_{jw}, \quad \forall j \in J \quad (10)$$

$$\sum_{w \in W} a_{jw} = 1, \quad \forall j \in J \quad (11)$$

$$a_{jw} \leq y_{tw}, \quad \forall j \in J, \forall w \in W \quad (12)$$

$$a_{jw} \geq y_{cj} + y_{tw} - 1, \forall j \in J, \forall w \in W \mid j = w \quad (13)$$

$$f_{jw} \leq M \cdot a_{jw}, \quad \forall j \in J, \forall w \in W \quad (14)$$

$$\sum_{j \in J} f_{jw} \leq \text{Cap}_t \cdot y_{tw}, \quad \forall w \in W \quad (15)$$

$$\sum_{k \in K} a_{jk} = 1, \quad \forall j \in J \quad (16)$$

$$a_{kl} \leq y_{pk}, \quad \forall k \in K, \forall l \in L \quad (17)$$

$$a_{jk} \leq y_{pk}, \quad \forall j \in J, \forall k \in K \quad (18)$$

$$a_{jk} \geq y_{cj} + y_{pk} - 1, \forall j \in J, \forall k \in K \mid j = t \quad (19)$$

$$\sum_{j \in J} f_{jk}(1 - \text{wast}_t) = \sum_{j \in J} f_{kl}, \quad \forall k \in K \quad (20)$$

$$f_{jk} \leq M \cdot a_{jk}, \quad \forall j \in J, \forall k \in K \quad (21)$$

$$\sum_{j \in J} f_{jk} \leq \text{Cap}_p, \quad \forall k \in K \quad (22)$$

$$\sum_{l \in L} f_{kl} = 1, \quad \forall k \in K \quad (23)$$

$$a_{li} \leq \text{Cad}_l, \quad \forall l \in L, \forall i \in I \quad (24)$$

$$a_{kl} \leq \text{Cad}_l, \quad \forall k \in K, \forall l \in L \quad (25)$$

$$a_{kl} \geq y_{pk} + \text{Cad}_l - 1, \forall k \in K, \forall l \in L \mid k = l \quad (26)$$

$$\sum_{k \in K} f_{kl}(1 - \text{wast}_{ad}) = \sum_{k \in K} f_{li} + \text{Inv}_l, \forall l \in L \quad (27)$$

$$f_{kl} \leq M \cdot a_{kl}, \quad \forall k \in K, \forall l \in L \quad (28)$$

$$\sum_{k \in K} f_{kl} \leq \text{Cap}_{ds}, \quad \forall l \in L \quad (29)$$

$$\sum_{l \in L} a_{li} = 1, \quad \forall i \in I \quad (30)$$

$$f_{li} \leq qd_i \cdot a_{li}, \quad \forall l \in L, \forall i \in I \quad (31)$$

$$a_{li} \geq \text{Cad}_l, \quad \forall l \in L, \forall i \in I \mid l = i \quad (32)$$

$$qd_i \cdot a_{li} \leq f_{li}, \quad \forall l \in L, \forall i \in I \quad (33)$$

$$\text{Inv}_l, f_{XX'} \geq 0; \; a_{XX'}, y_{bX} \in \{0, 1\} \quad (34)$$

The objective function minimize the total cost of the blood supply chain (Eq. (1)), considering facility installation and transportation costs. Constraints (2)–(33) ensure network feasibility by enforcing allocation rules, flow conservation, capacity limits, activation dependencies, and demand satisfaction across all stages (collection, testing, production, storage, and distribution). Finally, constraint (34) defines the domains of the variables.

4.3 Inventory Management

Inventory Management module operates at the tactical level and addresses the challenge of managing perishable blood products under uncertainty. It adopts a simulation-optimization approach (adapted from Magalhães et al. [19], as shown in Fig. 5) to support decisions on inventory size and replenishment points for each blood product and storage center.

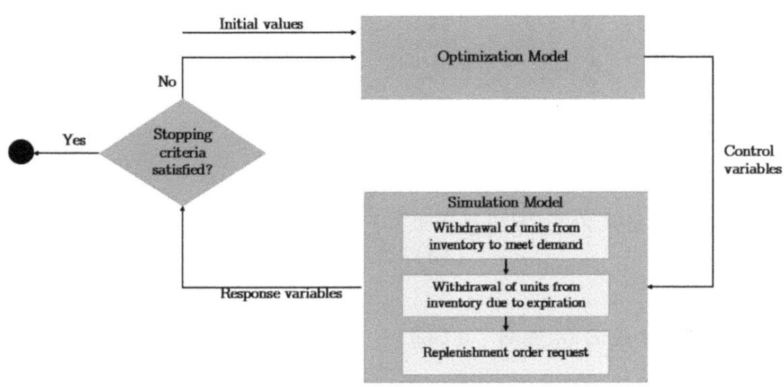

Fig. 5. Simulation-optimization approach for Blood Component Inventory Management. Source: Adapted from Magalhães et al. [19].

The simulation-optimization approach iteratively combines simulation and optimization models to define inventory size and replenishment points for each blood product and type. Simulation captures stochastic elements such as demand variability, wastage, and shortages, while optimization adjusts control variables to minimize these effects. The process continues until convergence (difference $<$ 0.01%) or a predefined iteration limit is reached.

The simulation model manages blood inventory through three steps: (i) demand fulfillment using probability distributions with blood type compatibility, (ii) inventory update with expiration removal and shelf-life reduction, and (iii) replenishment order generation when levels fall below reorder points, considering lead time distributions.

The optimization model determines the ideal inventory size and replenishment point for each blood product individually. Decision variables include inventory size (TE_{ij}) and replenishment point (PR_{ij}), while shortage ($PRup_{ij}$) and wastage ($PExp_{ij}$) rates depend on these variables and lack deterministic solutions. Therefore, these performance measures are estimated through simulation, enabling accurate evaluation of inventory policies under uncertainty. The optimization model uses the elements described in Table 2.

Table 2. Sets, parameters, decision variables, and performance measures.

Sets and indices
I blood types (AB^+, AB^-, A^+, A^-, B^+, B^-, O^+, O^-); J blood products
Parameters
α_{ij}, β_{ij}, γ_{ij} max percentages for wastage, shortage, and inventory;
LIS_{ij} lower bound of inventory size; δ_j max fraction for replenishment point
Decision variables
PR_{ij} replenishment point; TE_{ij} inventory size for blood type i and product j
Performance measures
VT_{ij} expired units; URE_{ij}, URE_j inventory levels; $PExp_{ij}$ % expired; $PRup_{ij}$ % shortages;
RS_{ij} shortage units; $PEst_{ij}$ % inventory proportion
Source: Based on Magalhães et al. [19]

The optimization mathematical model is defined by equations (35)–(43).

$$\min Z = \sum_{w \in W} TE_{ij}, \qquad \forall j \in J. \tag{35}$$

$$PExp_{ij} \leq \alpha_{ij}, \qquad \forall i \in I, \forall j \in J \tag{36}$$

$$PRup_{ij} \leq \beta_{ij}, \qquad \forall i \in I, \forall j \in J \tag{37}$$

$$PEst_{ij} \leq \gamma_{ij}, \qquad \forall i \in I, \forall j \in J \tag{38}$$

$$\sum_{i \in I} PEst_{ij} = 1, \qquad \forall j \in J \tag{39}$$

$$PR_{ij} \leq \delta_j \cdot TE_{ij}, \qquad \forall i \in I, \forall j \in J \tag{40}$$

$$TE_{ij} \geq LIS_{ij}, \qquad \forall i \in I, \forall j \in J \tag{41}$$

$$PExp_{ij}, PRup_{ij}, PEst_{ij} \in \mathbb{Q}^+. \tag{42}$$

$$PR_{ij}, TE_{ij} \in \mathbb{Z}^+. \tag{43}$$

The objective function (35) minimizes the total inventory size for each blood type and product. Constraints (36)–(38) impose upper limits on wastage, shortages, and inventory proportions. Constraint (39) maintains proportional distribution among blood types, while constraint (40) restricts replenishment points to recommended thresholds. Constraint (41) enforces minimum inventory levels. Finally, constraints (42)–(43) define variable domains.

4.4 Production and Collection Planning

Production and Collection Planning modules operate at the tactical level to define quantities that ensure a balanced blood supply within the logistics network. They use the locations and allocations established by the Network Design module (strategic level) to set collection targets and production policies. Inventory needs at storage centers drive production requirements, which in turn determine the number of whole blood units to collect, considering processing yields.

The Production Planning module calculates quantities to meet storage replenishment policies, account for production losses, and respect capacity limits. Equation (44) defines required production per center based on storage demand and waste rates, while Eq. (45) enforces capacity constraints. Conversely, the Collection Planning module converts product demand into whole blood collection targets, considering yields, losses, and capacity. Equations (46)–(47) ensure flow balance between collection and production.

$$QP_{zkt} = \sum_{l \in L} Q_{zlt} \cdot a_{kl} \cdot (1 + \gamma_{zkt}), \qquad \forall z \in Z, \forall k \in K, \forall t \in T. \tag{44}$$

$$\sum_{z \in Z} QP_{zkt} \leq \mathrm{Cap}_k \cdot \varphi_k, \qquad \forall k \in K, \forall t \in T. \tag{45}$$

$$SQ_{jt} = \left(\sum_{z \in Z, k \in K} \eta_z \cdot QP_{zkt} \cdot a_{jk} \right) \cdot (1 + \omega_j), \qquad \forall j \in J, \forall t \in T. \tag{46}$$

$$SQ_{jt} \leq \mathrm{Cap}_j \cdot \varphi_j, \qquad \forall j \in J, \forall t \in T. \tag{47}$$

where: QP_{zkt} quantity of product z at production center k in time t; Q_{zlt} demand at storage l; a_{kl} flow binary; γ_{zkt} waste rate; Cap_k capacity; ϕ_k efficiency; SQ_{jt} quantity for collection center j in time t; η_z yield coefficient for product z; a_{jk} flow binary; ω_j testing waste rate; Cap_j capacity; ϕ_j efficiency factor.

4.5 Distribution Planning

Distribution Planning uses a CVRPTW model to optimize delivery routes, minimizing transportation costs while respecting vehicle capacity and time windows. Table 3 summarizes model elements, followed by the modeling given by equations (46)–(53).

The objective function (46) minimizes transportation costs across selected routes. Constraints (47)–(51) ensure feasibility: each customer is visited once, vehicle capacity is respected, routes start and end at the origin, and flow continuity is maintained. Constraint (52) eliminates subtours and enforces temporal logic using the big-M method. Finally, constraint (53) defines variable domains.

Table 3. Sets, parameters, and decision variables for the distribution model.

Sets and indices
$V(v)$ vehicles; $C(n)$ demand points; $N(n+1)$ vertices incl. depot
Parameters
t_{ij} travel time; c_{ij} distance; Q vehicle capacity; d_i demand; M large constant
Decision variables
x_{ijk} binary: vehicle k travels from i to j; s_{ik} service start time at i by vehicle k

$$\min Z(\text{Total\_Cost}) = \sum_{k \in V} \sum_{i \in N} \sum_{j \in N} c_{ij}\, x_{ijk} \tag{46}$$

$$\sum_{j \in N} x_{ijk} = 1 \qquad \forall i \in C,\ \forall k \in V \tag{47}$$

$$\sum_{i \in C} d_i \left(\sum_{j \in N} x_{ijk} \right) \leq Q \qquad \forall k \in V \tag{48}$$

$$\sum_{j \in N} x_{0jk} = 1 \qquad \forall k \in V \tag{49}$$

$$\sum_{i \in N} x_{i,\,n+1,\,k} = 1 \qquad \forall k \in V \tag{50}$$

$$\sum_{i \in N} x_{ihk} - \sum_{j \in N} x_{hjk} = 0 \forall h \in C,\ \forall k \in V \tag{51}$$

$$s_{ik} + t_{ij} - M\,(1 - x_{ijk}) \leq s_{jk} \forall i, j \in N,\ \forall k \in V \tag{52}$$

$$x_{ijk} \in \{0,1\}, s_{ik} \geq 0 \qquad \forall i, j \in N,\ \forall k \in V \tag{53}$$

5 Integrated Planning System and Practical Application

The proposed decision-support system integrates strategic and tactical planning through five interconnected modules, achieving both vertical integration across planning levels and horizontal coordination among blood supply chain stages.

The system targets both regional and country-wide blood supply chain. Users are supply chain managers and operations directors with access to historical data. Strategic planning (Network Design) operates on 3–5 year horizons, reviewed annually; tactical modules employ 12-month rolling horizons, updated monthly.

The operational workflow comprises four phases: (i) Collect historical data and generate demand forecasts (Sect. 4.1); (ii) Determine facility locations and allocations (Sect. 4.2); (iii) Configure modules using strategic outputs as inputs: Collection (Sect. 4.4) sets targets, Inventory (Sect. 4.3) determines policies via simulation-optimization, Production (Sect. 4.4) defines quantities, Distribution

(Sect. 4.5) optimizes routes; (iv) Review of KPIs (service level, wastage, costs) with parameter adjustments. The system supports strategic decisions (facility location, capacity, allocation) and tactical decisions (collection targets, inventory levels, production quantities, delivery routes).

Network Design establishes the structural foundation by defining facility locations and interconnections. At the tactical level, integration follows a demand-driven approach: demand forecasting drives inventory management, determining optimal inventory levels and replenishment points while considering acceptable wastage and shortage thresholds. Inventory targets define distribution requirements, which drive production quantities through flow balance equations. Production requirements translate into collection targets, ensuring alignment. Distribution Planning optimizes delivery across all stages. This reverse flow perspective enables responsive adjustments that reduce waste and improve service.

Key challenges include harmonizing planning horizons, managing uncertainty, and real-world validation. Next steps involve implementing the modules (Sects. 3–4) to generate a functional system and conducting empirical validation in Blood Supply Chain contexts.

References

1. Beliën, J., Forcé, H.: Supply chain management of blood products: a literature review. Eur. J. Oper. Res. **217**(1), 1–16 (2012)
2. Williamson, L.M., Devine, D.V.: Challenges in the management of the blood supply. Lancet **381**(9880), 1866–1875 (2013)
3. Blood safety and availability. https://www.who.int/news-room/fact-sheets/
4. Osorio, A.F., Brailsford, S.C., Smith, H.K.: A structured review of quantitative models in the blood supply chain: a taxonomic framework for decision-making. Int. J. Prod. Res. **53**(24), 7191–7212 (2015)
5. WHO: Global status report on blood safety and availability (2021)
6. Meneses, M., Santos, D., Barbosa-Póvoa, A.: Modelling the blood supply chain. Eur. J. Oper. Res. **307**(2), 499–518 (2023)
7. Chopra, S., Meindl, P., A.: Supply Chain Management: Strategy, Planning, and Operation, 6th edn. Pearson (2016)
8. Imamoglu, G., Topcu, Y.I., Aydin, N.: A systematic literature review of the blood supply chain through bibliometric analysis and taxonomy. Systems **11**(3), 124 (2023)
9. Shokouhifar, M., Goli, A.: Designing a resilient–sustainable supply chain network of age-differentiated blood platelets using vertical–horizontal transshipment and grey wolf optimizer. Int. J. Env. Res. Public Health **20**(5), 4078 (2023)
10. Altunoglu, B., Batur Sir, G.D.: Multi-objective location-distribution optimization in blood supply chain: an application in Turkiye. BMC Public Health, 3181 (2024)
11. Şahinyazan, F.G., Kara, B.Y., Taner, M.R.: Selective vehicle routing for a mobile blood donation system. Eur. J. Oper. Res. **245**(1), 22–34 (2015)
12. Ghandforoush, P., Sen, T.K.: A DSS to manage platelet production supply chain for regional blood centers. Decis. Support Syst. **50**(1), 32–42 (2010)
13. Dillon, M., Oliveira, F., Abbasi, B.: A two-stage stochastic programming model for inventory management in the blood supply chain. Int. J. Prod. Econ., 27–41 (2017)

14. Ganesh, K., Narendran, T.T., Anbuudayasankar, S.P.: Evolving cost-effective routing of vehicles for blood bank logistics. Int. J. Logist. Syst. Manage., 381–415 (2014)
15. Osorio, A.F., Brailsford, S.C., Smith, H.K., Blake, J.: Designing the blood supply chain: how much, how and where? Vox Sang (2018)
16. Kazemi, S.M., Rabbani, M., Tavakoli-Moghaddam, R., Shahreza, F.A.: An exact solution for joint optimization of inventory and routing decisions in blood supply chains: a case study. Econ. Comput. Econ. Cybern. Stud. Res. **51**(4), 315–333 (2017)
17. Haijema, R., van der Wal, J., van Dijk, N.M.: Blood platelet production: optimization by dynamic programming and simulation. Comput. Oper. Res., 760–779 (2007)
18. Zahiri, B., Torabi, S.A., Mohammadi, M., Aghabegloo, M.: A multi-stage stochastic programming approach for blood supply chain planning. Comp. Ind. Eng., 1–14 (2018)
19. Magalhães, V.S., Pinto, L.R., Rodrigues, L.F., Blake, J.T.: Simulation-optimisation approach to support management of blood components inventory. J. Simul. **18**(4), 671–686 (2024)
20. Nagurney, A., Dutta, P.: Supply chain network competition among blood service organizations: a generalized Nash equilibrium framework. Ann. Oper. Res. (2019)
21. Yang, H., Yin, Y., Wang, D., Cheng, T.C.E., Zhang, R., Hu, H.: An integrated blood supply chain network design during a pandemic. Int. J. Prod. Res., 1–26 (2024)
22. Bishop, C.M., Nasrabadi, N.M.: Pattern Recognition and Machine Learning, vol. 4, no. 4, p. 738. Springer, New York (2006)
23. James, G., Witten, D., Hastie, T., Tibshirani, R.: An Introduction to Statistical Learning: with Applications in R, vol. 103. Springer, New York (2013)
24. Géron, A.: Hands-on machine learning with Scikit-Learn, Keras, and TensorFlow. In: Concepts, Tools, and Techniques to Build Intelligent Systems, 2nd edn. (2019)

Multicriteria Decision Support for Public Collective Transport System Selection Using the FITradeoff Method

Gustavo Alves Leite[1(✉)], Ana Paula Henriques de Gusmão[1,2], and Jônatas Araújo de Almeida[1,3]

[1] Universidade Federal de Pernambuco/PPGEP-CAA, Caruaru, Brazil
{gustavoalves.leite,jonatas.almeida}@ufpe.br,
anapaulahg@academico.ufs.br

[2] Universidade Federal de Sergipe/Management Engineering Department, Sao Cristovao, Brazil

[3] Universidade Federal de Pernambuco/MAPS group - Modelling and Alignment of Portfolio and Strategy Group, Recife, Brazil

Abstract. In Brazil, despite recent investments and policy advances in urban mobility, structural challenges such as low system productivity and unmet demand persist. Selecting an appropriate public collective transport system (PCTS) technology therefore constitutes a complex multicriteria decision problem, particularly in medium-sized cities subject to financial and operational constraints. This study proposes a structured decision-support framework to guide the selection of PCTS technologies for the municipality of Caruaru, Brazil. The framework combines technical guidelines from the Brazilian Development Bank (BNDES) with the Flexible and Interactive Tradeoff (FITradeoff) method. Five transport alternatives are evaluated against nine criteria related to spatial coverage, financial feasibility, operational performance, demand capacity, user attractiveness, socio-environmental impacts, flexibility, and investment requirements. A reference consequence matrix was developed based on technical documentation and transport planning literature and adapted to local conditions. The application demonstrates how the proposed approach structures the decision process, supports sensitivity analysis, and helps in identifying potential alternatives, highlighting its applicability to public transport planning.

Keywords: Public Transport Planning · Public Collective Transport Selection · Multicriteria Decision Analysis · FITradeoff

1 Introduction

According to [1], between 2019 and 2025 urban mobility in Brazil showed significant progress, particularly through the expansion of public subsidies for bus-based public transport, which increased from 120 municipalities in 2019 to 241 cities in 2025. In addition, investments in infrastructure also grew, with 725 active projects nationwide, including 237 exclusive bus lane projects and 123 bus corridor projects.

A. P. Cabral Seixas Costa et al. (Eds.): ICDSST 2026, LNBIP 582, pp. 366–379, 2026.
https://doi.org/10.1007/978-3-032-25146-6_24

However, despite these advances, structural challenges remain, such as low system productivity, an average fleet age of 6 years and 5 months in 2024, and demand levels that have not yet recovered to pre-pandemic volumes, reaching only 86% of the demand recorded in 2019. With the New Growth Acceleration Program (Novo PAC), which allocates R$ 44.2 billion to sustainable urban mobility, a recovery in investment levels and improvements in the efficiency and accessibility of public transport are expected [1].

A relevant issue in this context is the initial definition of the transport mode to be implemented. Based on the different methodologies available, the Brazilian Development Bank (BNDES) proposed the development of a guide intended to provide a comprehensive overview of the various urban public transport technologies adopted in Brazil [2]. This initiative was promptly welcomed, with an invitation extended to the National Secretariat for Urban Mobility of the Ministry of Cities (MCidades) to participate in the project, which was positively accepted due to recurring municipal demands related to the characterization and implementation of public collective transport systems (PCTS). As a result, the originally proposed scope was expanded to include guidelines covering all project phases, from initial formulation to service operation. The guide aims to equip public managers with consistent information to support more effective use of available resources.

Considering the issues discussed above and the Technical Cooperation Agreement signed with the Autarchy of Mobility, Traffic and Transport of Caruaru (AMTTC), aimed at improving the city's public transport services, the authors of this study propose the adaptation and application of the BNDES guide to define the most appropriate public transport technology or mode. The guide originally recommends the use of a multicriteria decision-making approach in the final stage of technology selection. Thus, this study seeks to propose the use of a more flexible and interactive tradeoff method (FITradeoff), proposed by [3, 4]. The analysis relies on a reference decision matrix constructed from technical guidelines and transport planning literature and adapted to the operational and demand characteristics of Caruaru, a Brazilian medium-sized city.

This study contributes to the literature by providing a structured and reproducible operationalization of public transport evaluation guidelines within an MCDA framework. It translates the BNDES assessment guidelines into a coherent value-based model, integrates the FITradeoff method into a public-sector transport planning context, and proposes a transparent decision-support structure adaptable to participatory settings. The contribution is primarily methodological and structural, as the application serves to demonstrate the framework's analytical consistency and capabilities rather than to deliver empirical findings.

2 Background

2.1 Public Transportation Service in Brazil

Brazilian public transportation has historically been affected by the lack of adequate policies, incentives for private vehicle use, and the inability to fully meet population needs. As a result, negative outcomes such as declining demand for public transport, poor service quality, increased congestion and pollution, reduced quality of life, and longer travel times have been observed [5].

In an effort to address these challenges, the National Urban Mobility Policy was established through Law No. 12,587/2012, which aims to integrate different transport modes and improve accessibility and mobility for people and goods, ensuring the right to access the city [5, 6]. Despite the administrative and political autonomy granted to states and municipalities by the 1988 Federal Constitution, these entities generally show unsatisfactory performance in public transport management, reinforcing the importance of quality management and the definition of indicators suited to regional realities [7, 8].

In the international context, different countries have adopted strategies to finance and modernize their public transport systems. In the United Kingdom, investments in large-scale infrastructure projects, such as the high-speed rail system (HS2), are combined with subsidy policies aimed at maintaining the affordability of bus services, along with substantial funding for railway modernization and electrification aligned with decarbonization targets [9, 10]. In France, the current strategy focuses on regenerating existing infrastructure through significant annual investments and a diversified financing model aimed at reducing emissions [11].

In Brazil, federal public investment in transport infrastructure reached R$ 14.90 billion in 2024, representing modest growth compared to the previous year but remaining below the increase observed between 2022 and 2023. At the same time, private investment showed a slight decline, indicating a slowdown in private sector participation in transport financing [12]. This scenario highlights the need for mechanisms that enable continuous evaluation, monitoring, and improvement of the quality of public transport services in the country.

Public Collective Transport Systems

The PCTS play a structuring role in urban mobility and are fundamental to promoting accessibility, social equity, and efficient use of urban space. The proper characterization of these systems represents an essential step in the planning and decision-making process, particularly in medium-sized cities such as Caruaru (PE), which experience accelerated urban growth, spatial dispersion, and increasing demand for motorized travel. As highlighted in the Guide for the Selection of Technologies and Implementation of Public Collective Transport Projects, prepared by the BNDES, there is no universally superior technology; instead, the analysis of local conditions and demand patterns is indispensable [2].

According to the guide, PCTS can be grouped into three major categories: bus-based systems, rail-based systems, and other public transport systems, which differ in terms of capacity, cost, operational complexity, and urban impact. This classification supports the comparative evaluation of technological alternatives, especially in municipalities seeking to improve public transport services without adopting solutions incompatible with their financial and institutional capacity, as is the case of Caruaru [2].

Bus-based public collective transport systems (PCTS) predominate in Brazil, particularly in medium-sized cities, due to their flexibility, lower implementation costs, and adaptability to existing road infrastructure. They may operate with varying levels of traffic priority, from mixed traffic to exclusive lanes and BRT systems. In Caruaru, where demand is concentrated along specific corridors and varies over time, bus systems with some degree of right-of-way priority are especially suitable. Exclusive lanes or structured corridors can significantly improve speed, regularity, and reliability without the high investments required for rail systems, and their performance depends on segregation level, operational management, and urban integration [2].

Rail-based systems, such as metro, urban rail, and LRT, offer high capacity and operational regularity, making them appropriate for dense, high-demand corridors. However, they involve substantial capital and maintenance costs, institutional complexity, and long-term planning requirements [2]. In Caruaru, their implementation faces economic and demand-related constraints, and their adoption would require careful assessment of urban structure and financing capacity.

In addition to bus- and rail-based systems, the [2] presents other PCTS modalities, such as monorails, urban cable cars, ferries, and automated people movers. These systems are generally intended for highly specific contexts in which topographical, environmental, or land-use constraints limit the applicability of conventional solutions. While such alternatives may offer localized advantages, their applicability in cities such as Caruaru is generally limited.

The characterization of PCTS demonstrates that the selection of the most appropriate technology for Caruaru should be addressed as a multicriteria decision-making problem, as proposed in the BNDES guide itself [2]. This decision requires the simultaneous consideration of technical, operational, economic, social, environmental, and institutional criteria, while accounting for local specificities and budgetary constraints.

In this regard, the application and adaptation of the BNDES guide to the context of Caruaru, combined with the use of more flexible and interactive multicriteria methods, can effectively support public managers in selecting the technological alternative most consistent with the operational reality and demand characteristics of the local public transport system, thereby contributing to improvements in efficiency, service quality, and urban accessibility.

2.2 Multicriteria Decision

The MCDA is employed when a decision maker (DM) faces at least two alternative courses of action and the choice among them involves the simultaneous pursuit of multiple objectives, which are often conflicting. These objectives are associated with the consequences of the alternatives and are represented through criteria, attributes, or dimensions that allow the evaluation of each alternative from different perspectives [13]. MCDA has been widely used as a structured process that supports DM and stakeholders in problem structuring, preference elicitation, and the development of consistent and transparent decision recommendations. Examples of applications in different contexts include operations management [14], data protection [15], power distribution systems [16], agriculture [17] and others.

According to [18], multicriteria decision problems can be classified into four fundamental problematics: choice or selection, ranking, classification, and description. This typology guides both problem structuring and the selection of an appropriate decision support method. In addition to these classical categories, the literature also recognizes the portfolio problematic, which focuses on selecting subsets of alternatives under constraints such as budget limitations [13, 19]. Thus, the correct definition of the decision problematic constitutes a central step in multicriteria modeling.

MCDA methods can generally be grouped into two main approaches: single-criterion synthesis and outranking. In the former, a single aggregation function is constructed, with methods based on Multi-Attribute Value Theory (MAVT) standing out, such as SMARTS, SMARTER, and FITradeoff [3, 20–22]. These methods are typically compensatory, allowing superior performance in one criterion to offset inferior performance in others [23]. In contrast, outranking methods, such as ELECTRE and PROMETHEE, are based on pairwise comparisons and do not allow full compensation between criteria [24, 25].

One of the main challenges in applying multicriteria methods concerns the elicitation of the DM's preferences, as these preferences may be imprecise, incomplete, or difficult to express due to cognitive limitations or lack of availability [3, 4]. In this context, methods based on partial information have gained prominence by reducing the cognitive effort required from DM and increasing the practical feasibility of multicriteria applications [26]. Within this framework, the FITradeoff method emerges as a flexible and interactive alternative capable of handling trade-offs and partial preference information, making it particularly suitable for complex decision problems in organizational and public environments characterized by multiple criteria, uncertainty, and constraints.

FITradeoff Method

The FITradeoff (Flexible and Interactive Tradeoff) method is a multicriteria decision support method developed within the scope of Multi-Attribute Value Theory (MAVT). This method was proposed to overcome the limitations of traditional tradeoff-based elicitation approaches, which often require complete and precise preference information and impose a high cognitive burden on the DM, making them less suitable for real-world selection contexts [3].

In selection problems, FITradeoff operates by progressively reducing the set of admissible alternatives through dominance relations derived from partial preference information. Instead of requiring exact tradeoff judgments, the method elicits simple pairwise comparisons between hypothetical consequences, which are translated into linear constraints on the criteria weights. These constraints define a feasible weight space that is iteratively refined, allowing the identification of alternatives that are dominated under all admissible weight vectors. As a result, the method supports the elimination of inferior options while preserving those that remain potentially optimal for the DM [3, 4]

The FITradeoff procedure for selection problems is structured as an interactive process comprising three main stages: ordering the criteria by the importance of their range of consequences, elicitation of partial tradeoff information, and recommendation of the most suitable alternative(s). At each iteration, linear programming models are solved to verify dominance relationships among alternatives and to update the set of candidates. This process continues until a single alternative is selected or until the DM considers the remaining alternatives sufficiently close in performance. The interactive nature of the method enables preference learning while maintaining consistency with the additive value model assumed in MAVT [4].

In organizational and public decision-making environments, FITradeoff offers significant advantages by accommodating incomplete and imprecise preferences, which are common when DMs face time constraints or complex evaluation contexts. The availability of a dedicated decision support system enhances transparency and facilitates the interpretation of results throughout the selection process. Empirical applications have demonstrated that FITradeoff provides robust and defensible recommendations for choice problems, particularly when the objective is to identify a preferred alternative under uncertainty and limited preference information [4, 27].

3 Proposed Framework

As a contribution of this study, the framework illustrated in Fig. 1 is proposed to support the selection of the most appropriate technology or PCTS for the city of Caruaru, based on BNDES guidelines and the criteria deemed relevant by public managers (DM).

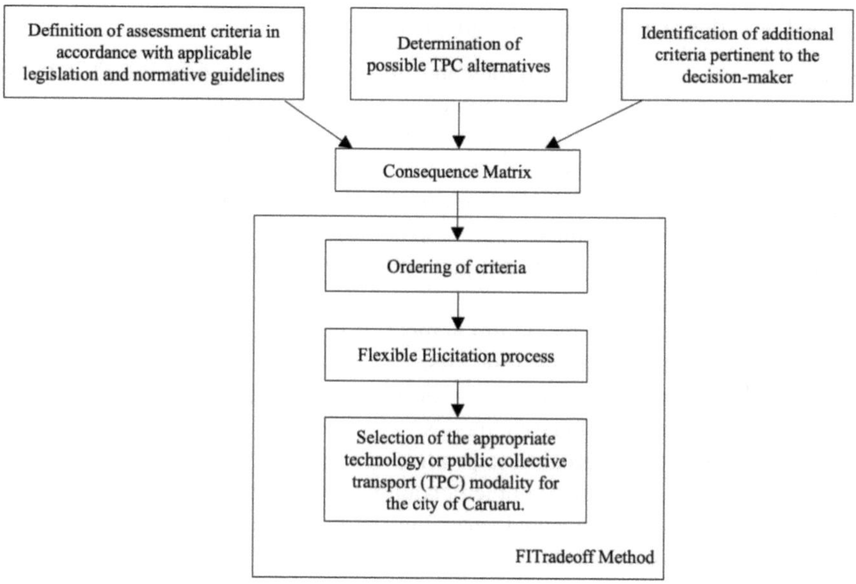

Fig. 1. Proposed framework for technology or PCTS selection

The proposed framework aims to present a decision-support framework that contributes to standardization and assists in the selection of the most appropriate technology/modality for the provision PCTS. To this end, a consequence matrix is constructed, comprising the decision alternatives and criteria. The alternatives are initially identified based on all systems described in sect. 2.2. Subsequently, an analysis is conducted to eliminate those alternatives that are not suitable for the local reality and the specific context evaluated by the public manager. By way of example, three conditioning factors can be highlighted to support this pre-assessment stage: (i) population size and municipal scale, (ii) urban density, and (iii) corridor length. As a result, a set of potentially applicable alternatives is obtained, which is then subjected to comparative evaluation using the proposed multicriteria method.

Regarding the definition of decision criteria, in this context, the BNDES guide, approached in [2], considers that three groups of stakeholders should be taken into account in the analysis of solutions, based on their governance perspectives and reflections on the public and social processes involved: users, public authorities, and society at large. Accordingly, evaluation criteria are suggested for each of these stakeholder groups based on their respective perspectives. However, since this research is linked to a technical cooperation agreement with service providers, only the stakeholder group associated with public authorities is considered, along with the criteria suggested for this group and additional criteria deemed relevant by the DM.

The attributes associated with public authorities are related to their role as those responsible for the provision of Public Collective Transport (PCT) services, either as direct operators or as granting authorities. In this role, public authorities are responsible for balancing the provision of attractive, sustainable, and high-quality services to society with resource demands and the costs inherent to service delivery. The main attributes associated with this stakeholder group are presented below [2]:

- Spatial coverage of the network/demand: territorial coverage of the PCT system, associated with demand attraction potential and social benefits.
- Cost/financial feasibility: relationship between operating costs, revenues, and economic sustainability indicators.
- Operational reliability: level of operational stability, considering risks of service interruptions and irregularities.
- Operational efficiency: system productivity, assessed through indicators relating supply and demand.
- Demand service capacity: maximum number of passengers served and the system's ability to absorb demand above forecasts.
- User attractiveness: characteristics that promote system acceptance, user attraction, and reduction in the use of private modes.
- Socio-environmental externalities: positive or negative social and environmental impacts resulting from system implementation and operation.
- Flexibility in response to demand variations: system capacity to adapt to changes in demand and urban expansion.
- Investments: volume of financial resources required for solution implementation.

Once the alternatives, criteria, and evaluations of alternatives for each criterion are defined, the consequence matrix is obtained, which is required to initiate the FITradeoff method. In this research, the FITradeoff Decision Support System (DSS) [28] used is freely available on the website www.fitradeoff.org. The DSS comprises three stages: criteria ordering, preference elicitation, and selection of the appropriate PCTS technology. In the following section – Application of the Proposed Framework – an application of the proposed framework is presented. The application, of the proposed multicriteria framework, is illustrative and aims to demonstrate its applicability in the context of public transport planning.

4 Application of the Proposed Framework

Considering the framework proposed in Sect. 3, an application was conducted in the context of the municipality of Caruaru. The objective is to select, among the PCTS modes, the alternative that best fits the city's urban, operational, and demand characteristics. For this purpose, the following alternatives were considered: A1 – Conventional Bus (mixed traffic); A2 – Bus with Exclusive Lanes; A3 – Bus Rapid Transit (BRT); A4 – Light Rail Transit (LRT/VLT); and A5 – Urban Rail or metro-like system. Alternative A5 was included for comparative purposes only, as it presents low suitability to the local context.

The decision criteria are presented in Table 1, together with a description of the Likert scale [29] adopted for the evaluation. The definition of the 1–5 scale was based on technical guidelines provided by [2], complemented by evidence from international transport planning literature.

Regarding Table 1, it is important to emphasize that (i) the 1–5 levels represent ordered performance categories rather than direct cardinal measurements; (ii) the additive aggregation model operates on marginal value functions derived through the FITradeoff elicitation procedure, not on the raw ordinal scores themselves; and (iii) the scaling assumptions are aligned with established MAVT modeling principles, reflecting a value-based transformation of performance levels instead of a literal numerical interpretation of the Likert scale.

The evaluations used to construct the consequence matrix were not directly elicited from a specific decision maker (DM). Instead, the assigned performance values were derived from a structured technical assessment grounded in the BNDES guidelines [2] and supported by national and international literature on public transport planning. These values were subsequently adapted to reflect the urban, operational, and demand characteristics of the municipality of Caruaru. Accordingly, the resulting matrix represents a reference scenario designed to illustrate the applicability and analytical potential of the adopted multicriteria method. In real-world implementations, however, the entries of the consequence matrix would correspond to the value judgments of a DM or a group of DMs regarding the performance of each alternative under each criterion Table 2.

Considering the objective of the decision problem, FITradeoff for choice was adopted, since the aim is to identify the most appropriate PCTS technology for the city.

After entering the data into the DSS, the criteria ordering stage was initiated according to the importance of their range of consequences. At this stage, DM is asked to express preferences between two fictitious alternatives, A and B, which present different performance levels with respect to two criteria. Once again, as the main purpose of this study was to illustrate the applicability of the proposed framework, real data were not employed and the preference judgments were simulated.

Subsequently, the ordering of the criteria was established and the preference elicitation process began using the decomposition procedure rather than a holistic elicitation approach. After four interactions, alternative A1 was identified as the most appropriate option according to the elicited preferences. A summary of the interactions, extracted from the DSS, is presented in Table 3.

Table 1. Description of decision criteria.

Criterion	Definition	Objective
Spatial Coverage of the Network/Demand (C1)	1. Very limited coverage, restricted to a single axis or specific area 2. Low coverage, serving few structuring corridors 3. Moderate coverage, serving a significant part of the urban area 4. Extensive coverage, with good territorial reach 5. Very extensive coverage, with high reach and adaptation to the urban fabric.	Maximize
Cost / Financial Feasibility (C2)	1. Very low financial viability; costs incompatible with the size of the municipality 2. Low viability; requires high continuous subsidies 3. Moderate viability; intermediate costs 4. High viability; costs compatible with the municipality's capacity 5. Very high viability; low costs and easy financial sustainability	Maximize
Operational Reliability (C3)	1. Highly unstable operation, subject to frequent interruptions 2. Low reliability, with high operational variability 3. Moderate reliability 4. High reliability, with few interferences 5. Very high reliability, with segregated and stable operation	Maximize
Operational Efficiency (C4)	1. Very low operational efficiency 2. Low efficiency 3. Moderate efficiency 4. High efficiency 5. Very high efficiency, with high productivity	Maximize
Demand Service Capacity (C5)	1. Very low capacity (low demand) 2. Low capacity 3. Medium capacity 4. High capacity 5. Very high capacity (high and continuous demand)	Maximize

(*continued*)

Table 1. *(continued)*

Criterion	Definition	Objective
User Attractiveness (C6)	1. Very low attractiveness 2. Low attractiveness 3. Moderate attractiveness 4. High attractiveness 5. Very high attractiveness	Maximize
Socio-environmental Externalities (C7)	1. Predominantly negative socio-environmental impacts 2. Significant negative impacts 3. Neutral or moderate impacts 4. Positive impacts 5. Highly positive impacts (low emissions, social inclusion)	Maximize
Flexibility in Response to Demand Variations (C8)	1. Very low flexibility 2. Low flexibility 3. Moderate flexibility 4. High flexibility 5. Very high flexibility	Maximize
Investments (C9)	1. Very low investment 2. Low investment 3. Medium investment 4. High investment 5. Very high investment	Minimize

Table 2. Consequence matrix.

Heading level	C1	C2	C3	C4	C5	C6	C7	C8	C9
A1	4	5	2	2	2	2	3	5	1
A2	4	4	3	3	3	3	3	4	2
A3	3	3	4	4	4	4	4	3	3
A4	2	2	5	4	3	4	5	2	4
A5	1	1	5	5	5	5	5	1	5

The FITradeoff DSS provides several graphs and tables for result analysis. One of these graphical outputs, presented in Fig. 2, allows the identification of the maximum and minimum values that the criteria weights may assume while maintaining the recommendation of alternative A1.

The graph presented in Fig. 2 also highlights the criteria for which the DM should seek greater precision regarding the weights and the evaluation of alternatives. For these criteria, any change may lead to the recommendation of a different alternative. To assess

Table 3. Summary of questions answered.

Cycle	Consequence A	Consequence B	Answer	Potentially optimal alternatives	Holistic Evaluation (HE) performed?
1	3 of C2	Best of C6 (5)	Consequence A	A1, A2, A3	no
2	3 of C2	Best of C4 (5)	Consequence A	A1, A2, A3	no
3	3 of C4	Best of C1 (5)	Consequence B	A1, A2, A3	no
4	3 of C1	Best of C5 (5)	Consequence A	A1	no

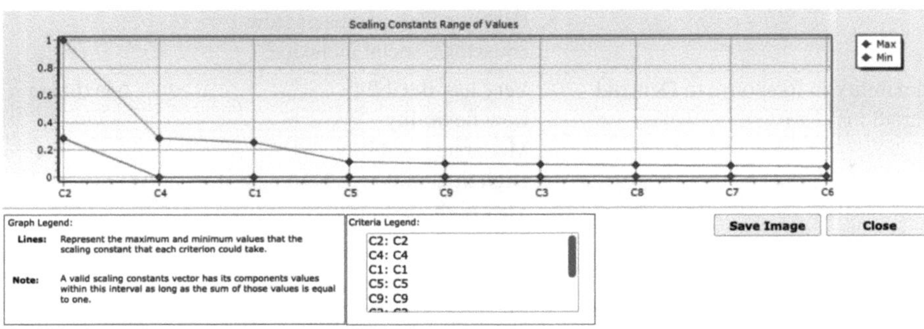

Fig. 2. Range of weights for final interaction recommending the alternative selected

this effect, a sensitivity analysis is conducted using the FITradeoff DSS, in which performance values (consequence) are varied while the elicited intercriteria preference information is kept fixed. For discrete criteria, variations occur between adjacent scale levels. The decision-maker defines which criteria will vary and the allowable bounds of variation, and the system performs 1,000 Monte Carlo simulations, solving a new linear programming problem in each run. The results indicate how often each alternative remains optimal or admissible, thereby providing a measure of decision robustness under uncertainty. For instance, Fig. 3 presents the results of a sensitivity analysis for a scenario in which possible maximum and minimum variations, considering the adopted scale for each criterion, are defined for the evaluation of the alternatives.

In this sensitivity analysis example, alternative A2 becomes a potentially optimal option in 67% of the iterations, while alternative A3 is identified as potentially optimal in 9.8% of the iterations.

The obtained results and the analyses provided by the FITradeoff method demonstrate that the problem under study can be addressed in a more structured manner, consistent with stated preferences, and in a transparent and flexible way. Therefore, the results should not be interpreted as a definitive recommendation for the municipality, but rather as an illustrative outcome that highlights the potential of the proposed decision-support framework. Future applications should involve the direct participation of public decision

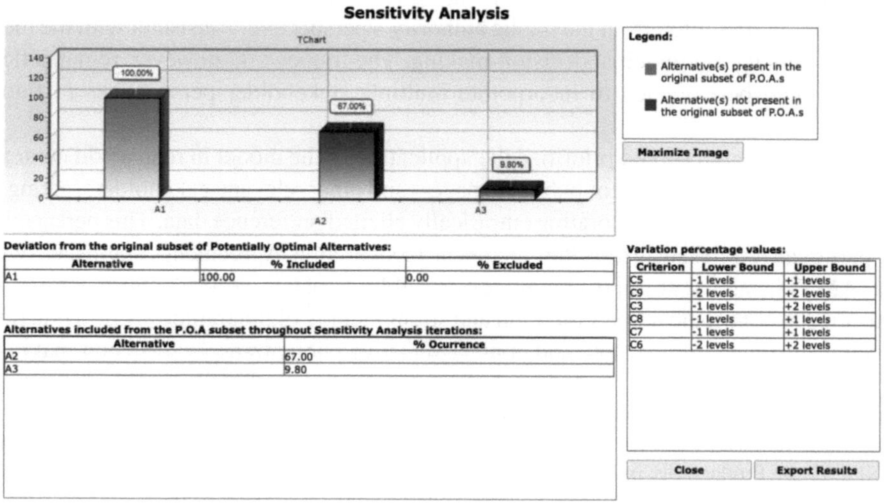

Fig. 3. Sensitivity analysis in a specific scenario

makers, enabling the elicitation of preferences and the validation of the framework under real decision-making conditions.

5 Conclusions

This study addressed the problem of selecting an appropriate PCTS technology for Caruaru, a medium-sized Brazilian city, proposing a structured multicriteria decision-support framework. By integrating technical guidelines from the BNDES with the FITradeoff method, the research demonstrates how complex transport planning decisions – characterized by multiple, often conflicting criteria – can be systematically organized and analyzed. The framework allows decision makers to explicitly consider technical, operational, economic, and socio-environmental aspects while accounting for local urban and demand conditions.

The illustrative application to the municipality of Caruaru demonstrated the model's ability to structure criteria, organize performance assessments, and support the comparison of alternatives under defined assumptions. By constructing a reference consequence matrix and applying FITradeoff for preference modeling and choice, the study identified potentially admissible alternatives and evaluated their robustness through sensitivity analysis. The results are not intended as a definitive recommendation, since the application was methodological and based on simulated evaluations and preferences. Instead, the findings emphasize the analytical robustness, transparency, and adaptability of the framework, as well as its capacity to clarify trade-offs within an additive value model.

Methodologically, the study confirms the suitability of FITradeoff for public-sector decisions characterized by incomplete or imprecise preference information. The partial elicitation procedure reduces cognitive burden while preserving consistency with MAVT principles, and the decision support system enhances result interpretation. Although transport mode selection is inherently multi-stakeholder and politically sensitive, the criteria were structured from the public authority's perspective, consistent with the institutional context of municipal decision-making. The framework, however, remains flexible and can be extended to incorporate multiple stakeholder perspectives in future applications.

Future research should prioritize the application of the model in real-world contexts with the direct involvement of public managers and other relevant stakeholders, acting as decision-makers and incorporating empirically elicited preference data. This perspective is particularly relevant given the Technical Cooperation Agreement established with the AMTTC, aimed at improving the city's public transport services. Such efforts are expected to contribute to the validation and strengthening of the framework as a practical tool for informed, transparent, and context-sensitive public transport decision making.

Acknowledgements. This study was financed in part by the Coordenação de Aperfeiçoamento de Pessoal de Nível Superior - Brasil (CAPES) – Finance Code 001. The second author thanks CNPq for the Productivity in Research Fellowship (Grant no. 311643/2023-0).

References

1. National Association of Urban Transport Companies NTU Yearbook of Urban Public Transport (2025)
2. Banco Nacional de Desenvolvimento Econômico e Social (BNDES) Guia para seleção de tecnologias e implementação de projetos de transporte público coletivo (2017)
3. De Almeida, A.T., De Almeida, J.A., Costa, A.P.C.S., De Almeida-Filho, A.T.: A new method for elicitation of criteria weights in additive models: flexible and interactive tradeoff. Eur. J. Oper. Res. **250**, 164–170 (2016). https://doi.org/10.1016/j.ejor.2015.08.058
4. de Almeida, A.T., Frej, E.A., Roselli, L.R.P.: Combining holistic and decomposition paradigms in preference modeling with the flexibility of FITradeoff. Cent. Eur. J. Oper. Res. **29**, 7–47 (2021). https://doi.org/10.1007/s10100-020-00728-z
5. Pavelski, L.M.: Gestão de sistemas de transporte público na atualidade, 1st edn. InterSaberes (2021)
6. Echechipia, R., Abdal, A.: Os dez anos da Política Nacional de Mobilidade Urbana. FGV RIC Revista de Iniciação Científica. **4**, 85969 (2023)
7. Tavares Da Silva RR, Felipe C, Loureiro G: CARACTERIZAÇÃO E DIAGNÓSTICO DA PROBLEMÁTICA DA MUNICIPALIZAÇÃO DO TRÂNSITO NO BRASIL (2009)
8. Cidade, R.B., de A, L.J.T.M.: O Direito ao Transporte como Direito Fundamental Social. Revista de Direito Urbanístico, Cidade e Alteridade. **2**, 196–216 (2016). https://doi.org/10.26668/indexlawjournals/2525-989x/2016.v2i1.509
9. Department for Transport: Annual Report and Accounts 2024–2025, London (2025)
10. Zhang, Y., Cheng, L.: The role of transport infrastructure in economic growth: empirical evidence in the UK. Transp. Policy (Oxf). **133**, 223–233 (2023). https://doi.org/10.1016/j.tranpol.2023.01.017
11. Ministère des Transports: Ambition France Transports: Financer l'avenir des mobilités. Rapport de la conférence Ambition France Transports, July (2025)
12. Observatório Nacional de Transporte e Logística ANUÁRIO ESTATÍSTICO DE TRANSPORTES 2015–2024 (2025)
13. De Almeida, A.T.: Processo de decisão nas organizações: construindo modelos de decisão multicritério. Atlas (2013)

14. Ferreira, D.V., de Gusmão, A.P.H.: Maturity assessment in the context of industry 4.0 - an application using FITradeoff method in a textile industry. In: Decision Support Systems XII: Decision Support Addressing Modern Industry, Business, and Societal Needs Lecture Notes in Business Information Processing (2022)
15. Vieira Dantas, N., Henriques de Gusmão, A.P., Teixeira Gonzaga, M.: A multicriteria model to support decisions regarding data protection compliance. J. Decis. Syst. **33**(4), 744–760 (2023). https://doi.org/10.1080/12460125.2023.2263683
16. Ferreira, D.V., de Oliveira, V., Ferreira, T.V., Henriques de Gusmão, A.P.: Decision model for selecting a portfolio of non-technical loss mitigation projects for power distribution systems. Util. Policy. **95**, 101967 (2025). https://doi.org/10.1016/j.jup.2025.101967
17. Biswas, T., Ishizaka, A., Majumder, A., et al.: The PROMETHEE-GAIA: A multi-criteria decision-making method for identifying best conservation agricultural practices. Soil Tillage Res. **245**, 106315 (2025). https://doi.org/10.1016/j.still.2024.106315
18. Pomerol, J.C., Roy, B., Rosenthal-Sabroux, C.: Developing an "intelligent" DSS for the multicriteria evaluation of railway timetables: problems and issues. J. Decis. Syst. **5**, 249–267 (1996). https://doi.org/10.1080/12460125.1996.10511691
19. Pessoa, M.E.B.T., Frej, E.A., de Almeida, A.T.: A novel information-driven approach for selecting a project portfolio: A benefit-to-cost based method combining decomposition and holistic elicitation paradigms. Omega (United Kingdom). **139**, 103456 (2026). https://doi.org/10.1016/j.omega.2025.103456
20. Saaty, T.L.: How to make a decision: the analytic hierarchy process. Eur. J. Oper. Res. **48**, 9–26 (1990). https://doi.org/10.1016/0377-2217(90)90057-I
21. Edwards, W., Barron, F.H.: Smarts and smarter: improved simple methods for multiattribute utility measurement. Organ. Behav. Hum. Decis. Process. **60**, 306–325 (1994). https://doi.org/10.1006/obhd.1994.1087
22. Barron, F.H., Barrett, B.E.: Decision quality using ranked attribute weights. Manag. Sci. **42**, 1515–1523 (1996). https://doi.org/10.1287/mnsc.42.11.1515
23. Greco, S., Figueira, J.R., Ehrgott, M.: Multiple Criteria Decision Analysis, vol. 233. Springer, New York (2016)
24. Roy, B.: The outranking approach and the foundations of electre methods. Theor. Decis. **31**, 49–73 (1991). https://doi.org/10.1007/BF00134132
25. Brans, J.P., Vincke, P., Mareschal, B.: How to select and how to rank projects: the Promethee method. Eur. J. Oper. Res. **24**, 228–238 (1986). https://doi.org/10.1016/0377-2217(86)90044-5
26. Frej, E.A., de Almeida, A.T., Costa, A.P.C.S.: Using data visualization for ranking alternatives with partial information and interactive tradeoff elicitation. Oper. Res. **19**, 909–933 (2019). https://doi.org/10.1007/s12351-018-00444-2
27. de Almeida, A.T., Frej, E.A., Roselli, L.R.P., Costa, A.P.C.S.: A summary on FITRADEOFF method with methodological and practical developments and future perspectives. Pesquisa Operacional. **43**, e268356 (2023). https://doi.org/10.1590/0101-7438.2023.043spe1.00268356
28. De Almeida, A.T., Frej, E.A., Almeida, J.A., et al.: FITradeoff-Web. 2021 (2021)
29. Likert, R.: A technique for the measurement of attitudes. Arch. Psychol. **140**, 1–55 (1932)

Academic Regulation Alignment in Transnational Education Partnerships: An Application of Multi-objective Optimisation Models

Yu-Ling Lin[1(✉)], Rui Yang[2], and Salem Chakhar[2,3]

[1] Goldsmiths, University of London, London, UK
Yuling.Lin@gold.ac.uk
[2] University of Portsmouth, Portsmouth, UK
{rui.yang,salem.chakhar}@port.ac.uk
[3] Centre of Operational Research and Decision Analysis, University of Portsmouth, Portsmouth, UK

Abstract. Transnational Education (TNE) partnerships, particularly dual-degree and joint programmes, require close alignment of academic regulations across institutions. Yet partners often operate under different governance systems—UK universities follow OfS and QAA standards, while Malaysian institutions comply with MQA requirements—creating regulatory friction that can delay programme approval and quality assurance. This study introduces a multi-objective optimisation model that reconciles such discrepancies by integrating regulatory clauses into a unified governance framework. Core academic criteria (curriculum design, assessment, staff qualifications, and quality assurance) are treated as decision variables weighted by institutional utility. Using multi-criteria decision analysis (MCDA) with linear and nonlinear optimisation, the model maximises total utility while ensuring compliance with both national systems. Two scenarios, single- and dual-institution recruitment, demonstrate how power distribution and decision autonomy shape optimal alignment. The model offers a quantitative decision-support tool that enhances transparency, efficiency, and fairness in cross-border academic governance.

Keywords: Transnational Education · Academic Regulation Alignment · Multi-Objective Optimisation · Quality Assurance Frameworks · Higher Education Governance

1 Introduction

Transnational Education (TNE) has become a defining feature of global higher education, encompassing diverse forms such as branch campuses, joint degrees, and dual-award partnerships [15,35]. Its expansion reflects both institutional strategies for internationalisation and the growing demand for cross-border access to higher education [1]. Yet, the governance of TNE partnerships is increasingly complex, requiring institutions to navigate overlapping and sometimes contradictory regulatory systems [9,11]. These

A. P. Cabral Seixas Costa et al. (Eds.): ICDSST 2026, LNBIP 582, pp. 380–394, 2026.
https://doi.org/10.1007/978-3-032-25146-6_25

challenges extend beyond administrative coordination, raising questions of legitimacy, sustainability, and equity in global partnerships [12,32].

Differences between national frameworks are particularly evident in curriculum design, assessment standards, staff qualifications, and programme validation. For example, UK universities operate under the Office for Students (OfS) and the Quality Assurance Agency (QAA), which emphasise outcomes-based accountability and institutional autonomy, while Malaysia's Malaysian Qualifications Agency (MQA) applies a more prescriptive, compliance-oriented approach [18,25,27]. These structural asymmetries create measurable friction in programme approval, recognition, and quality assurance [5,29]. Recent studies suggest that such frictions follow identifiable patterns, with regulatory clauses categorised as identical, similar, contradictory, or unique across partner institutions, providing a foundation for quantitative approaches to alignment [1].

At the same time, higher education governance is becoming more complex, involving multiple stakeholders and requiring new models of coordination [23,31]. Optimisation and decision-science methods offer promising tools to reconcile competing objectives. Multi-objective optimisation has been applied in related fields to balance efficiency and fairness [3,14], while Multi-Criteria Decision Analysis (MCDA) has proven effective in structuring decisions with multiple stakeholders [2,19]. Extending these approaches to TNE governance provides a novel way to integrate institutional preferences, bargaining power, and compliance thresholds into a unified framework.

The primary stakeholders of this research include university senior management, TNE partnership directors, quality assurance professionals, programme leaders, and national regulators. These actors are responsible for negotiating programme structures, ensuring compliance with multiple regulatory frameworks, and maintaining academic standards across jurisdictions. However, decision-making in TNE partnerships is often characterised by informal negotiation processes, fragmented governance structures, and limited analytical support tools, which can lead to delays in programme approval and inconsistencies in regulatory interpretation. Prior research has highlighted the growing complexity of cross-border higher education governance and the need for more systematic decision-support approaches to manage competing institutional and regulatory priorities [9,11,12]. By providing a quantitative optimisation framework, this study offers stakeholders a structured mechanism to evaluate trade-offs, anticipate regulatory conflicts, and identify mutually acceptable alignment strategies. Such an approach has the potential to improve transparency, reduce negotiation uncertainty, and support evidence-based decision-making in international higher education partnerships [2,20].

Against this backdrop, the present study develops an optimisation-based decision model to address regulatory misalignment in TNE partnerships. By conceptualising academic criteria as decision variables, regulatory requirements as constraints, and institutional preferences as utility functions, the model seeks to maximise joint utility while ensuring compliance with both national systems. This approach contributes to ongoing debates on how to balance efficiency, fairness, and sustainability in cross-border higher education governance [12,32].

The focus on UK and Malaysian regulatory systems serves as an illustrative analytical case rather than a limitation of the model's applicability. These two jurisdictions represent contrasting governance paradigms within transnational education. The

UK regulatory framework emphasises institutional autonomy, outcomes-based quality assurance, and risk-based oversight, whereas the Malaysian system adopts a more prescriptive and compliance-oriented approach to programme accreditation and delivery [18,25]. This contrast provides a theoretically meaningful context for examining regulatory misalignment and negotiation dynamics. Moreover, both countries are highly active in transnational education partnerships, particularly in dual-degree and validation arrangements, making them relevant empirical contexts for modelling regulatory alignment challenges [6,33]. Importantly, the optimisation framework developed in this study is country-agnostic and can be extended to other national systems by modifying regulatory thresholds and institutional parameters.

The remainder of this paper is structured as follows. Section 2 reviews the existing literature on transnational education governance, regulatory complexity, and optimisation approaches. Section 3 presents the conceptual framework and formal optimisation model, including the definition of decision variables, parameters, and the linearised formulation. Section 4 outlines the empirical implementation plan, detailing the extraction of regulatory minima, stakeholder elicitation of weights, and calibration procedures . Section 5 discusses the limitations of the proposed model and potential extensions to multi-institutional or dynamic contexts. Finally, Sect. 6 concludes by highlighting the theoretical contributions and practical implications for policy and institutional decision-making in transnational education partnerships.

2 Literature Review

2.1 Transnational Education and Regulatory Complexity

Transnational education (TNE) has become a core strategy in global higher education, encompassing branch campuses, dual-degree, and validation partnerships [15,35]. However, the governance of TNE partnerships is often complicated by overlapping and sometimes contradictory regulatory systems [5,9]. Institutions must adhere simultaneously to their home country's quality assurance frameworks, making it challenging to maintain coherence and academic integrity across jurisdictions [11].

Differences between national frameworks are most visible in assessment regulations, curriculum mapping, staff qualifications, and programme validation standards. For instance, the UK's Office for Students (OfS) and the Quality Assurance Agency (QAA) emphasise outcomes-based accountability, institutional autonomy, and evidence-led evaluation [25,27]. In contrast, the Malaysian Qualifications Agency (MQA) applies a more prescriptive, compliance-oriented approach, focusing on workload ratios, accreditation procedures, and curriculum uniformity [18]. These structural asymmetries between autonomy-driven and compliance-driven systems have created measurable regulatory friction that affects programme approval and recognition [29]. Recent studies indicate that these frictions follow identifiable patterns [1]. Regulatory clauses can be categorised as identical, similar, contradictory, or unique across partner institutions. This typology provides a foundation for quantitative approaches to align regulatory frameworks, forming the basis for the optimisation model developed in this research.

2.2 Partnership Dynamics and Institutional Bargaining Power

Transnational partnerships are shaped by power asymmetries and strategic dependencies [9, 15]. Power imbalances can arise from differences in institutional reputation, control over student recruitment, and national regulatory leverage. Studies show that the "dominant" partner often dictates quality control, branding, and programme structure, while the local partner assumes operational responsibilities [34, 36].

In decision science, such asymmetries are conceptualised as differences in bargaining power or preference weighting. The authors in [4, 14] formalised these dynamics through multi-objective optimisation, demonstrating how trade-offs between efficiency (total utility) and fairness (equity between parties) affect feasible solutions. Their work shows that fairness constraints often entail an efficiency loss, the "price of fairness" [3] indicates a principle directly applicable to TNE, where regulatory alignment (fairness) may reduce administrative efficiency (utility). This supports modelling institutional preferences and influence as explicit parameters within an optimisation framework, capturing how each partner's power shapes the resulting agreement structure.

2.3 Multi-criteria Decision Analysis (MCDA) in Higher Education

Multi-Criteria Decision Analysis (MCDA) effectively structures decisions involving multiple stakeholders and conflicting criteria [2, 19]. It translates qualitative judgments into quantitative importance weights, enabling rational evaluation of trade-offs [28]. Within higher education, MCDA has been applied to institutional ranking, resource allocation, and curriculum design [13]. MCDA enables compromise in multi-stakeholder decisions. The authors in [2] showed that integrating MCDA into group decision support helps reconcile divergent preferences. Similarly, the authors in [30] used MCDA to balance energy efficiency and health outcomes in UK schools, demonstrating its flexibility in contexts with multiple policy objectives. These applications justify using MCDA to elicit institutional weights (s_i) and regulatory priorities (w_i^A, w_i^B) in TNE partnership negotiations.

2.4 Optimisation and Multi-objective Decision Frameworks

Optimisation models identify efficient solutions when objectives conflict [22]. Combining MCDA with optimisation allows both qualitative engagement and quantitative precision [21]. Recent research demonstrates its effectiveness in education-related contexts [7, 17]. For instance, the authors in [17] developed a multi-objective model for educational resource allocation under equity constraints, showing how varying fairness altered Pareto-efficient frontiers. Similarly, the authors in [37] proposed a negotiation-based multi-party optimisation model to resolve stakeholder conflicts, illustrating the relevance of iterative optimisation to partnership management. These studies provide the methodological grounding for this research, which extends the multi-objective paradigm to regulatory negotiation in TNE by integrating institutional preferences, bargaining weights, and compliance thresholds.

2.5 Research Gap

Despite extensive work on internationalisation and partnership management, few studies have modelled the alignment of academic regulations quantitatively. The TNE literature has focused on descriptive governance frameworks [9, 11], while optimisation and decision-science approaches remain underexplored. This study bridges that gap by integrating regulatory alignment, institutional utility, and power asymmetry into a single optimisation framework. It links higher education management and operational research to reconcile national regulatory frameworks and institutional objectives in TNE partnerships.

3 Conceptual Framework

3.1 General Overview

Figure 1 illustrates the conceptual framework underpinning the proposed optimisation model for aligning academic regulations in transnational education (TNE) partnerships. The framework integrates regulatory inputs from two institutions—typically a UK university governed by the Office for Students (OfS) and the Quality Assurance Agency (QAA), and a Malaysian partner regulated by the Malaysian Qualifications Agency (MQA). Each institution's regulatory system contributes a set of extracted criteria (r_i^A and r_i^B) covering dimensions such as curriculum design, assessment standards, quality assurance mechanisms, and staff qualifications. These criteria are processed within a multi-objective optimisation model that balances institutional utility, compliance requirements, and misalignment penalties.

The optimisation process yields an equilibrium solution (a^*, b^*) representing the jointly optimal configuration of academic regulations that satisfies both partners' minimum thresholds while maximising shared utility. The output is a unified academic governance framework that supports regulatory coherence, transparency, and fairness in TNE programme delivery.

3.2 Optimisation Model

The optimisation model is designed to capture the negotiation dynamics between two universities engaged in a transnational education (TNE) partnership. At its core, the model balances institutional preferences, regulatory requirements, and bargaining power to identify feasible levels of alignment across multiple academic criteria. Let there be n regulatory criteria indexed by $i = 1, \ldots, n$. Examples include curriculum scope, assessment weighting, minimum staff qualification, credit volume, and moderation procedures. Each criterion represents a dimension of regulatory alignment that must be negotiated between the two institutions.

Decision Variables: The choices made by each university are represented by decision variables:

- $a_i \in [0, 1]$: the level of alignment chosen by University A on criterion i.
- $b_i \in [0, 1]$: the level of alignment chosen by University B on criterion i.

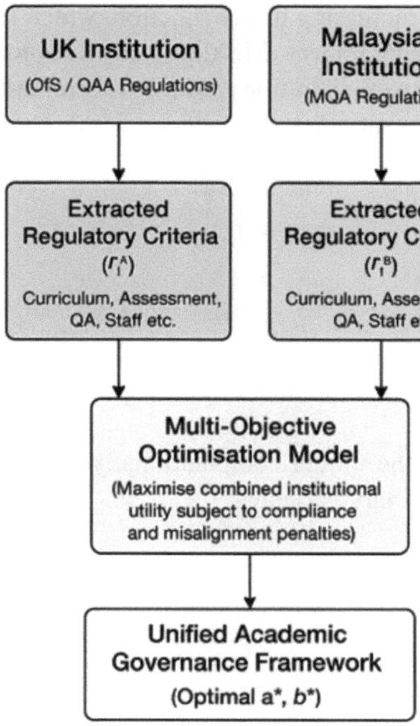

Fig. 1. Conceptual framework.

A value of 0 corresponds to non-alignment, while 1 indicates full alignment with the regulatory standard. Intermediate values capture partial alignment or compromise solutions.

Parameters: To reflect the complexity of decision-making, several parameters are introduced:

- $s_i \geq 0$: the intrinsic importance of criterion i, reflecting its weight in the overall partnership.
- $w_i^A, w_i^B \geq 0$: institutional weights, representing how strongly each university values criterion i relative to others.
- p_A, p_B: bargaining power coefficients, with $p_B = 1 - p_A$, capturing asymmetries in institutional influence.
- $\gamma_i \geq 0$: misalignment penalties, quantifying the cost of divergence between a_i and b_i.
- $r_i^A, r_i^B \in [0, 1]$: minimum regulatory requirements imposed by national authorities, ensuring compliance.

Objective Function: The model seeks to maximise joint utility, defined as the weighted sum of each university's satisfaction with its chosen levels of alignment, minus the penalties for misalignment:

$$\max_{a,b} U(a,b) = p_A \sum_{i=1}^{n} s_i w_i^A f(a_i) + p_B \sum_{i=1}^{n} s_i w_i^B f(b_i) - \sum_{i=1}^{n} \gamma_i |a_i - b_i| \quad (1)$$

Here, $f(\cdot)$ is a monotone increasing utility function, which may be linear (identity) or concave to reflect diminishing returns. Misalignment is captured by $\Delta_i = |a_i - b_i|$, ensuring that divergence between institutions reduces overall utility.

Constraints: The decision variables are bounded within the feasible space:

$$a_i \in [0, 1] \tag{2}$$

$$b_i \in [0, 1] \tag{3}$$

$$a_i \geq r_i^A \tag{4}$$

$$b_i \geq r_i^B \tag{5}$$

This ensures that both universities meet their respective regulatory minima while negotiating alignment levels.

Linearisation: To make the model computationally tractable, the absolute value terms are linearised using auxiliary variables:

$$z_i \geq a_i - b_i \tag{6}$$

$$z_i \geq b_i - a_i \tag{7}$$

$$z_i \geq 0 \tag{8}$$

$$\Delta_i = z_i \tag{9}$$

Final LP Formulation: The resulting linear programme is expressed as:

$$\max_{a,b,z} \; p_A \sum_{i=1}^{n} s_i w_i^A a_i + p_B \sum_{i=1}^{n} s_i w_i^B b_i - \sum_{i=1}^{n} \gamma_i z_i \tag{10}$$

Subject to:

$$z_i \geq a_i - b_i \tag{11}$$

$$z_i \geq b_i - a_i \tag{12}$$

$$r_i^A \leq a_i \leq 1 \tag{13}$$

$$r_i^B \leq b_i \leq 1 \tag{14}$$

Additional coupling constraints (e.g., staff qualification ratios, credit loads, and budgetary limits) can be incorporated linearly, allowing the model to reflect more complex institutional interdependencies. In this way, the optimisation framework provides a flexible yet rigorous tool for analysing regulatory alignment in TNE partnerships.

3.3 Managerial Interpretation of Decision Variables

From a practical perspective, the decision variables ai and bi represent negotiable policy levels that programme leaders and quality assurance managers can adjust during partnership discussions. For instance, ai may correspond to the proportion of locally delivered teaching, moderation intensity, or staff qualification thresholds. The optimisation results therefore provide decision-makers with recommended alignment levels that maximise joint institutional benefit while ensuring regulatory compliance.

3.4 Parameter Elicitation

The effectiveness of the optimisation model depends critically on the accurate elicitation of parameters. These parameters translate qualitative judgments and regulatory requirements into quantitative inputs, ensuring that the model reflects both institutional priorities and external compliance obligations. The elicitation process combines structured expert consultation, interpretive rubrics, and normalisation procedures to guarantee consistency and comparability across criteria.

- **Importance weights s_i:** Each criterion i carries a different level of intrinsic importance depending on its role in programme delivery and regulatory oversight. Importance weights are elicited through stakeholder surveys and Delphi panels, where experts rank or score criteria according to perceived significance. To ensure comparability, the weights are normalised so that $\sum_i s_i = 1$, reflecting a complete distribution of importance across all criteria [10,26].
- **Institutional weights w_i^A, w_i^B:** While s_i captures the general importance of a criterion, institutional weights reflect how strongly each university values that criterion relative to others. These are elicited through pairwise comparisons or structured Delphi rounds, allowing institutions to express preferences in a systematic way. Normalisation ensures that $\sum_i w_i^A = \sum_i w_i^B = 1$, so each institution's priorities form a coherent weighting scheme [19].
- **Bargaining Power p_A, p_B:** Bargaining power coefficients capture asymmetries in institutional influence. They can be derived from observable indicators such as student recruitment shares, revenue contributions, or composite multi-factor indices. For example, in single-institution recruitment scenarios, $p_A = 1.0$ and $p_B = 0.0$, while joint recruitment scenarios may allocate bargaining power proportionally to student or revenue shares. These coefficients ensure that the optimisation model reflects negotiation dynamics rather than assuming equal influence [12,32].
- **Misalignment penalties γ_i:** Not all criteria carry the same cost of divergence. For legally constrained items (e.g., minimum staff qualifications or credit volume), penalties are set higher to reflect the risk of regulatory non-compliance. For negotiable items (e.g., assessment weighting), penalties may be lower, allowing greater flexibility. These values are elicited through expert judgment and validated through sensitivity analysis [7].
- **Minimum regulatory requirements r_i^A, r_i^B:** Each institution must satisfy its national regulator's minimum thresholds. These requirements are mapped from formal regulator documents (e.g., OfS/QAA for the UK, MQA for Malaysia) to a $[0, 1]$ scale, ensuring that the optimisation model respects compliance boundaries while allowing negotiation above the minima [18,25,27].

Together, these elicited parameters provide the foundation for the optimisation model. By systematically translating qualitative judgments into quantitative inputs, the model achieves both methodological rigour and practical relevance in guiding TNE partnership negotiations.

3.5 Pareto Frontier Analysis

As illustrated in Fig. 2, the Pareto frontier demonstrates in detail how varying bargaining weights (p_A, p_B) influence the trade-off between regulatory alignment and total utility. The diagram plots three distinct curves corresponding to different bargaining scenarios: full dominance by University A ($p_A = 1.0$), equal bargaining power between partners ($p_A = 0.5$), and an intermediate case. Each curve shows that as regulatory alignment increases (i.e., misalignment penalties γ_i decrease and Δ_i approaches zero), total utility $U(a, b)$ tends to decline across all scenarios. This decline reflects the efficiency loss associated with enforcing fairness and compliance across both regulatory systems. The dominant partner scenario yields the highest utility, since the institution with greater bargaining power can prioritise its preferences with minimal compromise. By contrast, equal bargaining results in lower overall utility but promotes a more equitable distribution of influence, ensuring that both partners' regulatory requirements are respected. The intermediate case illustrates how partial bargaining power produces outcomes that balance efficiency and fairness, lying between the extremes of dominance and equality. Taken together, the figure complements the optimisation model by visually highlighting how shifts in bargaining power reshape the feasible solution space, clarifying the efficiency—equity dynamics central to transnational education governance. It provides stakeholders with a clear representation of the "price of fairness" in regulatory alignment, showing how institutional utility is traded off against compliance and partnership equity.

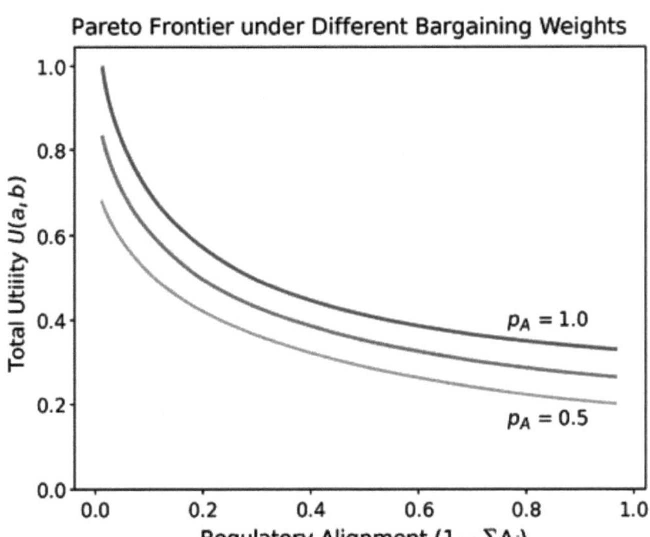

Fig. 2. Implementation workflow.

4 Empirical Implementation Plan

The empirical implementation involves several integrated steps to calibrate and validate the optimisation model. First, regulatory minima r_i^A and r_i^B are systematically extracted from official regulator manuals, ensuring that the model reflects the formal compliance thresholds of both partner countries [18,25,27]. This step guarantees that the optimisation framework is grounded in formal policy requirements rather than informal practice.

Second, Delphi panels comprising TNE managers, quality assurance officers, and curriculum leaders from both institutions are convened to elicit the importance weights s_i, institutional priorities w_i^A and w_i^B, and misalignment penalties γ_i. The Delphi technique is widely recognised for its ability to synthesise expert judgment and achieve consensus in complex decision contexts [10,26]. These expert evaluations are mapped to numerical scales using an interpretive rubric, with inter-rater reliability assessed to ensure consistency and validity of the inputs [19].

Third, the model is calibrated by solving the baseline linear program (P1) to generate initial optimal levels a^* and b^* for each regulatory criterion. Calibration ensures that the optimisation outputs are not only mathematically valid but also practically interpretable in the context of institutional governance. Results are iteratively reviewed in stakeholder workshops, where bargaining weights and other parameters can be adjusted to reflect negotiation dynamics or revised institutional priorities [12,32]. These workshops provide a participatory mechanism for embedding institutional legitimacy and transparency into the modelling process.

Forth, to demonstrate the practical application of the optimisation framework, consider a dual-degree partnership between a UK university and a Malaysian institution negotiating assessment weighting requirements. The UK partner requires at least 40% coursework assessment ($r^A = 0.4$), while the Malaysian regulator mandates 60% final examination weighting ($r^B = 0.6$). By incorporating institutional priorities and bargaining weights into the optimisation model, the solution identifies an equilibrium configuration that satisfies both regulatory minima while minimising misalignment penalties. In practice, this result translates into a blended assessment structure combining coursework, examination, and moderated components, providing a feasible governance specification that both institutions can adopt during programme validation.

Finally, the outcomes undergo a comprehensive validation, which integrates the calibrated model outputs with sensitivity analyses, alignment scenarios, and negotiation results. Sensitivity analysis is particularly important in multi-objective optimisation, as it reveals how changes in weights or penalties affect Pareto-efficient solutions [7,17]. Validation also includes scenario testing to explore alternative recruitment and revenue-sharing arrangements, ensuring that the model remains robust under different partnership configurations. This multi-stage validation provides actionable insights for achieving a unified regulatory specification while demonstrating the practical applicability and resilience of the optimisation framework in real-world TNE partnerships [6,24,33].

5 Discussion

5.1 Summary of Contributions

The optimisation model developed in this study offers several important contributions to the governance of TNE partnerships. By integrating regulatory requirements, institutional preferences, and bargaining power into a unified framework, the model provides a structured and transparent approach to reconciling cross-border academic regulations. This quantitative decision-support tool enhances efficiency by reducing delays in programme approval and validation, while simultaneously promoting fairness through explicit modelling of power asymmetries and misalignment penalties. In doing so, it advances the literature on international higher education governance, which has often been criticised for relying on descriptive or qualitative frameworks without offering operationalisable solutions [1,9,11]. The model also demonstrates the applicability of optimisation and MCDA in higher education contexts, extending prior work on resource allocation and institutional ranking [2,17,19].

The contributions of this study are demonstrated through the scenario analysis and Pareto-frontier results, which quantitatively illustrate how variations in bargaining power influence regulatory alignment outcomes. Unlike existing descriptive governance frameworks in transnational education research, the proposed optimisation model operationalises institutional negotiation dynamics by explicitly integrating regulatory constraints, institutional preferences, and misalignment penalties into a unified decision framework. This enables the identification of efficient and equitable alignment configurations under different partnership conditions. In particular, the Pareto-frontier analysis provides insight into the efficiency—equity trade-off, illustrating the measurable "price of fairness" associated with enforcing balanced regulatory compliance between institutions, thereby extending theoretical concepts from optimisation literature into the higher education governance domain [3,14]. Furthermore, while MCDA has been applied in higher education contexts such as institutional evaluation and resource allocation [2,20], its integration with optimisation modelling for regulatory negotiation remains underexplored. By combining MCDA-based weight elicitation with multi-objective optimisation, this study advances methodological approaches for analysing complex governance decisions involving multiple stakeholders and conflicting objectives. This contributes to both the transnational education literature and the broader field of decision sciences by demonstrating how quantitative modelling can support policy and managerial decision-making in international higher education partnerships [8,16].

5.2 Limitations and Extensions

Despite these strengths, the model has several limitations and opportunities for extension. First, the assumption of linear utility functions may oversimplify stakeholders' preferences; incorporating concave or other nonlinear functions could better capture diminishing returns and heterogeneous institutional priorities [21,22]. Second, the elicited weights and penalties rely heavily on expert judgment, which introduces subjectivity; robustness checks, multiple elicitation rounds, and triangulation with empirical data can enhance reliability [28,30]. Third, regulatory frameworks evolve over time, and static optimisation may fail to capture temporal changes in compliance requirements. Dynamic or rolling-horizon models could address this limitation by updating constraints as regulations shift [23,31]. Fourth, the current formulation is limited to bilateral partnerships; extending the model to multi-institutional or multi-country networks would require additional decision vectors and misalignment penalties, reflecting the complexity of global higher education consortia [12,32]. Finally, strategic behaviour by partners (such as misreporting preferences or manipulating bargaining weights) remains a challenge. Incorporating mechanism design principles could ensure incentive compatibility and mitigate potential manipulation, aligning with broader debates on fairness and efficiency in optimisation [3,14].

5.3 Policy Implications

The findings of this study carry several implications for policymakers, regulators, and institutional leaders engaged in transnational education. First, the optimisation framework offers regulators a transparent tool for identifying where compliance thresholds can be harmonised without undermining institutional autonomy, supporting evidence-based policy design [6,24]. Second, by explicitly modelling bargaining power, the framework highlights how recruitment strategies and revenue-sharing arrangements shape governance outcomes. This insight can inform the design of partnership agreements that balance efficiency with fairness, reducing the risk of dominance by a single institution [12,33]. Third, the Pareto-frontier analysis provides a visual mechanism for communicating trade-offs to stakeholders, enabling more inclusive and participatory decision-making processes. Such visualisations can be used in negotiations to demonstrate the "price of fairness" and to justify regulatory compromises in a transparent manner [32]. Finally, the model underscores the importance of dynamic regulatory adaptation. As global education landscapes evolve, regulators may use optimisation-based simulations to anticipate the impact of new compliance standards, ensuring that policies remain responsive and sustainable [31].

In operational terms, the framework can be embedded within institutional programme approval workflows. During early partnership negotiations, stakeholders can input regulatory requirements and institutional priorities into the model to simulate feasible alignment scenarios before formal validation begins. This can reduce approval delays, clarify negotiation boundaries, and support evidence-based decision-making during quality assurance reviews.

6 Conclusion

The optimisation model developed in this study illuminates the delicate balance between institutional fairness and regulatory efficiency in transnational education (TNE) partnerships. By formalising the negotiation process through decision variables, bargaining weights, and misalignment penalties, the framework provides a structured lens for understanding how institutional preferences and regulatory constraints interact. In Scenario 1, reflecting single-institution recruitment, administrative coherence and efficiency are prioritised, whereas Scenario 2, with dual-institution recruitment, promotes shared ownership and a more equitable distribution of decision-making influence. These contrasting scenarios highlight the inherent trade-offs between efficiency and fairness that characterise cross-border governance arrangements [15,32].

By formalising negotiation through specific bargaining weights and misalignment penalties, our framework ensures that TNE partnerships remain sustainable and equitable. The application of this 'Fairness Weighting' provides a structured path toward ethical alignment, ensuring that global education partnerships are mutually beneficial and responsive to the evolving regulatory landscapes of both host and home nations.

Pareto-frontier analysis further strengthens the model by offering a practical method for visualising trade-offs between total institutional utility and fairness under varying bargaining weights (p_A, p_B). This visualisation enables stakeholders to explore how different recruitment strategies reshape optimal regulatory configurations, providing a transparent basis for negotiation and policy dialogue [7, 14]. The framework also supports policy simulations, allowing universities to evaluate the impact of changes in regulatory thresholds or compliance standards. Such simulations facilitate data-informed and transparent decision-making, aligning with calls for evidence-based governance in international higher education [12,24].

By integrating multi-criteria decision analysis (MCDA) with multi-objective optimisation, the model quantitatively captures institutional preferences, bargaining power, and regulatory constraints. This integration offers both theoretical insight and practical guidance for harmonising cross-border academic regulations, extending prior applications of optimisation in education policy and resource allocation [2, 19]. The proposed scenarios demonstrate how recruitment arrangements influence the balance between efficiency and equity, providing a structured approach for negotiation and regulatory alignment.

Overall, this study contributes to higher education governance by offering a decision-support tool that enhances transparency, fairness, and strategic planning in TNE partnerships. Future research may extend this framework to multi-country or multi-partner networks, incorporate stochastic optimisation to address regulatory uncertainty, or validate the model empirically using data from actual TNE programmes [6,33]. Such extensions would further strengthen the robustness and policy relevance of optimisation-based approaches, ensuring that TNE partnerships remain sustainable, equitable, and responsive to evolving global education landscapes. Practically, the proposed model functions as a decision-support instrument that universities can apply during partnership negotiations, validation planning, and regulatory compliance reviews. By converting complex regulatory requirements into quantifiable alignment scenarios, the framework enables institutions to anticipate conflicts, evaluate trade-offs, and identify mutually acceptable governance configurations before implementation.

References

1. Altbach, P.G., Knight, J.: The internationalization of higher education: motivations and realities. J. Stud. Int. Educ. **24**(3), 203–222 (2020). https://doi.org/10.1177/1028315319891276
2. Belton, V., Hodgkin, J., Koulouri, A.: A multiple criteria decision support system for group decision-making. Eur. J. Oper. Res. **160**(3), 466–482 (2005). https://doi.org/10.1016/j.ejor.2003.07.019

3. Bertsimas, D., Farias, V.F., Trichakis, N.: The price of fairness. Oper. Res. **59**(1), 17–31 (2011). https://doi.org/10.1287/opre.1100.0888
4. Bertsimas, D., Farias, V.F., Trichakis, N.: On the efficiency-fairness trade-off. Manage. Sci. **58**(12), 2234–2250 (2012). https://doi.org/10.1287/mnsc.1120.1565
5. British Council: Transnational education in the UK: Policy overview (2021). https://www.britishcouncil.org/education
6. British Council: Transnational education strategy 2025-2027: Supporting UK transnational education through partnership (2025). https://www.britishcouncil.org
7. Ehrgott, M., Gandibleux, X.: Multiobjective optimization: state of the art. Eur. J. Oper. Res. **177**(2), 347–374 (2007). https://doi.org/10.1016/j.ejor.2006.09.032
8. Ehrgott, M.: Multicriteria Optimization, 2 edn. Springer (2005). https://doi.org/10.1007/978-3-540-27659-3
9. Healey, N.M.: Managing international branch campuses: what do we know? High. Educ. Q. **69**(4), 386–409 (2015). https://doi.org/10.1111/hequ.12087
10. Hsu, C.C., Sandford, B.A.: The Delphi technique: making sense of consensus. Pract. Assess. Res. Eval. **12**(10), 1–8 (2007). https://doi.org/10.7275/pdz9-th90
11. Huang, F.: Cross-border higher education: quality assurance and recognition. High Educ. Pol. **35**(2), 163–178 (2022). https://doi.org/10.1057/s41307-021-00253-8
12. Ilieva, J., Maringe, F., Foskett, N.: Global higher education partnerships: sustainability, equity, and governance. High Educ. Pol. **37**(2), 215–233 (2024). https://doi.org/10.1057/s41307-023-00321-9
13. Ishizaka, A., Nemery, P.: Multi-Criteria Decision Analysis: Methods and Software. Wiley (2013)
14. Karsu, Ö., Morton, A.: Inequity averse optimization in operational research. Eur. J. Oper. Res. **245**(2), 343–359 (2015). https://doi.org/10.1016/j.ejor.2015.02.022
15. Knight, J.: Transnational education remixed: balancing access, quality, and sustainability. Int. High. Educ. **87**(3), 2–4 (2016). https://doi.org/10.6017/ihe.2016.87.9342
16. Liu, S., Eckert, C., Earl, C.: A review of fuzzy AHP methods for decision-making with subjective judgements. Exp. Syst. Appl. **161**, 113738 (2020). https://doi.org/10.1016/j.eswa.2020.113738
17. Liu, Y., Zhao, J., Zhang, X.: Multi-objective optimization of educational resource allocation under fairness constraints. Comput. Ind. Eng. **140**, 106272 (2020). https://doi.org/10.1016/j.cie.2019.106272
18. Malaysian Qualifications Agency: Malaysian Qualifications Framework, 2nd edn. (2020). https://www.mqa.gov.my
19. Mardani, A., Jusoh, A., Zavadskas, E.K.: Fuzzy multiple criteria decision-making techniques and applications. Econ. Res.-Ekonomska Istraživanja **28**(1), 210–263 (2015). https://doi.org/10.1080/1331677X.2015.1100522
20. Mardani, A., Jusoh, A., Nor, K.M., Khalifah, Z., Zakwan, N., Valipour, A.: Multiple criteria decision-making techniques and their applications - a review of the literature from 2000 to 2014. Econ. Res.-Ekonomska Istraživanja **28**(1), 516–571 (2015). https://doi.org/10.1080/1331677X.2015.1075139
21. Marler, R.T., Arora, J.S.: Multi-objective optimization: concepts and methods for engineering. J. Mech. Des. **127**(4), 611–618 (2010). https://doi.org/10.1115/1.1391230
22. Miettinen, K.: Nonlinear Multiobjective Optimization. Springer (1999). https://doi.org/10.1007/978-1-4612-0572-4
23. OECD: Education Governance in Action: Lessons from Case Studies. OECD Publishing (2016). https://doi.org/10.1787/9789264262829-en
24. OECD: Trends shaping education 2025: Global conflict, cooperation, and futures thinking (2025). https://doi.org/10.1787/trends-edu-2025-en

25. Office for Students: Regulatory framework for higher education providers (2023). https://www.officeforstudents.org.uk
26. Okoli, C., Pawlowski, S.D.: The Delphi method as a research tool: an example, design considerations and applications. Inf. Manag. **42**(1), 15–29 (2004). https://doi.org/10.1016/j.im.2003.11.002
27. Quality Assurance Agency for Higher Education: Quality code for higher education (2022). https://www.qaa.ac.uk
28. Saaty, T.L.: The Analytic Hierarchy Process. McGraw-Hill (1980)
29. Shams, F., Huisman, J.: The role of institutional policies in international branch campus sustainability. Stud. High. Educ. **41**(9), 1598–1614 (2016). https://doi.org/10.1080/03075079.2014.1003386
30. Skidmore, M., Bell, S., Green, C.: Balancing energy and health outcomes in school infrastructure: an MCDA approach. Energy Policy **179**, 113125 (2023). https://doi.org/10.1016/j.enpol.2023.113125
31. Springer, S.: Complexity in Higher Education Governance: Global Perspectives. Springer (2025). https://doi.org/10.1007/978-3-031-12345-6
32. Tsiligkiris, V.: Transnational education partnerships and the governance of legitimacy. J. Stud. Int. Educ. **29**(1), 45–62 (2025). https://doi.org/10.1177/1028315324123456
33. Universities UK International: The scale of UK higher education transnational education 2023–24: Trend analysis and regional highlights (2023). https://www.universitiesuk.ac.uk
34. Wilkins, S.: Transnational higher education in the 21st century. J. Mark. High. Educ. **26**(1), 1–12 (2016). https://doi.org/10.1080/08841241.2016.1143831
35. Wilkins, S., Huisman, J.: The international branch campus as transnational strategy. High. Educ. **64**(5), 627–645 (2012). https://doi.org/10.1007/s10734-012-9540-1
36. Yang, R.: Global partnership governance and local legitimacy in higher education. Asia Pac. Educ. Rev. **22**(3), 421–434 (2021). https://doi.org/10.1007/s12564-021-09711-1
37. Zhao, J., Liu, Y., Zhang, X.: A negotiation-based multi-party decision model for inter-basin resource allocation. Water Resour. Manage **26**(12), 4029–4038 (2012). https://doi.org/10.1007/s11269-012-0115-5

Risk Identification and Control Strategies of Vaccine Supply Chain Based on SCOR Model

Manyi Yang[1] and Guoqing Zhao[2(✉)]

[1] School of Management Science and Engineering, Nanjing University of Information Science and Technology, Nanjing, China
[2] School of Management, Swansea University, Swansea, Wales, UK
guoqing.zhao@swansea.ac.uk

Abstract. From a whole-process perspective of the supply chain, this study identifies and analyses the risks existing in the vaccine supply chain (VSC) and proposes targeted risk control measures, aiming to provide directions and recommendations for building a safe VSC. Based on the Supply Chain Operations Reference model (SCOR), the risks in the operation process of the VSC are systematically identified, the risk factors in the VSC are summarized, and targeted governance countermeasures are proposed accordingly. The results show that the risks in the VSC mainly come from the processes of planning, procurement, production, distribution, and returns. Based on the SCOR model, a total of 27 specific risk factors are identified, and corresponding risk control countermeasures are proposed from the perspective of the five processes.

Keywords: SCOR model · vaccine supply chain · risk identification · risk control strategies

1 Introduction

At the onset of 2020, the COVID-19 pandemic rapidly spread worldwide, severely threatening human life and safety. In China, for example, by the end of September 2021, the country had reported a total of 96,162 confirmed COVID-19 cases and 4,636 deaths [1]. To mitigate the spread of the virus and minimize infections and fatalities, China initiated an emergency COVID-19 vaccination program. By the end of September 2021, more than 2.2 billion vaccine doses had been administered [2]. The vast demand for vaccines has not only exerted immense pressure on vaccine production but also presented significant challenges for cold chain logistics. Studies have shown that transportation is the most vulnerable part of the cold chain, and currently, China's cold chain logistics are relatively underdeveloped, with issues such as poor management. Given the current unique social context, the vaccine supply chain (VSC) is vital to the nation's well-being. A disruption in the VSC would inevitably impact the overall health of society and could even threaten public safety. Therefore, from a risk management perspective, it is essential to analyze the risks throughout the entire VSC and propose appropriate governance strategies to assist stakeholders in managing these risks comprehensively.

A. P. Cabral Seixas Costa et al. (Eds.): ICDSST 2026, LNBIP 582, pp. 395–406, 2026.
https://doi.org/10.1007/978-3-032-25146-6_26

This would help establish a safe and effective VSC. The findings of this research are of significant importance to pandemic prevention and control, the vaccine industry, and public health and safety.

Scholars' current research on vaccine risks primarily concentrates on two areas. The first area is risk analysis in the vaccine production stage. For example, Song et al. [3] analyzed risks during vaccine production using quality risk management theory and proposed targeted regulatory strategies. Xu [4] and Jian [5] employed FMEA/FMECA models to assess the relative importance of various risk factors in biological vaccine production. Based on their risk analysis and evaluation, they developed risk control measures and enhanced the internal quality management systems of vaccine manufacturing companies. The second area of focus is risk analysis in the vaccine transportation stage. Li [6], using the fault tree analysis method, evaluated the structural and probability importance of risk events in the pharmaceutical cold chain logistics process, identifying the significance of each basic event and suggesting corresponding control measures for critical risk factors. Ying et al. [7] reviewed the current status and challenges of vaccine cold chain management in China and offered improvement suggestions from the perspective of cold chain management. Du [8] analyzed issues related to COVID-19 vaccine transportation within cold chain logistics and proposed feasible strategies concerning supply chain models, the application of new technologies, and vaccine traceability systems.

A review of existing studies reveals that most research primarily focuses on risk identification in vaccine production and transportation, with limited attention to analyzing risks throughout the entire vaccine operation process from a supply chain perspective. Therefore, this paper aims to systematically identify risks across the entire operation process of the VSC by utilizing the supply chain operations reference (SCOR) model and total quality management (TQM) theory. It also offers targeted governance strategies based on the risk identification findings, aiming to assist stakeholders in managing potential risks within the VSC.

2 Supply Chain Operations Reference Model

The SCOR model, initially introduced by the Supply Chain Council, serves as a standard reference framework and diagnostic tool applicable across various industries' supply chains. The SCOR model consolidates concepts such as business process reengineering, benchmarking, and process assessment into a comprehensive cross-functional framework that includes business processes, performance metrics, best practices, and technology features. It provides a universal language for enterprises to describe and analyze their supply chains. The SCOR model is composed of five core management processes: Planning, Sourcing, Making, Delivering, and Returning, each of which is regarded as a critical internal function and a key inter-organizational process [9]. Commonly employed as a risk identification tool, the SCOR model facilitates systematic identification and evaluation of supply chain risks, thereby reducing them.

In this paper, building upon the general supply chain definition, the VSC is defined as a functional network focused on the vaccine as the core product. The process starts with the procurement of materials required for vaccine production, followed by manufacturing by vaccine producers, then distribution through logistics networks to vaccination centers, and finally, the administration of vaccines to individuals. This network integrates suppliers of vaccine materials, vaccine manufacturers, various levels of Centers for Disease Control (CDCs), epidemic prevention agencies, and the vaccinated individuals into a cohesive system. The structure of the VSC is clear: vaccine manufacturers are at the core, supported by upstream suppliers of vaccine materials, and downstream, various levels of CDCs and epidemic prevention departments. Logistics operations are typically managed either by the manufacturers themselves or through third-party logistics providers.

Compared to other supply chains, the VSC demands higher quality standards, is extremely time-sensitive, and involves significant costs in the event of losses. These unique characteristics of the VSC substantially increase the potential risk impacts and the complexity of risk management. If risks materialize in the VSC, they can not only compromise vaccine quality but also pose significant threats to the safety of those receiving the vaccine. The SCOR model provides a clear depiction of the VSC's structure and components while also enabling systematic identification of potential risks within the supply chain. This facilitates effective risk management by stakeholders involved in the VSC.

3 Risk Factors in the VSC Based on the SCOR Model

3.1 Risks in the Planning Process

The planning process involves strategizing the entire supply chain based on forecast outcomes, ensuring a balance between demand and supply. As the initial process at the highest level of the SCOR model, it is fundamental to the efficient operation of all subsequent processes. This section identifies and analyzes activities related to VSC risks by considering the main activities within the planning process and the unique characteristics of vaccine companies.

SCR1: Risk of Inaccurate Demand and Supply Forecasting

The presence of the bullwhip effect and information asymmetry between companies means that the accuracy of demand and supply forecasts significantly impacts the planning and decision-making processes of vaccine companies. This risk is identified as "Accuracy of Demand and Supply Forecasts".

SCR2: Risk from Inefficient Supply Chain Structure Design

China's vaccine distribution largely follows traditional supply chain structures. However, practical experience has shown that these traditional models, characterized by multiple hierarchical levels and slow response times, fail to meet the distribution needs of various vaccines. Inefficient supply chain structure design can lead to reduced vaccine supply efficiency and risks such as delivery delays. This risk is termed "Efficiency of Supply Chain Structure Design".

SCR3: Risk of Poor Management During Vaccine Product Decline

Vaccine companies must assess when a product is entering its decline phase using relevant indicators and data, and either innovate to meet market demands anew or decide to withdraw the product from the market. Poor management during this phase can not only harm the company's interests but also strain its relationship with the public [10]. This risk is referred to as "Management of Product Decline Phase".

VSCR4: Risk of Weak Risk Awareness Among Employees

Given the high quality and safety requirements for vaccines, it is essential for management and production staff in vaccine companies to maintain strong risk awareness throughout the process and to foster a deep understanding of risk culture. Weak risk awareness can easily lead to quality and safety issues with vaccines. This risk is labeled "Employee Risk Awareness".

3.2 Risks in the Procurement Process

The procurement process encompasses key activities such as supplier selection and raw material transportation. However, certain activities within this process overlap or are redundant. To identify the associated risks effectively, similar activities were consolidated, and the risks were analyzed based on the refined procurement process activities, resulting in the identification of the following risk factors.

SCR5: Risk from Inadequate Control Over Material Procurement Prices

For vaccine manufacturers, the cost of materials is a significant determinant of their overall production costs. Effective control over procurement prices is crucial for maintaining price stability. If procurement prices increase, it can reduce the company's profitability, impacting future production plans and material quality, potentially leading to risks within the VSC. This risk is termed "Procurement Price Management".

SCR6: Risk from Inappropriate Supplier Selection

Supplier selection criteria significantly influence the overall functioning of the supply chain. Evaluating suppliers should encompass aspects such as delivery speed, reputation, supply quality, and material costs. Poor supplier selection can impede the VSC's goals and may even introduce severe risks to the company. This risk is identified as "Supplier Selection Effectiveness".

SCR7: Risk from Non-Adherence to Material Quality Standards

Vaccine manufacturers must rigorously inspect raw materials during procurement to ensure compliance with quality standards. If the materials fail to meet these standards, the resulting vaccines will also be substandard. This risk is labeled "Quality Consistency".

SCR8: Risk from Delayed Material Supply

Without thorough assessment and continuous monitoring of suppliers, vaccine manufacturers may face increased instances of delayed material deliveries. Such delays can disrupt production schedules or result in the loss of market opportunities. This risk is referred to as "Supply Punctuality".

SCR9: Risk from Inaccurate Material Verification

During the material receipt process, if the staff does not meticulously verify the materials against procurement standards, substandard or incomplete materials may be accepted into the company. This risk is known as "Material Verification Accuracy".

SCR10: Risk from Poor Material Delivery Conditions

Once procurement is completed, materials must be transported from suppliers to the vaccine manufacturer. Mismanagement during transportation can result in significant material damage, potentially reducing the vaccine company's production capacity [11]. This risk is termed "Delivery Condition Integrity".

SCR11: Risk from Inadequate Warehouse Management

After materials are stored, improper management—such as the absence of labeling or exposure to a humid environment—can lead to issues like material mix-ups or quality degradation. This risk is called "Warehouse Management Efficiency".

3.3 Risks in the Production Process

To comprehensively identify risk factors in the vaccine production process, TQM theory is applied. The potential risks are categorized based on the five key elements: personnel, equipment, materials, methods, and environment.

3.3.1 Vaccine Production Personnel

SCR12: Risk of Contamination by Particles and Microorganisms

During production, it is challenging to fully disinfect and sterilize personnel entering sterile areas, leading to the risk of product contamination by particles and microorganisms. This risk is termed "Particle and Microbial Contamination".

SCR13: Risk from Non-Standard Operations

The vaccine production process requires human involvement, and improper operations by production staff can result in quality issues with vaccine products. This risk is identified as "Operational Standardization".

3.3.2 Vaccine Production Equipment

SCR14: Risk of Cross-Contamination

During the use of production equipment, containers may come into direct contact with the vaccine solution, leading to the risk of microbial adhesion and cross-contamination. This risk is labeled "Cross-Contamination".

SCR15: Risk from Improper Equipment Usage

If vaccine production equipment is not properly installed, designed, or used in accordance with specified procedures, it may adversely affect the quality of the vaccines. This risk is termed "Equipment Usage Compliance".

3.3.3 Vaccine Production Materials

SCR16: Risk from Substandard Production Materials

The quality of materials used in vaccine production directly impacts the final product, and substandard materials can compromise vaccine safety and efficacy [12]. This risk is referred to as "Material Quality Compliance".

SCR17: Risk from Material Mismanagement

Inadequate management and categorization of production materials can lead to confusion and the incorrect use of materials, resulting in significant risk events. This risk is identified as "Material Management Accuracy".

3.3.4 Vaccine Production Methods

SCR18: Risk from Inaccurate Production Processes and Methods

Successful vaccine production requires meticulous adherence to manufacturing methods and complex processes. Any deviation at any stage can result in substandard vaccine quality. This risk is labeled "Process and Method Accuracy".

SCR19: Risk from Insufficient Research Prior to Process and Method Changes

Changes in production processes and methods without thorough research can negatively impact the safety and efficacy of vaccine products. This risk is referred to as "Research Sufficiency Before Changes".

3.3.5 Vaccine Production Environment

SCR20: Risk of Environmental Contamination

Production personnel, equipment, and the manufacturing environment can all contribute to the contamination of vaccine products. This risk is labeled "Environmental Contamination".

SCR21: Risk from Improper Disinfection Methods and Low Frequency of Disinfection

If the disinfection methods used in the production environment do not meet standards or if disinfection is infrequent, the risk of environmental contamination increases. This risk is termed "Disinfection Practices and Frequency".

3.4 Risks in the Distribution Process

The distribution process includes managing product inventory and transportation. Based on the primary activities of distribution and the specific characteristics of vaccine companies, the following risks are identified.

SCR22: Risk from Improper Control of Finished Products

After production, vaccines must be stored under strict conditions, including specific temperature and cleanliness requirements. Poor inventory management can lead to vaccine damage, delaying delivery times. This risk is identified as "Finished Product Management".

SCR23: Risk from Poor Decision-Making in Distribution Methods

Temperature control during vaccine distribution is critical, especially in maintaining the cold chain. Poor decisions in transportation methods can lead to frequent changes in handling, risking improper temperature control. This risk is termed "Distribution Method Selection".

SCR24: Risk from Equipment Malfunction

Vaccines require cold chain transportation, which involves storing them in refrigerated equipment at controlled temperatures. Equipment malfunction can lead to temperature excursions, rendering vaccines unusable. This risk is identified as "Equipment Reliability".

SCR25: Risk from Low Employee Competence

The quality of distribution personnel is crucial in mitigating risks. Distribution staff must be skilled drivers, capable of handling emergencies, and knowledgeable about vaccine handling. This risk is labeled "Human Resource Competence".

3.5 Risks in the Return Process

The return process in the supply chain represents reverse logistics, involving the return of products and materials. This includes handling product returns from customers and material returns to suppliers, requiring communication, documentation, and the physical return of items.

SCR26: Risk from Low Product Quality

In the vaccine industry, product returns often occur due to issues such as deviations in vaccine efficacy or the presence of counterfeit products [13]. This risk is termed "Product Quality Compliance".

SCR27: Risk from Low Vaccine Distribution Satisfaction

Errors in vaccine distribution can lead to reduced customer satisfaction, prompting return requests from recipients. This risk is labeled "Distribution Satisfaction".

Table 1 summarises the risk factors at each stage of the VSC.

Table 1. The risk factors of VSC

No.	Risk category	Risk factors
SCR1	Risks in the planning process	Risk of inaccurate demand and supply forecasting
SCR2		Risk from inefficient supply chain structure design
SCR3		Risk of poor management during vaccine product decline
SCR4		Risk of weak risk awareness among employees
SCR5	Risks in the procurement process	Risk from inadequate control over material procurement prices
SCR6		Risk from inappropriate supplier selection
SCR7		Risk from non-adherence to material quality standards
SCR8		Risk from delayed material supply
SCR9		Risk from inaccurate material verification
SCR10		Risk from poor material delivery conditions
SCR11		Risk from inadequate warehouse management
SCR12		Risk of contamination by particles and microorganisms
SCR13	Risks in the production process	Risk from non-standard operations
SCR14		Risk of cross-contamination
SCR15		Risk from improper equipment usage
SCR16		Risk from substandard production materials
SCR17		Risk from material mismanagement
SCR18		Risk from inaccurate production processes and methods
SCR19		Risk from insufficient research prior to process and method
SCR20		Risk of environmental contamination
SCR21		Risk from improper disinfection methods and low frequency of disinfection
SCR22	Risks in the distribution process	Risk from improper control of finished products
SCR23		Risk from poor decision-making in distribution methods
SCR24		Risk from equipment malfunction

(*continued*)

Table 1. (*continued*)

No.	Risk category	Risk factors
SCR25		Risk from low employee competence
SCR26	Risks in the return process	Risk from low product quality
SCR27		Risk from low vaccine distribution satisfaction

4 VSC Risk Management Strategies

4.1 Risk Control in the Planning Process

Risks in the planning process include the accuracy of demand and supply forecasting, the reasonableness of supply chain structure design, the management of product decline, and the level of risk awareness. To address these, companies should consider integrating demand and supply planning with supply chain structure design to propose coordinated control measures. For instance, transitioning from a CDC-centric supply chain model to a logistics-centric one could help overcome the inefficiencies of traditional models, such as excessive layers, slow information flow, and challenges in vaccine quality control. This transition could also alleviate the bullwhip effect, reducing risks associated with inaccurate demand and supply forecasts (SCR1). Furthermore, leveraging modern technologies like the Internet of Things (IoT) can enable real-time visibility and information sharing within the VSC, enhancing the accuracy of planning and ensuring the supply chain's safety and efficiency (SCR2). Additionally, establishing a dedicated risk management department within the organization can facilitate rapid risk detection and response. Implementing safety principles, such as DuPont's top ten safety tenets, across the company can foster a culture of risk awareness, ensuring that all employees recognize safety as their responsibility (SCR4). Lastly, for vaccines entering the decline phase (SCR3), companies should develop robust product withdrawal plans, strictly adhere to regulations for recalls or market exits, and conduct thorough evaluations to provide valuable insights for future product improvements.

4.2 Risk Control in the Procurement Process

Risks in the procurement process encompass procurement pricing (SCR5), supplier selection capability (SCR6), quality consistency (SCR7), supply punctuality (SCR8), material approval (SCR9), delivery integrity (SCR10), and warehouse management (SCR11). Vaccine companies should begin by gathering comprehensive data on suppliers, evaluating their performance in areas like delivery speed, reputation, quality, and pricing, to select the most reliable partners. Additionally, maintaining relationships with alternative qualified suppliers can serve as a safeguard against potential disruptions. When procuring materials, companies should deploy a specialized team to establish and enforce stringent quality standards while negotiating prices to ensure cost-effectiveness.

Contracts should clearly define deadlines, and continuous monitoring should ensure suppliers meet their obligations until the transaction is finalized. Upon receiving materials, companies must conduct thorough inspections to verify quantities and compliance with standards. Post-delivery, enhancing warehouse management through regular inspections and strict labeling can prevent mishandling. Companies might also implement a tiered management system, categorizing materials based on their impact on vaccine quality, to ensure higher levels of control where needed.

4.3 Risk Control in the Production Process

Managing risks in the production process is vital to the effective operation of the entire VSC, so companies should treat it as a priority in risk governance. First, because vaccine quality is foundational, substandard materials can directly undermine product quality. Risk control should therefore start at the source, with rigorous management of procurement, transportation, and storage to ensure material quality. Second, to prevent material mix-ups, companies should strengthen oversight of production materials through routine inspections and clear labeling, and verify all materials before use to confirm they meet production requirements. Third, environmental contamination risks arise from personnel, the workshop environment, and equipment, and should be managed in an integrated manner. For personnel-related risks, firms should require pre-employment health screening and enforce strict entry protocols, including cleaning, disinfection, and appropriate clothing. For risks related to the workshop and equipment, regular disinfection and appropriate sterilization methods are needed to maintain a sterile setting, alongside routine equipment inspection and timely cleaning and disinfection after use to reduce cross-contamination and extend equipment life. Fourth, operational risks should be controlled by establishing strict standards, procedures, and change-management rules. Companies should recruit skilled staff, ensure new hires receive comprehensive training, and allow only those who pass assessments to work. Equipment should be properly designed, selected, and operated according to guidelines, and any changes to methods or processes should be supported by thorough research and validation to avoid arbitrary modifications.

4.4 Risk Control in the Distribution Process

Risks in the distribution process involve finished-product storage (SCR22), transportation decisions (SCR23), equipment reliability (SCR24), and personnel capability (SCR25). To strengthen risk management in the VSC, companies should first ensure vaccines are stored at the required temperature and in a clean environment, supported by routine cleaning, disinfection, and continuous temperature monitoring. Second, before distribution, companies should assess conditions at both the current storage site and the destination, then select appropriate routes and transport methods to reduce temperature fluctuations caused by frequent mode changes. Third, cold chain equipment should be inspected and maintained regularly, and key devices such as temperature monitors should be checked before each shipment to lower the likelihood of in-transit failures and vaccine damage. Finally, all distribution personnel should complete training and certification before work begins, covering vaccine handling, cold chain equipment maintenance, and driving skills, so they can respond to emergencies and ensure safe, efficient delivery.

4.5 Risk Control in the Return Process

To address the risks in the return process, which arise from poor product quality (SCR26) and low satisfaction with vaccine distribution (SCR27), companies should give this area careful attention. By enhancing the quality control of vaccines and managing deviations during the distribution process, it is crucial to ensure that vaccines meet quality standards before they are dispatched and that the distribution process remains error-free until safe delivery to vaccination centers. Moreover, in maintaining product quality, companies should rigorously enforce vaccine release procedures, including detailed information logging and verification, to prevent basic errors in distribution. Although the risks associated with the return process are of relatively lower priority compared to other risks, any such event can still disrupt the efficiency of the VSC and lead to increased operational costs. Therefore, vaccine manufacturers should still focus on this aspect to ensure that the supply chain achieves its intended objectives.

5 Conclusion

This study systematically identifies risks within the VSC operations by applying the SCOR model and integrating TQM theory to identify risks specifically in the vaccine production phase, identifying a total of 27 distinct risk factors. The study then proposes targeted risk management strategies from the perspective of planning and other processes. By analyzing risks across the entire supply chain, this approach addresses the shortcomings of previous studies that only focused on individual stages of vaccine risk identification. This comprehensive approach enriches the theoretical framework of VSC risk management, enabling stakeholders to better manage potential risks and establish a secure and effective VSC. This has significant implications for pandemic prevention, the vaccine industry, and public health and safety.

References

1. National Health Commission.: http://www.nhc.gov.cn
2. Yang, M., Qu, S., Ji, Y., et al.: Risk management of the vaccine supply chain: interactions of risk factors and control strategies. Socio-Econ. Plan. Sci. **98**, 102153 (2025)
3. Song, M., Huang, T.: Research on the Risk Supervision Strategy during Vaccine Production. China Pharm. **24**(5), 385–387 (2013)
4. Zhang, X.: Risk Management in Biological Vaccine Production Process. University of Chinese Academy of Sciences (2015)
5. Zhang, J.: The Ouality of the Vaccine Production Process of Risk Assessment and Control. University of Chinese Academy of Sciences (2014)
6. Li, L.: Research on Drug Cold Chain Logistics Process Risk Management. Beijing Jiaotong University (2021)
7. Ying, X., Liu, Q., Zhang, M., et al.: Current situation and development prospect of vaccine cold chain management in china. Chin. J. Mod. Appl. Pharm. **37**(5), 636–640 (2020)
8. Du, X.: Research on COVID-19 vaccine transportation issues and countermeasures based on cold chain logistics. China Storage Transp. Magazine. **4**, 96–98 (2021)

9. Abolghasemi, M., Khodakarami, V., Tehranifard, H.: A new approach for supply chain risk management: mapping SCOR into Bayesian network. J. Ind. Eng. Manag. **8**(1), 280–302 (2015)
10. Sankarankutty, S.: Application of risk assessment in product quality lifecycle management. Rapports De Pharmacie. **1**(2), 46–53 (2015)
11. Faisal Li, L., Su, Q., Chen, X.: Ensuring supply chain quality performance through applying the SCOR model. Int. J. Prod. Res. **49**(1), 33–57 (2011)
12. Moktadir, M.A., Ali, S.M., Mangla, S.K., et al.: Decision modeling of risks in pharmaceutical supply chains. Ind. Manag. Data Syst. **118**(7), 1388–1412 (2018)
13. Vishwakarma, V., Prakash, C., Barua, M.K.: A fuzzy-based multi criteria decision making approach for supply chain risk assessment in Indian pharmaceutical industry. Int. J. Logist. Syst. Manag. **25**(2), 245–265 (2016)

Tax Decision-Making: Methods to Increase the Financial Sustainability of Ecuadorian Rural Enterprises

Dolores Ortiz Guevara(✉), Elvira Elizabeth Ortega Decimavilla, and Fernando Rodolfo Orellana Intriago

University of Guayaquil, Guayaquil 090505, Ecuador
{Dolores.ortizgue,Elvira.ortegad,fernando.orellanai}@ug.edu.ec

Abstract. This research aimed to analyze the impact of tax decision-making methods on the sustainability of rural businesses in Ecuador. A mixed-methods approach, both descriptive and analytical, was employed, using the synthetic method. A sample of 100 businesses was studied, using a structured survey of seven closed-ended questions as the analytical instrument, supplemented by a four-question open-ended interview with three managers from these businesses. As part of the quantitative analysis, the chi-square test was used to confirm the statistically significant association between the variables investigated. Simultaneously, the statistical analysis was combined with qualitative results and theoretical corroboration through methodological triangulation, thus strengthening the validity of the conclusions. The results indicate that businesses lacking structured decision-making processes are hampered by a lack of tax planning, inadequate financial analysis, and weaknesses in their evaluation systems. It is concluded that the proper structuring of fiscal decision-making methods has a positive impact on strengthening financial sustainability in rural businesses.

Keywords: decision-making process · rural entrepreneurship · methods · business sustainability · tax management

1 Introduction to Decision-Making Within Rural Enterprises

Today, tax decisions have become a strategic part of financial and administrative management in the business world. Beyond simply complying with tax requirements, these decisions directly impact liquidity, profitability, financial planning, and the sustainability of organizations, becoming an essential tool that contributes to the economic and rural transition of businesses in the rural sector toward sustainability [1].

In this sense, the methods used in tax decision-making are mechanisms that improve a company's ability to respond to changing laws and policies. When the implementation of these approaches is not carried out rationally, technically, and strategically, companies often face unplanned financial crises, involuntary non- compliance, and lost growth opportunities [2]. This is even more frequent in rural businesses, which, despite representing a key element within the economic system, face structural barriers and constant

A. P. Cabral Seixas Costa et al. (Eds.): ICDSST 2026, LNBIP 582, pp. 407–423, 2026.
https://doi.org/10.1007/978-3-032-25146-6_27

challenges due to limitations in technical resources and scarce access to specialized advice [3].

In the case of rural enterprises in Ecuador, these have emerged as an alternative to improve the country's economic and social conditions. By combining traditional practices with technology and innovation, they have directed their efforts toward transforming food security and territorial equity [4].

This type of enterprise represents 29% within the different economic activities in the country, classifying the rural sector as a contributor of great relevance to regional and national growth, highlighting that these companies "rely on the integration of local resources, traditional practices and economic sustainability strategies [4].

However, in most rural areas of Ecuador, it is evident that many of these initiatives face major challenges and structural limitations that prevent them from achieving a solid competitive position, remaining behind the rest of the companies [5].

One of the most concerning deficiencies that has become evident is related to tax management, which has been affected not only by the poor financial and accounting organization of these entrepreneurs, but also by the lack of appropriate methods and processes that would allow for sound decision-making in this area. Furthermore, as Herruzo et al. mention in [5], these types of businesses lack innovative tools to facilitate accounting activities and have limited access to personal training and technical assistance.

This lack of instruments and mechanisms for decision-making has led these ventures to develop incomplete tax management, where strategic actions are insufficient to make sound decisions in the face of regulatory changes, take advantage of tax benefits, develop their financial resources and plan their tax obligations, which has led to the closure of many of these companies and most of them do not last 3 years [6].

Therefore, this study seeks to analyze the impact of tax decision-making methods on the sustainability of rural businesses in Ecuador, with the aim of providing theoretical and empirical tools for strengthening the tax management of this economic sector.

2 Making and Its Impact on Business Consolidation: Theoretical Elements

Globally, changes and transformations have occurred, where business decisions face a competitive system that requires systematic models that allow the effective management of related information and data.

Thus, in the evolution of business, decision-making has become a process of paramount importance, especially when it comes to entrepreneurship, since the viability of this type of business depends on sound actions in all its constituent areas, and on which its success or failure will depend [7].

This process is relevant in the construction of key strategies for the consolidation of any undertaking, defining its organizational, administrative, operational, strategic and financial direction, determining its structure and functioning, which would allow it to achieve its objectives, goals and the resolution of conflicts, both internal and external [8].

According to Imran in [1] Decision-making in the business environment represents a comprehensive and fundamental process for recognizing the risks and advantages that these companies face in the external context. This process allows them to achieve the expected results and resolve problems.

Furthermore, this author mentions that decision-making is based on several stages that include setting objectives, obtaining relevant information and data, reviewing the alternatives presented, and selecting the strategy that best suits the expected goal.

In this sense, this process depends on a series of elements that are essential for making effective decisions: (a) every undertaking must have clearly defined objectives and goals so that decisions are aligned with the vision and strategic vision of the business; (b) Having accurate and precise information; (c) Evaluating the risks and benefits, considering the consequences of each action; (d) Considering the legal and ethical implications of the decisions and their impact on organizational principles; and (e) Implementing a system for monitoring the decisions made, through a plan [3].

Therefore, considering the theoretical approaches of Velázquez et al. [3], the decision-making process is based on a series of stages that guarantee its efficiency (Fig. 1), including: (1) Situational analysis, which promotes a review of the business environment to establish appropriate decisions; (2) Situational diagnosis, which seeks to identify the determining factors for successful actions; (3) The identification of indicators and criteria that serve as references; (4) Selection of the alternative that allows addressing the reality of the business; and finally, (5) The monitoring and follow-up stage that allows measuring the long-term impact of the decisions.

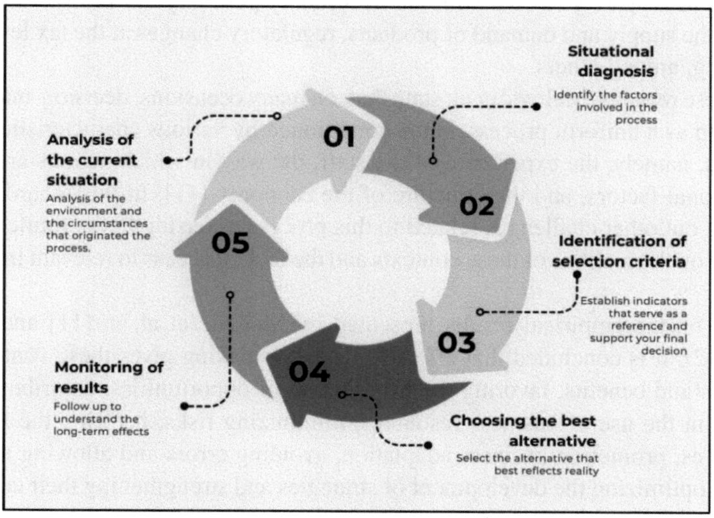

Fig. 1. Tax decision-making process

2.1 Benefits of Decision-Making for Rural Ventures

The sustainable development approach, which has gained momentum in recent years, has led to the growth and establishment of various business initiatives in rural areas, representing a fundamental part of strengthening local communities through new economic components and the integration of innovation. This type of entrepreneurship has experienced significant growth over the last few decades, considered a strategy for revitalizing economies, whose main focus has been on combating poverty and unemployment, contributing to new productive activities, and generating opportunities for consolidation in rural communities [9].

Within this scenario, these ventures must make daily decisions related to the various operational processes they manage, where the quality and availability of information becomes indispensable to guide effective actions that contribute to the growth of the business.

Various scientific and academic contributions, such as Cardona's in [9], demonstrate that these ventures have been defined as a business system of social transformation; however, the structural challenges and those of the economic, political, and technological system itself generate a set of challenges that affect the survival and success of this type of company.

It has been shown that one of the biggest problems affecting its functioning is the difficulty in making decisions. On many occasions, this process is usually done based on the intuition of directors or administrators, without taking into account key information. Other times, decisions are made with a short-term vision without measuring subsequent risks, exposing themselves to operational errors that can reduce competitiveness and solvency [10].

Furthermore, the lack of efficient, structured, and innovative methods has limited its organizational capacity to deal with the uncertainty posed by the markets themselves, related to the supply and demand of products, regulatory changes at the tax level, access to financing, among others.

For these reasons, Melović et al. state that, on many occasions, decision-making does not develop as a uniform process, but is conditioned by various characteristics of these companies, namely, the experience of the staff, the way in which actions are decided, organizational factors, and the structure of the economy [11]. In this regard, Krasniqi et al. point out other challenges related to this process, referring to the limited research conducted on the realities of these contexts and the lack of access to relevant information [12].

Based on the empirical results presented by Melović et al. in [11] and Krasniqi et al. in [12], it is concluded that effective decision-making gives these ventures great advantages and benefits, favoring the exploitation of opportunities, contributing to the efficiency in the use of business resources, minimizing risks, favoring the fulfillment of objectives, promoting business adaptation, avoiding errors and allowing them to be corrected, optimizing the development of strategies and strengthening their capacity for innovation and growth.

3 Empirical Analysis of Tax Decision-Making in Rural Businesses in Ecuador

The following is an empirical analysis conducted in various rural areas of Ecuador, where a convenience sample of 100 businesses was randomly selected to assess their current situation regarding the use of decision-making methods and their impact on financial sustainability. These businesses were identified through the Business Council for Sustainable Development in Ecuador and contacted via email.

3.1 Results of the Application of the Structured Survey

First, a structured survey questionnaire consisting of seven (7) closed questions was prepared, the questionnaire was designed in Google forms and sent to the business emails of each studied venture, which allowed the identification of the tax decision-making process in these companies.

Initially, in relation to the indicators related to decision-making methods, it was found that the majority of the studied ventures, that is, 61%, do not have established procedures (neither internally nor externally) (Fig. 2), that allow them to make effective decisions to comply with their tax responsibilities, due to the lack of information or lack of knowledge of the taxpayers and the low tax culture.

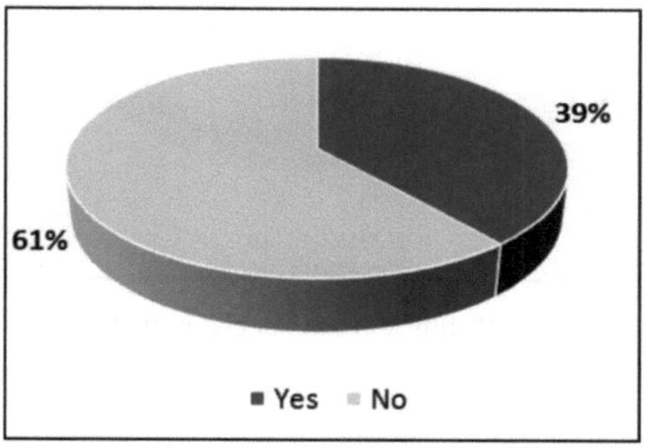

Fig. 2. Established procedures for tax decision-making

Regarding the availability of information for tax decisions, 58% stated that they do not have access to the data necessary to consolidate their tax management (Fig. 3).

This group of entrepreneurs stated that they have little knowledge of regulations, deadlines, penalties, and other actions related to tax processes. They lack in-house specialists in this area and receive very little support from the relevant agencies. This leads to poor compliance with these obligations and negatively impacts their economic performance.

Other aspects assessed through the research instrument revealed that 64% of the rural businesses interviewed (Fig. 4) stated that their tax decisions are not based on coordinated and organized planning that facilitates strategic decisions, but rather on pressures imposed by the tax system, unexpected situations, and information received from the environment in which they operate.

Within this scenario, it was also determined that this group of entrepreneurs, for the most part, 69% (Fig. 5), do not have digital tools or other types of technological systems that facilitate tax decision-making, arguing the lack of access to digital systems, the little knowledge they have about the management of these instruments and the digital

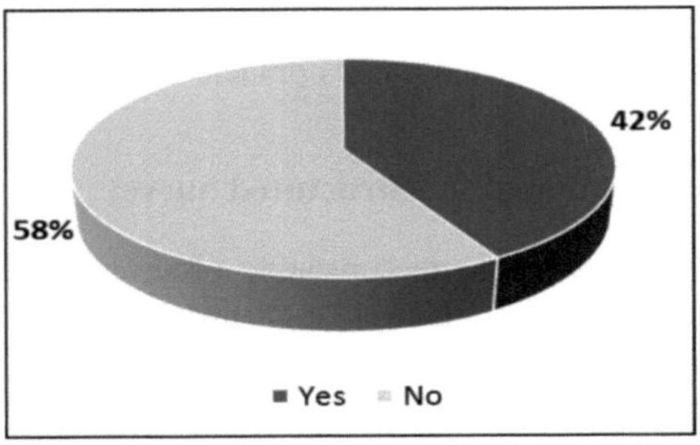

Fig. 3. Availability of information for tax decisions

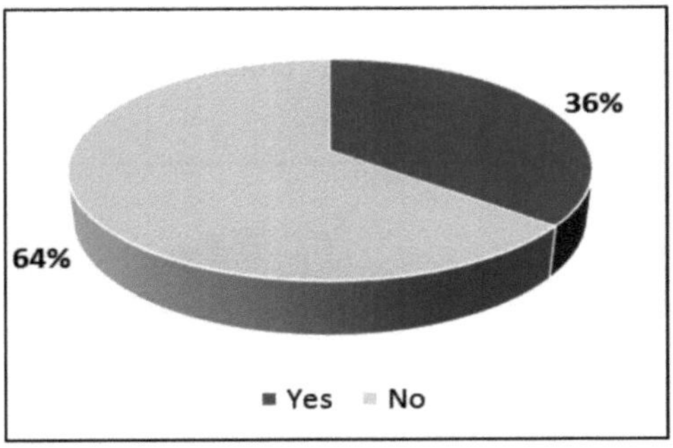

Fig. 4. Planning for tax decision-making

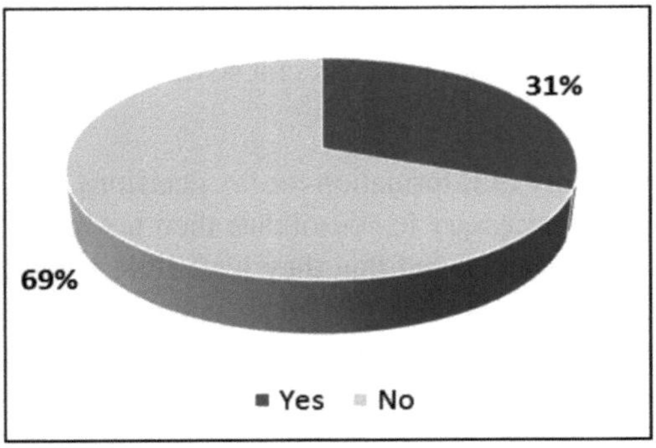

Fig. 5. Digital tools and technological systems for tax decision-making

divide that persists in Ecuador, especially in these rural areas where a true technological inclusion that promotes connectivity for this population has not yet been developed.

Furthermore, the findings show that these businesses, in 59% (Fig. 6), rarely evaluate whether their tax decisions were correct for the development of their businesses, which has generated economic, administrative and even organizational difficulties, because they do not take precautions about their tax management, given the inadequate communication that exists both internally and from the State.

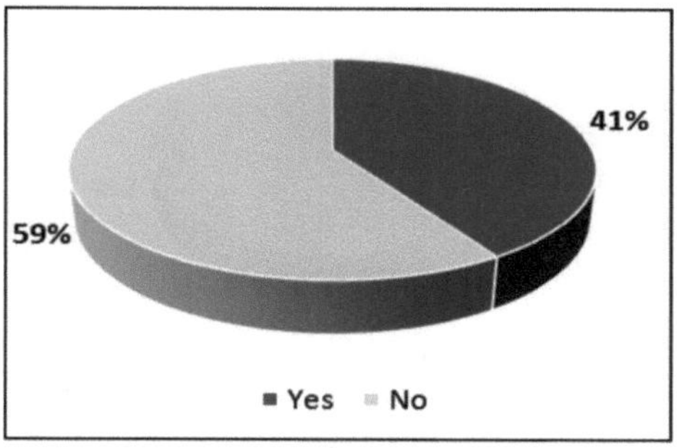

Fig. 6. Evaluation of successful tax decisions

Furthermore, it was established that 62% of the entrepreneurs who participated in the study (Fig. 7) indicated that they do not have effective methods for making tax decisions, which highlights that they do not establish tax planning procedures, do not have external technical advice, do not develop market analyses that allow them to identify the legal dimensions, and do not perform cost-benefit comparisons of their tax obligations.

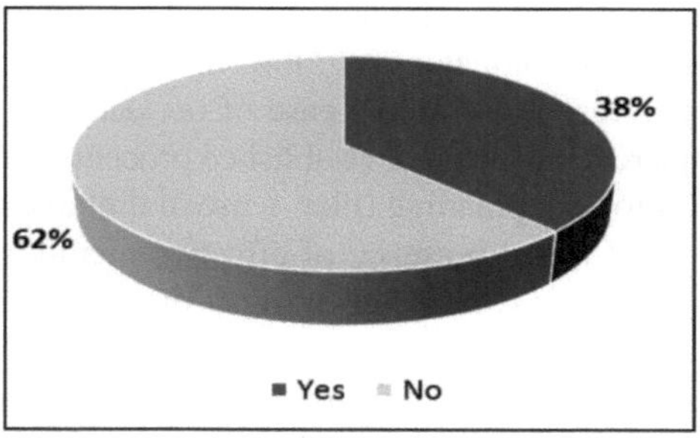

Fig. 7. Methods for tax decision making

Finally, when evaluating the financial sustainability variable of rural enterprises in Ecuador (Fig. 8), it was found that 54% established that it is low (frequent difficulties in covering costs and obligations), 28% indicated that it is medium (covers costs and obligations, but with limitations on growth), only 18% mentioned that the financial sustainability achieved by the company is high (maintains financial stability and growth capacity).

In addition to the problems and shortcomings identified in the tax decision-making process of rural businesses in Ecuador, some authors agree that other difficulties arise when making appropriate decisions [8]. These include the complexity of the Ecuadorian tax system, characterized by high tax rates, a lack of clear communication between the tax administration and entrepreneurs, a lack of transparency that has fostered institutional distrust, and the complexity of the regulations and laws.

Therefore, rural businesses in Ecuador require the implementation of methods and processes for making sound tax decisions. This will enable them to develop their activities, overcome the identified structural, administrative, organizational, and institutional

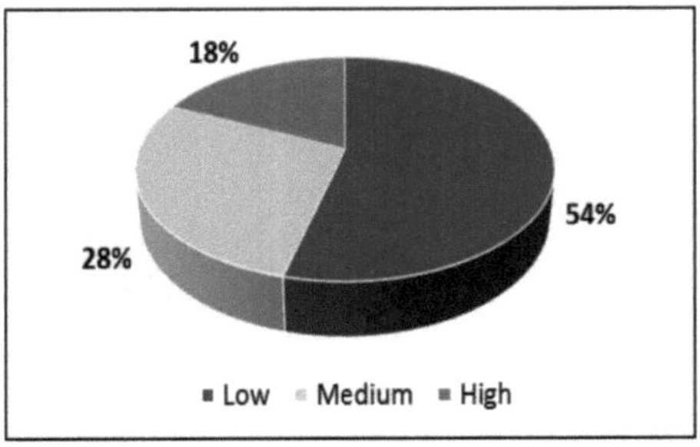

Fig. 8. Level of financial sustainability

deficiencies, achieve sustainability, and efficiently integrate tax management into the development and growth of their businesses.

Now, in order to delve deeper into the analysis and go beyond the percentage description of the results obtained, the statistical test of Chi-square (χ^2) of independence will be applied, in order to determine if there is a statistically significant association between the level of tax decision-making methods and the level of financial sustainability of the rural enterprises surveyed.

To this end, composite variables were first constructed from the indicators evaluated in the instrument, classifying companies into low, medium, and high levels according to the sum of affirmative responses. In the case of tax decision-making methods, six dimensions were considered: existence of established procedures (Proc.), availability of tax information (Disp. Info), tax planning (Plan.), use of digital tools (Herr.), evaluation of decisions adopted (Eval.), and presence of effective methods for tax management (Met.). Each of these items was measured using a dichotomous response (Yes/No), assigning a value of 1 point to affirmative responses and 0 points to negative responses.

Subsequently, the scores obtained by each company were summed, generating a total possible range between 0 and 6 points. Based on this cumulative score, three ordinal classification categories were established: low level (0 to 2 points), corresponding to companies with little structure in their tax methods; medium level (3 to 4 points), which identifies businesses with partially consolidated practices; and high level (5 to 6 points), which groups those companies that present organized and systematic tax management. This recoding allowed the transformation of multiple individual indicators into a single categorical variable, suitable for inferential analysis using the Chi-square test (Table 1).

Similarly, financial sustainability was classified into low, medium, and high levels, according to the survey results (Table 2). This standardization facilitated the construction of the contingency table and the analysis of the possible association between the two variables.

Based on these results, the information was cross-referenced company by company, counting how many coincided in each combination of levels. That is, it was identified how many companies with low-level methods exhibited low, medium, or high sustainability; how many with medium-level methods were located at each sustainability level; and similarly for those with high-level methods. This procedure yielded the joint distribution presented in Table 3, where the rows represent the levels of decision-making methods and the columns the levels of financial sustainability.

Table 1. Classification levels of the surveyed companies in relation to the decision-making methods variable

Total Score	Level	Number of companies
0–2	Low	60
3–4	Half	25
5–6	High	15
Total		100

Table 2. Classification levels of the surveyed companies in relation to the financial sustainability variable

Total Score	Level	Number of companies
0–2	Low	54
3–4	Half	28
5–6	High	18
Total		100

Table 3. Contingency Table: Decision-making methods vs. Financial sustainability

Methods / Sustainability	Low	Half	High	Total
Low (60)	40	15	5	60
Medium (25)	10	10	5	25
High (15)	4	3	8	15
Total	54	28	18	100

Table 3 shows that, of the 100 companies surveyed, 60 have a low use of decision-making methods, 25 of these ventures use them at a medium level and the remaining 15 at a high level.

Based on this information, the Chi-square test is calculated, and for this it is necessary to establish the following hypotheses:

Null Hypothesis (H_0)
There is no statistically significant association between tax decision-making methods and the financial sustainability of rural businesses in Ecuador.

Alternative Hypothesis (H_1)
There is a statistically significant association between tax decision-making methods and the financial sustainability of rural enterprises in Ecuador.

In this sense, the Chi-square is calculated, from the following formula [13]:

$$X^2 = \frac{(O - E)^2}{E} \tag{1}$$

where:

O: is the observed frequency
E: is the expected frequency

In this way, the data was processed in the Excel program, obtaining the following results:

The sum of all cells, constructed from the data in Table 3, was equal to:

$$X^2 \approx 20.17$$

The next step was to calculate the degrees of freedom (df):

$$gl = (filas - 1)(columnas - 1) \tag{2}$$

$$gl = 4$$

Furthermore, the critical value must be established, which is obtained from the degree of freedom: 4 and the value corresponding to the significance level α = 0.05, obtaining:

$$X^2_{crítico} = 9.49$$

Therefore, to measure the probability of association between the variables obtained, the following assumptions are made:

If $X^2 > X^2_{crítico}$: the null hypothesis is rejected
Yeah $X^2 \leq X^2_{crítico}$ The null hypothesis is accepted.

In this sense,

$$X^2 = 20.17 \text{ and } X^2_{crítico} = 9.49$$

$$18.6 > 9.49$$

This indicates that the null hypothesis is rejected, demonstrating a significant association between tax decision-making methods and the financial sustainability of rural businesses. This suggests that better structuring of tax methods is associated with higher levels of business sustainability.

3.2 Results of the Application of the Structured Interview

In addition to the above analysis, a structured interview was conducted with three (3) administrators of these rural enterprises, who were selected through an intentional convenience sampling, based on the researcher's access and the availability of the interviewees,

Table 4. Interview Results

N°	Pregunta	Entrevistado 1	Entrevistado 2	Entrevistado 3
1	How do you currently make decisions related to tax compliance in your business?	They are made with the support of an accountant and we review our obligations monthly.	We use external consulting and keep digital records to avoid errors.	Based on the company's accounting results.
2	Do you believe that a poor decision-making process can affect the sustainability of the business?	Of course, it can generate unexpected debts and affect liquidity.	Yes, it can result in penalties that directly impact profits.	Absolutely, it can jeopardize the continuity of the business.
3	Do you believe that businesses that use sound decision-making methods have a greater chance of financial sustainability?	Yes, because they generate confidence and financial stability.	Yes, they are more formal and have easier access to credit.	Yes, orderly management strengthens business sustainability.
4	From your experience, what relationship do you observe between correct tax decision-making and the growth of entrepreneurship?	It's straightforward: better planning, greater growth	Of course, good communication management opens up opportunities for financial consolidation and business growth.	There is a clear relationship between decision-making methods and financial sustainability, since tax decisions.

who gave their consent for their participation, but requested that their personal data be treated confidentially.

For this purpose, a 4-question open-ended interview questionnaire was designed, which was applied via WhatsApp video calls, the results of which are presented in Table 4.

The interview findings reveal a notable trend toward the formalization and structuring of tax decision-making in rural businesses. Overall, the responses indicate that tax compliance decisions are not made spontaneously but are informed by accounting advice, regular review of financial information, and accounting tools. Therefore, it is established that open dialogue and sound tax decisions are generally considered crucial factors in determining the financial sustainability of the entity.

3.3 Methodological Triangulation of the Results

To support the proposed analysis, methodological triangulation is applied, combining theoretical frameworks and empirical results with statistical analysis. Theoretically, approaches such as those proposed by Velázquez et al. in [3] on financial management and tax planning suggest that the use of decision-making methods reduces economic risks and promotes business longevity.

Empirically, surveys and interviews indicate that businesses adopting structured decision-making approaches have demonstrated relatively stronger financial stability and projections.

This correspondence between empirical evidence and theory is statistically supported by the chi-square test, which allowed for the rejection of the null hypothesis. This demonstrates a significant relationship between the tax decision-making method and the financial sustainability of rural businesses.

As a result, the findings show that better structuring of decision-making methods is positively associated with sustainable businesses, thus confirming the correlation between theoretical validation, the stakeholder's perspective, and statistical validation.

4 Necessary Decision-Making Methods: A Proposal to Strengthen the Financial Sustainability of Rural Enterprises in Ecuador

In any company, decision-making requires the implementation of methods aligned with the needs of the business environment [14], which help to maintain order, planning and administration, especially in rural ventures, a sector that is underdeveloped and requires strengthening its tax management.

Canco et al. in [14] state that "the methods applied in decision-making vary over time depending on the objectives, situations and expected results. This evolution is a direct reflection, especially, of social, economic and scientific advances."

In this sense, the theory offers a series of methods or models that allow for appropriate decisions for business development [14], as shown in Fig. 9.

These traditional models or methods, according to Aedo et al. [15], offer a wide variety of alternatives for companies, as is the case with rural businesses in Ecuador, and are characterized as follows (Table 5):

Therefore, given the results obtained in this research, it is relevant that the studied businesses adopt the strategies offered by these methods within their tax management, as follows.

The traditional methods mentioned can be adapted to tax decision-making in rural businesses. For example, in the case of the rational model, it offers basic practices that would allow the entrepreneur to gather necessary and informed information about their tax obligations (such as compliance schedules, tax regimes, changes in regulations, among others, which would help in making appropriate decisions) [15].

Regarding the bounded rationality model [16], this becomes relevant for these ventures, since they lack the technical skills to process the necessary information. Therefore, this method promotes the search for information from state or public organizations, as well as from experts in this area who would provide specialized advice.

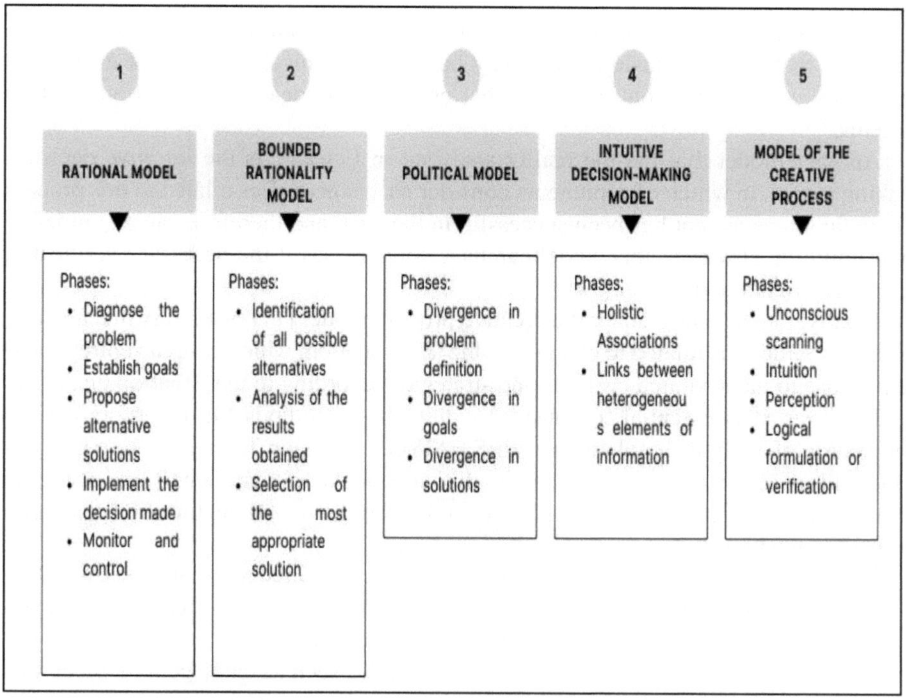

Fig. 9. Models to optimize the tax decision-making process

Table 5. Methods or models for decision making

Method or model	Description
Rational model	It is based on several assumptions from which the decision-maker can access various alternatives to analyze the variables involved
Bounded rationality model	This involves information processing which allows one to think, analyze, or understand related data and thus make appropriate and correct decisions.
Political model	This type of model is very useful, as it allows decisions to be made when conditions are uncertain
Intuitive decision-making model	This consists of decisions that are made based on previous experiences, which can be observed in the external market, and all participants have the possibility of intervening in these decisions
Model of the creative process	He argues that creativity is not a spontaneous or chaotic act, but a structured process involving experience, analysis, reflection, and critical evaluation

The political model also offers alternatives for rural entrepreneurs [14], since the decisions that must be made are not only based on technical aspects, but also on the regulatory context and control bodies that establish the deadlines, sanctions and tax benefits.

Another model that fits the reality analyzed in Ecuador is the intuitive decision-making model, in which entrepreneurs consider past experiences related to this process, where tax management has been successful in the past, and therefore can be guided by their intuition. However, this model can have some biases if the necessary verification processes are not carried out [14].

Finally, they can also adapt the creative process model [17], when tax management faces new situations related to changes in the tax system, in which case the entrepreneur can choose to incorporate new tools and strategies according to the situation that arises.

However, it is essential to consider within these methods the incorporation and application of analytical tools such as PEST, SWOT, Decision Tree and Pareto analysis, which allow for clear conclusions to be drawn to make valuable decisions at the tax level, guiding entrepreneurs towards areas for improvement and the financial evaluation of their companies.

Within this context and in light of the deficiencies identified in the analysis, the Decision Support System (DSS) emerges as an instrument to articulate, offering a systematic tool that complements the proposed theoretical models [11].

This type of system integrates a series of elements such as data, technological tools and evaluative methods that would be recommended for rural businesses in Ecuador to support their tax processes, based on the collection of financial, administrative and economic data.

It offers a comprehensive tool that allows for the selection, analysis, review, and decision-making of tax matters based on established legal stipulations, as well as the implementation of strategies that guide effective actions to achieve effective and transparent tax management.

In this way, a theoretical model of DSS is proposed for the strengthening of the financial sustainability of rural enterprises in Ecuador, which is detailed in Fig. 8, where the simplification of the information required to guarantee an adequate tax process is shown (Fig. 10).

This model begins with the external environment, considering fluctuations in the national economy and all regulations, rules, and control measures implemented by the State for tax administration. It then considers the database, which integrates internal information about businesses, as well as historical administrative, financial, and organizational data, forming the core structure of the DSS (Data Service System).

It also contains a model base, which outlines the methods needed to make timely decisions, incorporating models for tax planning simulation, cost-benefit analysis, and others.

The analysis engine is responsible for integrating the information obtained in each of the activities related to tax management, generating knowledge about the current situation of the company, allowing the selection of strategies in accordance with the current situation, reducing uncertainty.

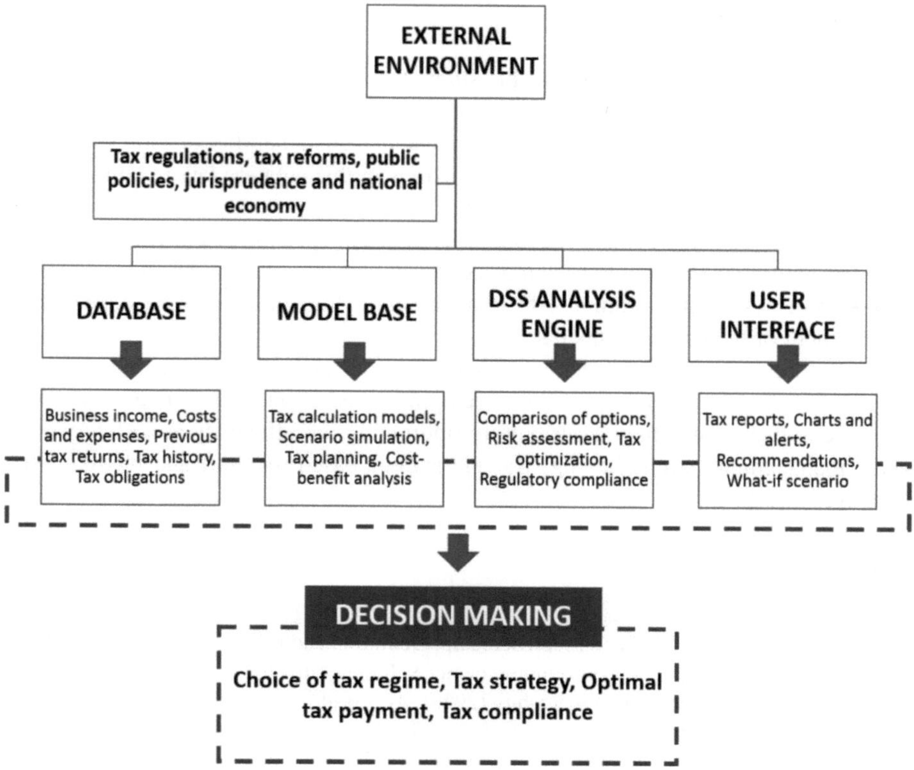

Fig. 10. Models to optimize the tax decision-making process

The user interface provides data that allows interaction between entrepreneurs and the entire system, facilitating the interpretation and understanding of the results achieved throughout this process.

In this sense, the theoretical formulation of the DSS suggested as a solution has a direct relevance to the findings presented, since this model operationalizes the recognized importance of organizing fiscal decision-making processes that could potentially improve the financial sustainability of rural businesses.

Qualitative results indicated that when information, technical guidance, and planning are incorporated into decisions, risks, penalties, and liquidity effects decrease, and the chi-square test indicates a significant finding that there is a relationship between decision-making methods and financial sustainability.

The DSS integrates databases, analytical models, and scenario evaluation tools that enable the comparison of options, risk estimation, and optimization of tax compliance, thus becoming systematic and data-driven in decision-making. In this way, the model not only aligns with theoretical and empirical evidence but also serves as a tangible tool to support the claim that the effective application of decision-making methods leads to greater financial sustainability for the business.

5 Conclusions

The analysis carried out throughout this study, through methodological triangulation, has made it possible to verify that there is a relevant association between the effective use of tax decision-making methods and the financial sustainability of rural enterprises in Ecuador, demonstrating that the right decisions in this area are key to consolidating adequate fiscal management.

Thus, it has become evident that the process of tax decision-making in rural enterprises or any other organization cannot be developed in isolation, but must focus on an integrated system that promotes the application of strategies to achieve consolidated management of financial and tax data.

Therefore, it is concluded that systematic methods and decision-making tools represent an essential alternative for the development and financial sustainability of this type of company, since they contribute to optimizing the efficiency and effectiveness of tax management, strengthening strategic actions based on short and medium-term objectives and goals.

References

1. Imran, H.: Decision making entrepreneurship: Definition and tips – TimeTrack Blog. Timetrack App-Zeiterfassung für Einzelunternehmer & Teams. https://www.timetrackapp.com/en/blog/decision-making-entrepreneurship/ (2023). Accessed 29 Jan 2026
2. Simon, H.A.: Making Management Decisions: The Role of Intuition and Emotion. AMP. **1**, 57–64 (1987). https://doi.org/10.5465/ame.1987.4275905
3. Velázquez, M.D.C.R., Chirinos, A.A.N., Brito, A.V.: Decision-making styles developed by commercial enterprises in the municipality of Barrancas. Management (Montevideo). **2**, 35–35 (2024)
4. Sevillano-Aguilar, C.J., Zambrano-Farías, I.: Characterization of rural agricultural enterprises on San Cristóbal Island in Galápagos. Sci. J. Episteme Praxis. **4**, 63–70 (2026)
5. Herruzo, E., Hernández-Sánchez, B., Cardella, G.M., Sánchez-García, J.: Entrepreneurship and Innovation: Opportunities for All. Dykinson (2019)
6. Navarrete, C.M., Boné-Andrade, M.F., Arboleda-Salazar, C.S.: Innovation and entrepreneurship in Ecuador as key factors for sustainable economic development. J. Econ. Soc. Sci. Res. **5**, 177–188 (2025)
7. Bravo Bravo, M.J.: Tax and Accounting Digitalization in the Informed Decision Making of Entrepreneurs of the Productive Development Directorate of the Gad of Jipijapa. Bs thesis. Jipijapa-Unesum (2024)
8. Robbins, S., Coulter, M.: Management. Pearson Education Mexico Edition (2005)
9. Cardona-Giraldo, L.A., R-G, C.J., V-M, J.: Explaining successful rural entrepreneurship from a multidimensional analysis: a case study in Colombia. Rev. Econ. Sociol. Rural. **63**, e291857 (2025)
10. Bartlett, D.: Management and business ethics: a critique and integration of ethical decision-making models. Br. J. Manag. **14**, 223–235 (2003). https://doi.org/10.1111/1467-8551.00376
11. Melović, B., Veljković, S.M., Ćirović, D., Vulić, T.B., Dabić, M.: Entrepreneurial decision-making perspectives in transition economies – tendencies towards risky/rational decision-making. Int. Entrep. Manag. J. **18**, 1739–1773 (2022). https://doi.org/10.1007/s11365-021-00766-2

12. Krasniqi, B.A., Berisha, G., Pula, J.S.: Does decision-making style predict managers' entrepreneurial intentions? J. Glob. Entrepr. Res. **9**, 68 (2019). https://doi.org/10.1186/s40497-019-0200-4
13. Montaña, R.M., Roco-Videla, Á., Nieves, A.R., Flores, S.V.: Chi-square test of homogeneity in clinical studies: a tool to analyze differences between treatments. Med. Fam. SEMERGEN. **51**, 102332 (2025)
14. Canco, I., Kruja, D., Iancu, T.: AHP, a reliable method for quality decision making: a case study in business. Sustainability. **13**, 13932 (2021)
15. Aedo, R.E., Hoyos, N., Herrera, D.: Models for business decision making. Semillas Saber J. **1**, 101–108 (2022)
16. Geng B, Li Q, Varshney PK. Human decision making with bounded rationality. ICASSP 2022–2022 IEEE International Conference on Acoustics, Speech and Signal Processing (ICASSP) IEEE 2022, p. 5493–5497.
17. Tamm, T., Hallikainen, P., Tim, Y.: Creative analytics: towards data-inspired creative decisions. Inf. Syst. J. **32**, 729–753 (2022). https://doi.org/10.1111/isj.12369

Author Index

A. P. Cabral Seixas Costa et al. (Eds.): ICDSST 2026, LNBIP 582, pp. 425–426, 2026.
https://doi.org/10.1007/978-3-032-25146-6

MIX
Papier aus verantwortungsvollen Quellen
Paper from responsible sources
FSC® C105338

If you have any concerns about our products,
you can contact us on
ProductSafety@springernature.com

In case Publisher is established outside the EU,
the EU authorized representative is:
Springer Nature Customer Service Center GmbH
Europaplatz 3, 69115 Heidelberg, Germany

Printed by Libri Plureos GmbH
in Hamburg, Germany